A Guide
To Introductory
Physics Teaching

A Guide to Introductory Physics Teaching

Arnold B. Arons
UNIVERSITY OF WASHINGTON

JOHN WILEY & SONS
New York · Chichester · Brisbane · Toronto · Singapore

Copyright © 1990, by John Wiley & Sons, Inc.

All rights reserved. Published simultaneously in Canada.

Reproduction or translation of any part of
this work beyond that permitted by Sections
107 and 108 of the 1976 United States Copyright
Act without the permission of the copyright
owner is unlawful. Requests for permission
or further information should be addressed to
the Permissions Department, John Wiley & Sons.

Library of Congress Cataloging-in-Publication Data:

Arons, A. B. (Arnold B.)
    A guide to introductory physics teaching / Arnold B. Arons.
       p. cm.
    Includes bibliographical references.
    ISBN 0-471-51341-5
    1. Physics–Study and teaching.   I. Title.
QC30.A76   1990
530'.0712–dc20   89-22653
                                                                                  CIP

Printed in the United States of America

10 9 8 7 6 5 4 3 2

Preface

Starting approximately twenty years ago, members of the physics teaching community began conducting systematic observations and research on student learning and understanding of physical concepts, models, and lines of reasoning. Some of these investigations began with, or subsequently spilled over into, research on more general aspects of the development of the capacity for abstract logical reasoning. In this book, I have tried to bring together as many as possible of the relevant insights into the teaching of the most basic aspects of introductory physics—covering high school through first year college level, including basic aspects of the course aimed at physics and engineering majors, without penetrating the full depth of the latter.

Very little that I present is based on conjecture. I have invoked and referred to most of the systematic research of which I am aware, and I have drawn on my own observations, which have been under way for more than forty years and have been extensively replicated over that time. One of my sources has been the direct interview in which one asks questions and listens to the individual student response; the other has been the analysis of students' written response to questions on tests and examinations. It is impossible to give all of the protocols of student interviews and all of the detailed supporting evidence without producing a book of impossible length. Although I give specific examples of student response from time to time, some of the insights are asserted without the full support they deserve. I can only ask the careful and critical reader to bear with these gaps, test them as opportunity arises, or turn to the more detailed literature for deeper penetration.

It is also impossible to include, in a book of reasonable length, all of the insights emerging from research on teaching, learning, and cognitive development. The literature is rich, varied, and rapidly increasing. I have been selective and have tried to include observations having the most direct bearing on classroom practice at the most basic levels of subject matter; the list of references will open the door to those wishing to pursue greater detail and explore primary evidence. Where a significant reference at this level is missing, the fault is in my judgment or in my not having fully encompassed the extensive literature.

Both the *American Journal of Physics* and *The Physics Teacher* are rich in articles discussing the logic and epistemology of various laws and concepts, outlining improved modes of presentation, suggesting demonstrations and other ways of making abstractions clearer and more concrete, describing ways of engaging students in direct activities, criticizing loose and faulty approaches, introducing new derivations, new laboratory experiments, and so forth. Every one of these functions is valuable and important to our community, and I wish someone, more competent than I, would undertake to bring together the heritage that has accumulated over the years in these areas into another book on physics teaching.

It is necessary for me to make clear, however, that my own purpose is different. I have undertaken to discuss some of the elements that I believe *underlie* and *precede* a great many of the ideas and presentations appearing in the journals. In fact, many of the excellent suggestions appearing in the journals turn out to be ineffective with large numbers of students, not because of anything wrong with the suggestions, but because the students have not had a chance to master the necessary *prior* concepts and lines of abstract logical reasoning. It is to this end that I have elected to concentrate on some of these prior aspects of cognitive development and on underlying problems of learning and understanding that have been commanding increasing attention in recent years. In doing this, I in no way disparage the valuable materials and modes of presentation that are described in the journals and that enter in full force at the points where I leave off.

It must further be emphasized that I am *not* formulating prescriptions as to how items of subject matter should be presented to the students or how they should be taught, nor am I suggesting that there is one single way of getting any particular item "across to the student." There is tremendous diversity in style and method of approach among teachers, and such diversity should flourish. My objective is to bring out as clearly and explicitly as possible the conceptual and reasoning difficulties many students encounter and to point up aspects of logical structure and development that may not be handled clearly or well in substantial segments of textbook literature. With respect to modes of attack on these instructional problems (avenues of explanation and presentation, balance of laboratory versus classroom experience, use of computers and of audiovisual aids), I defer to the style and predilections of the individual teacher.

I have endeavored to cover the range from high school physics through college and university calculus-based courses. Some of the material, therefore, goes well beyond high school level, and high school teachers should draw appropriate lines, limiting the more sophisticated material to their front running students if invoking it at all. At the other end of the spectrum, teachers in elite colleges, dealing with highly selected students, or teachers with a highly selected student body in calculus-based engineering-physics courses will find less relevance in the discussions of some of the more mundane underpinnings. However, it is necessary to issue a warning: there is much more overlap between the disparate populations than most teachers realize, and it is frequently startling to find how many students, at a presumably fairly high level, have the same difficulties, preconceptions, and misconceptions as do much less sophisticated students. It is only the *percentage* of students having a certain difficulty that changes as one goes up or down the scale; there is not an abrupt drop to zero at some intermediate level. Also, students at higher levels of scholastic ability, especially verbal skills, can usually remediate or overcome such initial difficulties at a more rapid pace than do other students, and a teacher needs to calibrate each of the classes with which he or she must deal.

Some of the chapters contain appendixes giving illustrations of possible test questions or homework problems. To keep down discursive length, I have not included detailed discussions of these questions, but they are all designed to implement some of the knowledge gained in the research protocols. They illustrate the kinds of questions that might be *added* to the normal regimen of quantitative end-of-chapter problems to confront the mind of the learner with aspects otherwise not being made explicit. The examples being given are an invitation more than an end point. The pool of such questions must be greatly expanded to enhance variety and flexibility. Such expansion will take place not through the output of one individual, whose imagination gives out at some finite point, but through the superposition of effort on the part of numerous interested individuals, each of whom brings a new imagination to the effort. I long to see my limited set of examples greatly expanded.

Finally I point to the following unwelcome truth: much as we might dislike the implications, research is showing that didactic exposition of abstract ideas and lines of reasoning (however engaging and lucid we might try to make them) to passive listeners yields pathetically thin results in learning and understanding—except in the very small percentage of students who are specially gifted in the field. Even in the calculus-based course, many students have the difficulties, and need all of the help, outlined in these pages. In expressing this caveat, I am, of course, *not* advocating *un*clear exposition. I am pointing to the necessity of supplementing lucid exposition with exercises that engage the mind of the learner and extract explanation and interpretation in his or her own words.

It is obvious that ideas and information such as I have summarized here cannot be developed in seclusion. I am deeply indebted to the hundreds of students who have submitted to my questioning, accepting the tension that goes with my shutting up and waiting for their answers. I am indebted also to the many colleagues and associates with whom I have discussed physics, prepared test questions, and worried about the meaning of learning and understanding. Among these are my former colleagues at Amherst College: Bruce Benson, Colby Dempesy, Joel Gordon, Robert Romer, Theodore Soller, and Dudley Towne; at the University of Washington: David Bodansky, Kenneth Clark, Ronald Geballe, James Gerhart, Patricia Heller, Lillian McDermott, James Minstrell, and Philip Peters. Robert Romer, Kenneth Clark, and Phillip Peters have read extensive sections of this book and have supplied me with valued criticism, corrections, and suggestions.

ARNOLD B. ARONS
Seattle, Washington

Contents

CHAPTER 1 **Underpinnings** — 1
    1.1 Introduction, 1
    1.2 Area, 1
    1.3 Exercises with "Area", 2
    1.4 Volume, 3
    1.5 Mastery of Concepts, 3
    1.6 Ratios and Division, 3
    1.7 Verbal Interpretation of Ratios, 4
    1.8 Exercises in Verbal Interpretation, 4
    1.9 Comment on the Verbal Exercises, 5
    1.10 Arithmetical Reasoning Involving Division, 6
    1.11 Coupling Arithmetical Reasoning to Graphical Representation, 8
    1.12 Scaling and Functional Reasoning, 10
    1.13 Elementary Trigonometry, 13
    1.14 Horizontal, Vertical, North, South, Noon, Midnight, 13
    1.15 Interpretation of Simple Algebraic Statements, 14
    1.16 Language, 15
    1.17 Why Bother with the Underpinnings?, 17
    Appendix 1A   Sample Homework and Test Questions, 18

CHAPTER 2 **Rectilinear Kinematics** — 20
    2.1 Introduction, 20
    2.2 Misleading Equations and Terminology, 20
    2.3 Events: Positions and Clock Readings, 21
    2.4 Instantaneous Position, 22
    2.5 Introducing the Concept of "Average Velocity", 23
    2.6 Graphs of Position Versus Clock Reading, 24
    2.7 Instantaneous Velocity, 26
    2.8 Algebraic Signs, 27

2.9   Acceleration, 28
2.10 Graphs of Velocity Versus Clock Reading, 30
2.11 Areas, 30
2.12 Top of the Flight, 31
2.13 Solving Kinematics Problems, 32
2.14 Use of Computers, 33
2.15 Research on Formation and Mastery of the Concept of Velocity, 34
2.16 Research on Formation and Mastery of the Concept of Acceleration, 35
2.17 Implications of the Research Results, 38
2.18 Galileo and the Birth of Modern Science, 38
2.19 Observation and Inference, 42
Appendix 2A   Sample Homework and Test Questions, 43

CHAPTER 3   *Elementary Dynamics*   49

3.1   Introduction, 49
3.2   Logical Structure of the Laws of Motion, 50
3.3   An Operational Interpretation of the First Law, 51
3.4   Operational Definition of a Numerical Scale of Force, 52
3.5   Application of the Force Meter to Other Objects: Inertial Mass, 54
3.6   Superposition of Masses and Forces, 55
3.7   Textbook Presentations of the Second Law, 56
3.8   Weight and Mass, 57
3.9   Gravitational Versus Inertial Mass, 58
3.10 Understanding the Law of Inertia, 59
3.11 What We Say *Can* Hurt Us: Some Linguistic Problems, 63
3.12 The Third Law and Free Body Diagrams, 64
3.13 Logical Status of the Third Law, 67
3.14 Distributed Forces, 68
3.15 Use of Arrows to Represent Force, Velocity, and Acceleration, 68
3.16 Understanding Terrestrial Gravitational Effects, 69
3.17 Strings and Tension, 74
3.18 "Massless" Strings, 75
3.19 The "Normal" Force at an Interface, 76
3.20 Objects Are Not "Thrown Backwards" When Accelerated, 78
3.21 Friction, 79
3.22 Two Widely Used Demonstrations of "Inertia", 80
3.23 Different Kinds of "Equalities", 81
3.24 Solving Problems, 83
Appendix 3A   Sample Homework and Test Questions, 86

CHAPTER 4 *Motion in Two Dimensions* 91

    4.1 Vectors and Vector Arithmetic, 91
    4.2 Defining a "Vector", 92
    4.3 Components of Vectors, 93
    4.4 Projectile Motion, 94
    4.5 Phenomenological Thinking and Reasoning, 96
    4.6 Radian Measure and π, 99
    4.7 Rotational Kinematics, 100
    4.8 Preconceptions Regarding Circular Motion, 101
    4.9 Centripetal Force Exerted by Colinear Forces, 103
    4.10 Centripetal Force Exerted by Noncolinear Forces, 106
    4.11 Frames of Reference and Fictitious Forces, 108
    4.12 Revolution Around the Center of Mass: The Two-Body Problem, 108
    4.13 Torque, 111
    Appendix 4A Sample Homework and Test Questions, 114

CHAPTER 5 *Momentum and Energy* 115

    5.1 Introduction, 115
    5.2 Developing the Vocabulary, 116
    5.3 Describing Everyday Phenomena, 116
    5.4 Force and Rate of Change of Linear Momentum, 118
    5.5 Heat and Temperature, 118
    5.6 The Impulse–Momentum and Work–Kinetic Energy Theorems, 121
    5.7 Real Work and Pseudowork, 123
    5.8 The Law of Conservation of Energy, 124
    5.9 Digression Concerning Enthalpy, 126
    5.10 Work and Heat in the Presence of Sliding Friction, 128
    5.11 Deformable System with Zero-Work Force, 130
    5.12 Rolling Down an Inclined Plane, 132
    5.13 Inelastic Collision, 134
    5.14 Some Illuminating Exercises, 136
    5.15 Spiraling Back, 138
    Appendix 5A Sample Homework and Test Questions, 141

CHAPTER 6 *Static Electricity* 144

    6.1 Introduction, 144
    6.2 Distinguishing Electric, Magnetic, and Gravitational Interactions, 145
    6.3 Frictional Electricity, Electrical Interaction, and Electrical Charge, 145
    6.4 Electrostatics Experiments at Home, 146

- 6.5 Like and Unlike Charges, 147
- 6.6 Positive and Negative Charges; North and South Magnetic Poles, 150
- 6.7 Polarization, 152
- 6.8 Charging by Induction, 154
- 6.9 Coulomb's Law and the Quantification of Electrical Charge, 154
- 6.10 Electrostatic Interaction and Newton's Third Law, 156
- 6.11 Sharing Charge Between Two Spheres, 157
- 6.12 Conservation of Charge, 158
- 6.13 Electrical Field Strength, 159
- 6.14 Superposition, 160

CHAPTER 7 Current Electricity 162

- 7.1 Introduction, 162
- 7.2 Which Should Come First, Static or Current Electricity?, 163
- 7.3 How Do We Know That Current Electricity Is "Charge in Motion?", 163
- 7.4 Batteries and Bulbs (I): Formation of Basic Circuit Concepts, 167
- 7.5 Batteries and Bulbs (II): Phenomenology of Simple Circuits, 170
- 7.6 The Historical Development of Ohm's Law, 172
- 7.7 Teaching Electrical Resistance and Ohm's Law, 175
- 7.8 Is Electric Current in Metals a Bulk or Surface Phenomenon?, 176
- 7.9 Building the Current–Circuit Model, 177
- 7.10 Conventional Current Versus Electron Current, 178
- 7.11 Not Every Load Obeys Ohm's Law, 180
- 7.12 Free Electrons in Metals: The Tolman–Stewart Experiment, 181

Appendix 7A Sample Homework and Test Questions, 184

CHAPTER 8 Electromagnetism 188

- 8.1 Introduction, 188
- 8.2 Oersted's Experiment, 188
- 8.3 Force Between Magnets and Current Carrying Conductors, 191
- 8.4 Ampère's Experiment, 192
- 8.5 Mnemonics and the Computer, 193
- 8.6 Faraday's Law in a Multiply Connected Region, 194
- 8.7 Faraday's Criticism of Action at a Distance, 195
- 8.8 Infancy of the "Field" Concept, 198
- 8.9 Laboratory Measurement of a Value of **B**, 200

CHAPTER 9 **Waves and Light** 201

    9.1 Introduction, 201
    9.2 Distinguishing Between Particle and Propagation Velocities, 201
    9.3 Graphs, 202
    9.4 Transverse and Longitudinal Pulse Shapes, 203
    9.5 Reflection of Pulses, 204
    9.6 Derivation of Propagation Velocities, 208
    9.7 Velocity of Propagation of a Kink on a String, 208
    9.8 Propagation Velocity of a Pulse in a Fluid, 210
    9.9 Propagation Velocity of Surface Waves in Shallow Water, 213
    9.10 Transient Wave Effects, 215
    9.11 Sketching Wave Fronts and Rays in Two Dimensions, 216
    9.12 Periodic and Sinusoidal Wave Trains, 217
    9.13 Two-Source Interference Patterns, 218
    9.14 Two-Source Versus Grating Interference Patterns, 219
    9.15 Young's Elucidation of the Dark Center in Newton's Rings, 220
    9.16 Specular Versus Diffuse Reflection, 221
    9.17 Images and Image Formation: Plane Mirrors, 222
    9.18 Images and Image Formation: Thin Converging Lenses, 223
    9.19 Novice Conceptions Regarding the Nature of Light, 226
    9.20 Phenomenological Questions and Problems, 226

CHAPTER 10 **Early Modern Physics** 228

    10.1 Introduction, 228
    10.2 Historical Preliminaries, 229
    10.3 Prelude to Thomson's Research, 233
    10.4 Thomson's Experiments, 234
    10.5 Thomson's Inferences, 236
    10.6 Homework Assignment on the Thomson Experiment, 238
    10.7 The Corpuscle of Electrical Charge, 239
    10.8 From Thompson's Electron to the Bohr Atom, 240
    10.9 The Photoelectric Effect and the Photon Concept, 245
    10.10 Quotations from Einstein's Paper on the Photon Concept, 248
    10.11 Bohr's First Quantum Picture of Atomic Hydrogen, 250
    10.12 Introducing Special Relativity, 257
    Appendix 10A Written Homework on Thomson Experiment, 264
    Appendix 10B Written Homework on the Bohr Atom, 269

CHAPTER 11 **Miscellaneous Topics** 274

    11.1 Introducing Kinetic Theory, 274
    11.2 Assumptions of the Kinetic Theory of the Ideal Gas, 276

11.3 Hydrostatic Pressure, 281
11.4 Visualizing Thermal Expansion, 282
11.5 Estimating, 283
11.6 Examples of Mathematical Physics for Gifted Students, 283
11.7 Chaos, 286

CHAPTER 12 *Achieving Wider Scientific Literacy* 288

12.1 Introduction, 288
12.2 Marks of Scientific Literacy, 289
12.3 Operative Knowledge, 290
12.4 General Education Science Courses, 292
12.5 Illustrating the Nature of Scientific Thought, 295
12.6 Illustrating Connections to Intellectual History, 299
12.7 Variations on the Theme, 302
12.8 Aspects of Implementation, 303
12.9 The Problem of Cognitive Development, 305
12.10 The Problem of Teacher Education, 305
12.11 A Role for the Computer, 308
12.12 Learning from Past Experience, 309

CHAPTER 13 *Critical Thinking* 313

13.1 Introduction, 313
13.2 A List of Processes, 314
13.3 Why Bother with Critical Thinking?, 319
13.4 Existing Level of Capacity for Abstract Logical Reasoning, 320
13.5 Can Capacity for Abstract Logical Reasoning Be Enhanced? 321
13.6 Consequences of Mismatch, 323
13.7 Ascertaining Student Difficulties, 325
13.8 Testing, 325
13.9 Some Thoughts on Faculty Development, 326

Bibliography   328
Index   337

CHAPTER 1

Underpinnings

1.1 INTRODUCTION

Several fundamental gaps in the background of students may seriously impede their grasp of the concepts and lines of reasoning that we seek to cultivate from the beginning of an introductory physics course. These gaps, having to do with understanding the concepts of "area" and "volume" and with reasoning involving ratios and division, are often encountered, even among students at the engineering-physics level.

In principle, these gaps should not exist because the ideas are dealt with, and should have been mastered, at earlier levels in the schools. It is an empirical fact, however, that such mastery has not been achieved, and ignoring the impediment is counterproductive.

Unfortunately, it is illusory to expect to remediate these difficulties with a few quick exercises, in artificial context, at the start of a course. Most students can be helped to close the gaps, but this requires *repeated* exercises that are spread out over time and are integrated with the subject matter of the course itself. This statement is *not* a matter of conjecture; it reflects empirical experience our physics education research group at the University of Washington has encountered repeatedly [Arons (1976), (1983b), (1984c)].

This chapter describes some of the learning difficulties that are involved in the development of a number of underpinnings, including arithmetical reasoning, and suggests exercises that can be made part of the course work.

1.2 AREA

The concept of area is the foundation of many of the other basic physical concepts, such as pressure, stress, energy flux, and coefficients of diffusion and heat conduction. It underpins all the ratio reasoning associated with geometrical scaling. Furthermore, it is essential to the interpretation of velocity change as area under the graph of acceleration versus clock reading, to the interpretation of position change as area under the graph of velocity versus clock reading, to the definitions of work and impulse, and to the interpretation of integrals in general.

If you ask students how one arrives at numerical values for "area" or "extent of surface," many—if they have any response at all—will say "length times width." If you then sketch some very irregular figure without definable length or width and ask about assigning a numerical value to the area of the figure, very little response of any kind is forthcoming. Students who respond in this way have not formed a clear operational definition of "area."

The reason for this is fairly simple: Although the grade school arithmetic books, when they introduce the area concept, do have a paragraph about selecting a unit square, imposing a grid on the figure in question, and counting the squares within the figure, virtually none of the students have ever gone through such a procedure themselves in homework exercises. They have never been asked to define "area." All they have ever done is calculate areas of regular figures such as squares, rectangles, parallelograms, or triangles, using memorized formulas that they no longer connect with the operation of counting the unit squares, even though this connection may have been originally asserted.

Furthermore, virtually none of the students have had any significant exposure to the notion of operational definition. They have had little or no practice in defining a term by reference to shared experience or by describing, in simple words of prior definition, the actions through which one goes to develop the numerical value being referred to in the name of a technical concept.

1.3 EXERCISES WITH "AREA"

In introductory physics teaching, it is desirable to invoke the area concept at the earliest possible opportunity. Students should be led to articulate the operational definition in their own words—and to do so on tests. (This is an excellent opportunity to introduce the concept of operational definition in a context that is familiar and relatively unthreatening.) The fact that they had been using the technical term "area" without adequate mastery of the concept behind it makes a salutory impression on many students.

Homework and test problems should give students opportunity to execute the operations they describe in the definition, right through the selection of the unit square, superposition of the grid on the figure in question, and actually counting the squares. The operation of counting must involve the estimation of squares contained around the periphery of the figure. To many students the necessity of estimating the fractions appears in some sense "sinful," since it involves "error" and is not "exact," as seems to be the value obtained from a formula. The actual experience of counting and estimating should begin with "pure" areas, that is, surface extent of arbitrarily and irregularly shaped geometrical figures. Then, as soon as it becomes appropriate, the exercises should be extended to measurement and interpretation of areas under v versus t and under a versus t graphs. (This, of course, adds the arithmetical reasoning associated with the dimensionality of the coordinates.)

In calculus-physics courses, the latter exercises should be explicitly linked with the mathematical concept of "integral." Although this might seem so obvious as to be not worthy of mention, many students have not actually established this connection even though they may be taking, or may have completed, a calculus course. Although they have been *told*, perhaps many times, that the integral can be interpreted as an area, the idea has not registered because it has not been made part of the individual student's concrete experience, and they have never had the opportunity to articulate the idea in their own words.

Such exercises should be repeated still later when the context begins to involve "work" and "impulse." It is only such recycling of ideas over fairly extended periods of time, reencountered in increasingly rich context, that leads to a firm assimilation in many students.

In noncalculus-physics courses, the concept of "integral" is not at hand and is not necessary. Dealing with the areas, however, breaks the shackles to eternally constant quantities and shows the students how physics can easily and legitimately deal with continuous change. "Capturing the fleeting instant" was one of the great intellectual triumphs of the seventeenth century, and students can be given some sense of this part of their intellectual heritage through calculations that they can easily make without the necessity of a formal course in the calculus.

1.4 VOLUME

Initially, most students have the same difficulty with "volume" as with "area." They grasp for formulas without having registered an operational definition of the concept. As a result, quite a few students do not, in fact, discriminate between area and volume; they use the words carelessly and interchangeably as metaphors for size.

Once the operational definition of "area" has been carefully developed and anchored in the concrete experience of counting squares, however, the operational definition of "volume" can be elicited relatively easily. The analogy to "area" is readily perceived, and the counting of unit cubes is quickly accepted.

1.5 MASTERY OF CONCEPTS

It should be emphasized at this point that mastery of the operational definitions of "area" and "volume" up to the point of recognizing the counting of unit squares or cubes is only a beginning; it is still far short of the ability to use the concepts in more extended context. At this stage, for example, some students (particularly those who have had little or no prior work in science) do not discriminate between mass and volume.[1] Many students, including those in engineering-physics courses, are, at this stage, still unable to compare final with initial areas or volumes when the linear dimensions of an object have been scaled up or down.

The problem of scaling is a particularly important one. It involves ratio reasoning and will be discussed in more detail in Section 1.12.

1.6 RATIOS AND DIVISION

One of the most severe and widely prevalent gaps in cognitive development of students at secondary and early college levels is the failure to have mastered reasoning involving ratios. The poor performance reproducibly observed on Piagetian tasks of ratio reasoning has become well known during the past 15 years [McKinnon and Renner (1971); Karplus, et al. (1979); Arons and Karplus (1976); Chiappetta (1976)]. This disability, among the very large number of students who suffer from it, is one of the most serious impediments to their study of science.

For convenience, I separate reasoning with ratios and division into two levels or stages: (1) verbally interpreting the result obtained when one number is divided by another; (2) using the preceding interpretation to calculate some other quantity.

---

[1] For evidence concerning this assertion and for strategies that help students achieve such discrimination see McDermott, Piternick, and Rosenquist (1980); McDermott (1980); McDermott, Rosenquist, and van Zee (1983).

1.7 VERBAL INTERPRETATION OF RATIOS

Reasoning with ratios and division requires, as a first step, the capacity to interpret verbally the meaning of a number obtained from a particular ratio. The verbal interpretations are somewhat different in different contexts. Many students are deficient in this capacity and need practice in interpreting ratios in their own words.

In the primitive case in which the numbers have not been given specific physical meaning, we interpret the result of, say 465/23, as the number of times 23 is contained in 465. This may sound like a trivial statement, but it is not. Most students have memorized (successfully or unsuccessfully, as the case may be) the algorithm of division but have never been given the opportunity to recognize it as a shorthand procedure for counting successive subtractions of 23 from 465. Thus they do not see the operation of division in perspective or translate it into simpler prior experience. The phrase "goes into" is memorized without relation to other contexts. Those who have not developed this perspective should be given the opportunity to count the successive subtractions and to begin to see what they are doing in the memorized algorithm. They should finally have to tell the whole story in their own words.

At a next higher level of sophistication, we may be dealing with a ratio of dimensionally identical quantities, for example, L_2/L_1, the ratio, say, of the heights of two buildings, or of distances from a fulcrum in balancing, or the linear scaling of a geometrical figure. Here the numerical value of the ratio serves as a *comparison*: it tells us how many times larger (or smaller) one length is compared to the other.

Next we encounter division of dimensionally *inhomogeneous* quantities: mass in grams divided by volume in cubic centimeters; position change in meters divided by a time interval in seconds; dollars paid divided by number of pounds purchased. Here the result of division tells us how much of the numerator is associated with *one* unit of whatever is represented in the denominator.

Finally, if we have 500 g of a material that has 3.0 g in each cubic centimeter, the numerical value of 500/3.0 tells us how many "packages" of size 3.0 g are contained in the 500 g sample. Since each such "package" corresponds to one cubic centimeter, we have obtained the number of cubic centimeters in the sample.

1.8 EXERCISES IN VERBAL INTERPRETATION

Many students have great difficulty giving verbal interpretations like those illustrated in the preceding section, since they have almost never been asked to do so. Without such practice in at least several different contexts, students do not think about the meaning of the calculations they are expected to carry out, and they take refuge in memorizing patterns and procedures of calculation, manipulating formulas, rather than penetrating to an understanding of the reasoning. As a consequence, when they find themselves outside the memorized situations, they are unable to solve problems that involve successive steps of arithmetical reasoning.

Explaining or telling students who are in such difficulty the meaning of particular ratios, however frequently or lucidly this may be done, has very little effect. It is necessary to ask questions that lead the students to articulate the interpretations and explanations in their own words. Here are some typical excerpts from such conversations:

Suppose students having difficulty with a problem involving the use of the density concept are asked: "We took the measured mass (340 g) of an object and divided it by the volume (120 cm^3). How do you interpret the number 340/120? Tell what it means, using the simplest possible words." Some will answer "That is the density." These students have not separated the technical term, the *name* of the resulting number, from the verbal interpretation of its meaning. (This involves an important cognitive process that will be discussed in another chapter.)

When it is pointed out that the name is not an interpretation, some students will say "mass per volume"; others might say "the number of grams in 120 cubic centimeters." (Exactly parallel statements are likely to be given if the ratio is position change divided by time interval.) Very few students having trouble with the original problem will give a simple statement to the effect that we have obtained the number of grams in *one* cubic centimeter of the material.

One can now adopt the strategy of going back to some more familiar context: "Suppose we go to a store and find a box costing $5.00 and containing 3 kg of material. What is the meaning of the number 5.00/3?" Some students will still say "That is how much you pay for 3 kg" but, in this more familiar context, many will recognize that we have calculated how many dollars we pay for *one* kilogram. (The former group is in need of further dialog, using more concrete examples, before a correct response is found.) One can now try to get the students to the generalization that in such situations the resulting number tells us "how many of these (in the numerator) are associated with *one* of those (in the denominator)."

If one then asks: "In the case of the box costing $5.00 and containing 3 kg, suppose we now consider the number 3/5.00. In light of what we concluded in the previous example, does this number have an interpretation?" Many students, including some who gave the correct interpretation of 5.00/3, now encounter difficulty. Some revert to earlier locutions such as "how many kilograms you get for 5.00"; many consider the number meaningless or uninterpretable.

In such instances there seem to be two difficulties superposed: (1) although the students may have previously been given some opportunity to think about or calculate "unit cost" (how much we pay for one kilogram), they rarely, if ever, have been asked about the inverse (how much one gets for one dollar). (2) 5.00/3 involved the division of a larger number by a smaller one. To many students this is more intelligible and less frightening than the fraction 3/5.00.

After students have been led through the parallel interpretation of *both* ratios, one can usually go back to a case such as mass divided by volume or change of velocity divided by time interval and elicit a correct interpretation of the new ratio and its inverse. Then one can elicit the generalization being sought, namely, that such a ratio tells us how much of the numerator is associated with *one* unit of whatever is represented in the denominator. It is essential, however, to elicit the word "one"; use of the word "per" by the student is no assurance that he or she understands the concept (see the discussion in the next section).

1.9 COMMENT ON THE VERBAL EXERCISES

Note the strategy being employed in the dialogs suggested in the preceding section: although some students have responded previously to problems such as "calculate the cost of one kilogram if 3 kg cost $5.00," very few students have ever been confronted with the ratio and asked to interpret it in words, that is, they have never reversed the line of thought, traversing it in the direction opposite to that previously experienced.

In Piagetian terminology, the term "operations" denotes reasoning processes that can be reversed by the user. Thus students who can calculate the unit cost but do not recognize the interpretation of the ratio are not reversing the reasoning and have not brought it to the "operations" level. Leading them to reverse the direction of reasoning turns out to be a useful tool for helping them master the reasoning. (This idea will be discussed in more general terms in a subsequent chapter.)

Complete control of the interpretation of ratios is rarely attained with just one short sequence of exposure as outlined above. Many students must have the experience of carrying through the same kind of reversible reasoning in several additional contexts (e.g., what is the meaning of the number obtained in dividing the circumference of a circle by its diameter? If 16 g of oxygen combine with 12 g of carbon, what is the meaning of 16/12? Of 12/16? If a laboratory cart travels 180 cm in 2.3 s, what is the meaning of 180/2.3? Of 2.3/180? etc.) before they fully assimilate it.

A word of warning: If a teacher accepts casual use of the word "per"—particularly the incorrect and meaningless "mass per volume," which was quoted in the preceding section—he or she falls into a trap. Even though it contains only three letters, "per" is a technical term, and very few of those students who are having trouble with arithmetical reasoning know what it means. They inject it into a response only because they have a vague memory that "per" frequently turns up for some obscure reason in division, but they do not explicitly translate it into simpler words such as "in," "for each," "corresponds to," "goes with," "combines with," "is associated with," and so on.

Even if students correctly say "mass per *unit* volume" rather than "mass per volume" in interpreting M/V, there is no conclusive assurance that they really understand the meaning. Some do, but others have merely memorized the locution. It is important to lead all students into giving simple interpretation in everyday language before accepting a regular use of "per".[2]

Many students do not know what the word "ratio" means. Those having difficulty with reasoning and interpretation should always be asked, at an early stage, for the meaning of the word if they, the text, or the teacher invoke it.

It is also worth noting that the interpretations of division being illustrated underlie many of the manipulations of elementary algebra and are particularly relevant to the translation of verbal problems into the corresponding algebraic equations and vice versa. Remediating student difficulties with verbal interpretation of ratios eventually enhances students' ability to use elementary algebra.

1.10 ARITHMETICAL REASONING INVOLVING DIVISION

Verbal interpretations like those illustrated in the preceding section (how much of the numerator is associated with *one* unit of whatever is represented in the denominator) are only the first step in a sequence and involve only one of the several interpretations of the meaning of a result of dividing one number by

---

[2]Tobias (1988) notes a similar problem, stemming from inattention on the part of teachers, in connection with the word "of":

> A number of [students] reported getting lost during lessons on multiplication and division of fractions, and as they talked about this, I began to notice an ambiguity in use of the word "of." They had been instructed that the word "of" in expressions like one-third of three-quarters always means multiply. But this, they remembered, felt wrong or confusing. "Of" felt more like division. Indeed they were right. Words connote as well as denote. The word "of," in fact, means multiply only in one narrow context within mathematics.

another. The next fruitful step is made through such questions as: "We have 800 g of material having a density of 2.3 g/cm³. What must be the volume occupied by the sample?"

The first impulse of many students is to manipulate the density formula $\rho = M/V$. (In fact, if the word "density" is not used in the statement of the question and one merely says that the material has 2.3 g in each cubic centimeter, quite a few students are completely lost, not knowing what to do when they have not been cued as to a formula.) An investigation of what is happening in manipulation of the formula reveals what Piaget would characterize as an essentially "concrete operational" response. In many instances, the students are not reasoning either arithmetically or algebraically but are simply rearranging the symbols, as though they were concrete objects, in patterns that have become familiar. Obtaining a correct answer to the initial question does not necessarily indicate a grasp of the attendant arithmetical reasoning.[3]

Students should be led to articulate something like the following story: What does 2.3 g/cm³ *mean*? The quantity 2.3 is the number of grams in *one* cubic centimeter. We can think of 2.3 g as a clump or package. If we find how many such packages there are in 800 g of the material, we obtain the total number of cubic centimeters because each package corresponds to one cubic centimeter.

Similarly, when asked to find the diameter of a circle having a circumference of 28 cm, students should be led to argue that, since each "package" of 3.14 cm in the circumference corresponds to *one* centimeter in the diameter, we must find how many packages of size 3.14 are contained in 28. Manipulation of the formula $C = \pi D$, however correctly, does not testify to understanding of the meaning of π or to grasp of the underlying arithmetical reasoning.

One such exposure does *not* usually provide full remediation to students who have this difficulty. Repetition is essential, but repetition without some alteration of the context simply encourages memorization. One way of altering the context sufficiently to make the repetition nontrivial is as follows: "We have a block consisting of 5000 g of material having a density of 2.3 g/cm³. Suppose we add 800 g of the same material to the block. By how much have we *increased* the volume of the block?" (Similarly, one alters the circle problem by adding 28 cm to the circumference of a circle having some arbitrary initial diameter, large or small, and asking for the *increase* in diameter.)

Many students initially see these problems as entirely different from the original versions. They painstakingly calculate, for example, the volume of a 5800 g block and subtract the volume of a 5000 g block. When they are led to realize that 800/2.3 gives the answer to both versions, they make a significant stride toward mastery of the underlying reasoning, especially when they additionally recognize that the circle problems are exactly the same as the density problems.

To summarize: linguistic elements play an essential, underlying role in the development of the capacity for arithmetical reasoning with ratios and proportion. This observation is explicitly supported by Lawson, Lawson, and Lawson (1984) who remark that "a necessary . . . condition for the acquisition of proportional

---

[3]It should be pointed out that classical "proportional reasoning" (e.g., object A has a height of 8 measured in units of length of a small paper clip. Object B has a height of 12 in the same units. Object A has a height of 6 measured in units of length of a larger paper clip. What would be the height of B measured in large paper clips?) suffers from similar problems. Many students memorize the "this-is-to-this-as-that-is-to-that" routine and manipulate the given numbers as concrete objects in a spatial arrangement, frequently doing so incorrectly. Again, a correct result is not firm evidence of understanding the line of reasoning.

reasoning during adolescence is the prior internalization of key linguistic elements or argumentation." Failure to provide this linguistic experience in the schools underlies much of the difficulty students experience, and much of the "fear of mathematics" that we observe, at high school and college levels. The pace at which verbal security can be conveyed at the latter levels is no greater than the pace required at earlier age. This problem will not be rectified until we, in the colleges and universities, produce elementary school teachers who have mastered arithmetical reasoning sufficiently thoroughly to lead their pupils into articulating lines of reasoning and explanation in their own words. This is not currently being achieved in sufficiently large measure.

1.11 COUPLING ARITHMETICAL REASONING TO GRAPHICAL REPRESENTATION

A powerful way of helping students master a mode of reasoning is to allow them to view the same reasoning from more than one perspective. In the case of arithmetical reasoning, a very useful alternative perspective is that of graphical representation. Consider, for example, the different situations illustrated in the graphs of Fig. 1.11.1.

Students should *not* be confronted with these graphs all at once in some remedial orgy. They should be led into building up the representations in homework problems whenever the situations arise in the normal sequence of the course work. This allows for spiralling back to the modes of thinking and spreads the encounters out over weeks of time; both the spiralling back and the time spread are essential for effective assimilation. In each encounter, they should have to interpret the representations in their own words.

For example:

1. In Fig. 1.11.1*a* each line represents a different substance; the steepness (or slope) of the line is the number M/V and is interpreted as the number of grams in *one* cubic centimeter if the units are grams and cubic centimeters respectively; in any straight line relationship the amount added along the vertical axis is always the same for equal steps along the horizontal axis; when the graph is *not* a straight line, the steps along the vertical axis are *not* equal under such circumstances.

2. The steepness of such straight lines is frequently a *property* of the object or system being described. In Fig. 1.11.1*a* the property is called "density of the substance"; in (b) it is called "concentration of the solution"; in (c) it is called "inertial mass of the object"; in Fig. 1.11.1*d* it is called "coefficient of friction between the two surfaces," and so on.

3. In most of the graphs, different systems or objects possess their own different numerical values of the property in question, and there are different straight lines for different objects. Figure 1.11.1*g*, however, illustrates the remarkable fact that the steepness 3.14, to which we give the name π, is a property that *all* circles have in common, and there is only *one* straight line!

4. The problems in Section 1.10 that involve arithmetical reasoning with the concept of density and π can be represented and interpreted on Figs. 1.11.1*a* and 1.11.1*g*, and students should be led to do so. In order to calculate the total volume of a sample of known mass and known density, or the diameter of a circle of known circumference, one can use the straight line from the origin to the mass

1.11 COUPLING ARITHMETICAL REASONING TO GRAPHICAL REPRESENTATION

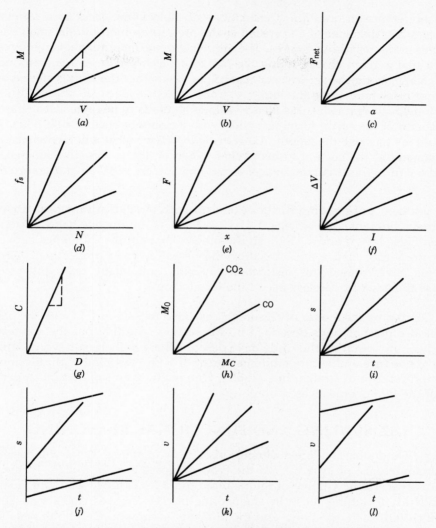

Figure 1.11.1 Linear relations and arithmetical reasoning. (*a*) Total mass M versus total volume V of three different homogeneous substances. (*b*) Total mass M of solute versus total volume V of solvent for three different solution concentrations using the same solute and the same solvent. (*c*) Net force F_{net} versus acceleration a for three different objects (having different inertial masses) in rectilinear motion. (*d*) Maximum static frictional force f_s versus normal force N for sliding involving three different pairs of surfaces. (*e*) Applied force F versus resulting extension x from relaxed condition for three different springs obeying Hooke's Law. (*f*) Potential difference ΔV versus current I for three different electrical conductors obeying Ohm's Law. (*g*) Circumference C versus diameter D for *all* circles. (*h*) Total mass of oxygen M_O versus total mass of carbon M_C in samples of carbon dioxide and carbon monoxide. (*i*) Rectilinear motion: position s versus clock reading t for three different objects all having position $s = 0$ at $t = 0$. (*j*) Rectilinear motion: position s versus clock reading t for three different objects all having different values of s at $t = 0$. (*k*) Rectilinear motion: instantaneous velocity v versus clock reading t for objects having zero velocity at $t = 0$. (*l*) Rectilinear motion: instantaneous velocity v versus clock reading t for objects having different velocities at $t = 0$.

or circumference in question. Calculations of the volume *added* to a sample, or the *increase* in diameter of a specified circle, are represented by the small dashed triangles in the respective figures. The graphical representation helps reinforce the insight that a given change along the horizontal axis produces a corresponding, fixed change along the vertical axis regardless of whether the shift is started at the origin or elsewhere along the line.

It helps to dramatize this idea by asking students to imagine a string around the equator of the earth, forming a circle with a circumference of 40,070 km. Now suppose we increase the length of the string by 6.0 m; what will be the *increase* in the diameter of the circle it forms? What would be the *increase* in diameter if we added 6.0 m to the circumference of a circle having an initial circumference of 8.0 cm?

5. In addition to providing further exercises with parallel arithmetical reasoning in entirely different context, Figs. 1.11.1i and 1.11.1k, on the one hand, juxtaposed against Figs. 1.11.1j and 1.11.1l, on the other, illustrate the difference between a direct proportion and a linear relation that is *not* a direct proportion. Very few students have formed this distinction explicitly, and many texts and teachers confuse the issue by careless use of the terminology.

Combining the modes of reasoning described in Section 1.10 with the parallel graphical representations described in this section, pointing out the connections explicitly, and requiring the students to describe them in their own words strongly serve to enhance and secure students' grasp of both reasoning with division and the interpretation of straight line graphs. One might even say that the superposition of the two perspectives is nonlinear.

1.12 SCALING AND FUNCTIONAL REASONING

Suppose we double the linear dimensions of, say, a statue: What will happen to the circumference of an arm? To the cross-sectional area of a leg? To the total surface area? To the volume of the required mold for casting? The great majority of students, including those in engineering-physics courses, have very serious difficulty with such questions, and the difficulty is compounded if the scale factor has a noninteger value. Many will guess, without thinking or analysis, that areas and volumes will increase by the given linear factor. They find themselves helpless in confronting the scale ratio alone without the actual initial dimensions of the object. They have no idea what to do in the absence of formulas for the relevant areas and volumes.

There are two principal difficulties behind this deficiency. The first has been discussed in Sections 1.2 and 1.3 above: the fact that the students have not been helped to form explicit operational definitions of "area" and "volume." The second difficulty resides in the fact that very few students have formed any conception of the basic *functional* relation between area and linear dimensions, on the one hand, or between volume and linear dimensions, on the other. Memorizing and using formulas for regular figures does *not* help form this conception. Hence students are unaware that all areas vary as the square of the length scale factor, and that volumes vary as the cube, regardless of regularity or irregularity of shape and regardless of existence or nonexistence of a formula.

Even if they are vaguely aware of the functional relations, they are unable to deal with them in terms of ratios, that is, they do not think in terms of what mature scientists and engineers call "scaling." Remediation must come by first

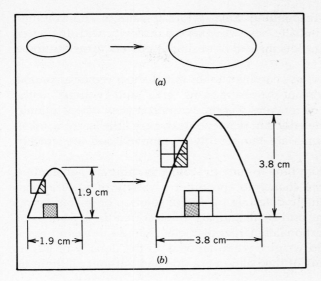

Figure 1.12.1 Two different plane figures scaled up by a factor of 2 in linear dimensions. In (*b*) it is shown that any one unit square in the smaller figure expands into *four* such squares in the larger figure and that this happens throughout the entire figure, including the periphery.

filling the gaps outlined in the preceding sections. Then students can be led to visualize what happens to unit squares as the dimensions of an arbitrary plane figure are doubled as illustrated, for example, in Fig. 1.12.1: any one unit square in the smaller figure expands into *four* such squares in the larger, whether in the interior of the figure or along the periphery. The reverse takes place when scaling down rather than up. Students should then sketch for themselves what happens when the scale factor is 3 or 4 rather than 2.

Those students, and there are many, who have difficulty extending the idea to noninteger scale factors should be led to sketch Fig. 1.12.2 in which dimensions are increased by a factor of 1.5, and one can readily confirm, by actually counting the squares, that the area increases by the factor $(1.5)^2/1$ since there are 2.25 unit squares in the larger figure.

Then one must extend the thinking to three dimensions and lead students to generalize the cubic functional relation for volume. Exercises can then be given in which areas and volumes are scaled up or down, as well as exercises in which the reasoning must be reversed, that is, given the ratio by which area has been scaled

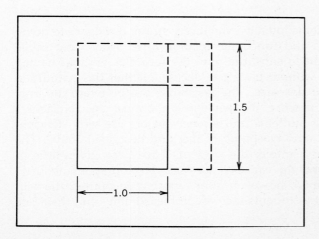

Figure 1.12.2 When the linear dimensions of a square are scaled up by a factor of 1.5, the new square contains 2.25 original squares.

up or down, what are the corresponding scale factors for length and volume? The great majority of students initially have very severe difficulty with the latter question; the necessity of taking roots instead of raising to powers turns out to be a formidable obstacle.

If these exercises, however, are confined to an initial short remedial period and are stated exclusively in terms of the abstractions "area" and "volume" without connection to visualization of concrete objects, without review of operational definition, and without being embedded in richer context, very little learning takes place; the calculational procedures are temporarily memorized and are quickly forgotten.

It is important to return from time to time to scaling in different substantive contexts, giving the students the chance to encounter a variety of applications: the role of surface-to-volume ratio in determining rate of solution or in comparing metabolic rates in cells or in large and small animals; the fact that the leg bones of elephants must have a disproportionately larger diameter than do those of horses in order to sustain the increased weight; what happens to the density of gas in a balloon if the linear dimensions of the balloon are doubled without addition or escape of gas?

Then, as more physics subject matter is developed, such thinking in terms of ratios should be extended to other and more abstract functional relations:

> If we have a bob on a string in horizontal circular motion, what happens to the centripetal force acting on the bob if the angular velocity is increased by a factor of 1.6, other quantities being held constant? What must be done to the tangential velocity in order to decrease the centripetal force by a factor of 2? What happens to the centripetal force if the mass of the bob and the radius of the circle are both tripled without change in angular velocity? If the tangential velocity is doubled, what must be done to the radius to keep the centripetal force unchanged?
>
> If the magnitude of the force acting on a certain lever arm is decreased by a factor of 2.3, what must happen to the length of the lever arm to keep the torque unchanged?
>
> If, in an interaction between point charges, one charge is increased by a factor of 3.5, what must be done to the separation between the charges to keep the force of interaction unchanged?

In all these examples, students initially exhibit *very* strong resistance to doing the thinking in terms of ratios and functional relationships. They want initial numerical values, and they want to substitute into the formulas without having to think through the ratios and without having to decide whether the quantity in question is going to increase or decrease. The resistance can be overcome only through repeated exposure and practice. It is well known to most college teachers that upper division engineering students and science majors are very deficient in ability to estimate and to do ratio reasoning of the kind described above. The reason for this deficiency is very simple: the students have been given little or no practice in such thinking, and the capacity does *not* develop spontaneously. When the breakthrough is attained, however, after repeated encounter, the self-confidence and self-respect of the students increase immeasurably, and their rate of progress is clearly enhanced.

Some teachers may remember the beautifully written Part I of the first two editions of the high school course *PSSC Physics*, with its fine overview of the science to be developed in more detail in the subsequent parts. Part I, which happened to be deeply infused with scaling and ratio reasoning, was deemed a "failure" and was removed in subsequent editions in the belief that the overview was premature and too sophisticated. In retrospect, I have come to believe that the problem with Part I was not so much in its subject matter as in the fact that neither the students nor the teachers were ready for the ratio reasoning, scaling, and estimating that permeated the sequence. The content was obscured by the impenetrability of the ratio reasoning.

If we do not help our students penetrate this obstacle, we shall never get them to the point of willingness to estimate or to make order of magnitude analyses and predictions, since such reasoning usually involves ratios, scaling, and functional relation. One hears frequent complaints that even physics majors and graduate students are gravely deficient in these skills. They are indeed deficient in this respect, and the reason is that they have had virtually no practice. (See Section 11.5 for references to papers giving problems and exercises on estimating.)

1.13 ELEMENTARY TRIGONOMETRY

Although in the more mathematically sophisticated sense sine, cosine, and tangent of an angle are to be regarded as functions, the students first encounter and use them as simple ratios of lengths of sides in right triangles. They laboriously memorize the standard definitions and use them as formulas to be rearranged by algebraic manipulation whenever a calculation on a right triangle is to be made.

The functional generalization is not necessary at this juncture and is not likely to be helpful. Students should first be led to see sines and cosines as simple *fractions*. If one multiplies the hypotenuse by the fractions, one obtains the lengths of the sides opposite and adjacent to the angle, respectively. This broadens the perspective by giving the students an alternative view of what the names "sine" and "cosine" stand for; it helps them think directly and concretely about the lengths—thinking that they are not doing when they mechanically and abstractly rearrange the standard formulas. The broadened perspective, however, rarely arises spontaneously; it must be deliberately induced by the teacher. This is clearly a matter of drill and practice that could readily be delivered via microcomputer.

Instructors should be explicitly aware of another basic aspect of trigonometry in which students are markedly deficient, even if they have had exposure in high school, namely that of radian measure. They have rarely, if ever, used radian measure in any significant context. They may have temporarily memorized a definition and used it in trivial conversion exercises, but they have not been shown *why* this dimensionless angular measure is useful, important, and even necessary. This deficiency is best remedied not by launching into a "review" at the beginning of a course but by showing the need for radian measure when an appropriate context is encountered. Hence the approach to radian measure will be discussed in more detail in Chapter 4 on two-dimensional motion.

1.14 HORIZONTAL, VERTICAL, NORTH, SOUTH, NOON, MIDNIGHT

Very few students can give intelligible operational definitions of the terms appearing in the title of this section. If one asks students, "What is meant by the term 'vertical'? How would you proceed to establish the vertical direction right here in this place?", a frequently occurring response is, "Perpendicular to the ground." If one then suggests going over to the steep slope nearby and erecting a perpendicular to the ground, the student recognizes an inconsistency but rarely sees any way out. It takes some minutes of hinting and questioning to draw out a proposal to hang a weight on a string and make a plumb bob. Relatively few students in this day and age have heard the term "plumb bob" or know what it means; nor do they know the meaning of the word "plumb" by itself.

Another acceptable, albeit more cumbersome, approach would be to establish the horizontal by means of a carpenter's level and then erect the perpendicular, but this suggestion very rarely emerges.

Similar discussions need to be conducted with respect to the other terms cited above. If asked how the local north direction is defined and established, most students refer to the magnetic compass as though this were a primary definition. They do not connect "north" with either the direction of the celestial pole or the shortest shadow cast by a vertical stick. If asked about the meaning of "local noon," most students are likely to refer to the sun being "directly overhead" without awareness that in the latitude at which most of them live the sun never passes through the zenith. When they are led to realize that the sun does not pass through the zenith, they can be led to the shadow of the vertical stick as a simple device for determining highest elevation and the like.

Again, such discussions are ineffective in an *a priori* review. They register most effectively if the student is challenged on the meaning of each term when it first arises in some specific context of problem or reading or discourse. The terms are so familiar and frequently invoked that the student has lost all sense of the fact that he or she does not really know what they mean. The necessity of groping for a simple operational definition of such familiar terms is, at first, embarrassing but provides a very salutary intellectual experience.

1.15 INTERPRETATION OF SIMPLE ALGEBRAIC STATEMENTS

Lochhead and Clement and their co-workers at the University of Massachusetts, Amherst, have studied the difficulties many individuals have with the translation of simple algebraic statements from words to an equation and from an equation to words [see Clement, Lochhead, and Monk (1981); Rosnick and Clement (1980)]. A typical exercise runs: "Write an equation using the variables S and P to represent the following statement: 'There are six times as many students as professors at this university.' Use S for the number of students and P for the number of professors."

Clement, Lochhead, and Monk report that "On a written test with 150 calculus-level students, 37% missed this problem and two-thirds of the errors took the form of a reversal of variables such as $6S = P$. In a sample of 47 nonscience majors taking college algebra, the error rate was 57%."

It is tempting to jump to the conjecture that these failure rates result from quick and careless misinterpretations of the wording of the problem. The investigators show, however, through detailed interviews and through altering the form of the problem, that the reversal is systematic and highly persistent. For example, the reversal is observed in problems that call for translation from pictures to equations or from data tables to equations.

Two principal patterns of incorrect reasoning emerged in the interviews: (1) Some students appeared to use a word order matching strategy by simply writing down the symbols $6S = P$ in the same order as the corresponding words in the text. (2) In the second approach, students were fully aware of the fact that there were more students than professors and even drew pictures showing six S's and one P. They still believed, however, that the relationship was to be represented by $6S = P$, apparently using the expression $6S$ to indicate the larger group and P to indicate the smaller. In other words, they did not understand S as a *variable* representing the *number* of students but rather treated it as a label or unit attached to the number 6 as in 6 feet or 6 meters; that is, they were reading the equation as they would read the statement 6 m = 600 cm, a statement which, incidentally, should be sedulously avoided for this as well as other reasons (cf. Section 3.23).

The very widespread occurrence of this difficulty is confirmed by Lochhead (1981) in his report of results of giving such tasks to university faculty members and high school teachers. Again this was not a matter of quick and careless misinterpretation. The task was administered in written form, and the subjects gave written explanations of their reasoning. The task in this instance was "Write one sentence in English that gives the same information as the following equation: $A = 7S$. A is the number of assemblers in a factory; S is the number of solderers in the factory."

Among university faculty members, 12% of a group in the physical sciences, 55% of a group in behavioral and social sciences, and 51% in a category "other" gave incorrect interpretations, reversing the meaning of the equation. Among the high school teachers, error rates in the same categories were 28%, 67%, and 47%, respectively. Although this was not a controlled or randomized experiment, the results testify eloquently to the persistence of the difficulty and to the fact that many individuals are not helped to overcome it in the course of their schooling.

This is a disability that should not be brushed off or treated casually, nor can one expect to remediate it by a short preliminary exercise. The most effective procedure is to give exercises in which the interpretations are traversed in both directions (words to symbols and symbols to words), and such questions should then be included on tests. The exercises should be given whenever the opportunity arises in subject matter being covered in the course, not as artificial episodes divorced from the course content.

1.16 LANGUAGE

Many aspects of the development and use of language play a deep underlying role in teaching and learning in all disciplines, not just in science. This is a huge subject attended by its own huge literature, and it is impossible to do it justice in this monograph. A few basic aspects, however, are so fundamental to our teaching that they will be mentioned here in the hope that some teachers may pursue them further in more sophisticated sources.

One aspect is that of operational definition of basic concepts. Few students, even at college level, have had direct experience, making them self-conscious about examining how words acquire meaning through shared experience. They tend to think that words are defined by synonyms found in a dictionary, and, when it comes to concepts such as velocity and acceleration or force and mass are completely unaware of the necessity of describing the actions and operations one executes, at least in principle, to give these terms scientific meaning. Since the words, to begin with, are metaphors, drawn from everyday speech, to which we give profoundly altered scientific meaning, only vaguely connected to the meaning in everyday speech, the students remain unaware of the alteration unless it is pointed to explicitly many times—not just once. Students must be made explicitly aware of the *process* of operational definition and must be made to tell the "stories" involved in generating numbers for velocity, acceleration, and so forth in their own words. This aspect is alluded to repeatedly in subsequent chapters.

The failure of many students to be aware when they do not fully comprehend the meaning of words and phrases in the context in which they occur underlies substantial portions of the "illiteracy" that we find currently decried in many disciplines, not science alone.

Still another linguistic aspect, crucial to understanding scientific reasoning and explanation as opposed to recall of isolated technical terms, resides in the use of words such as "then" and "because." A perceptive description of the difficulties exhibited by many students is given by Shahn (1988). In connection with "then," he remarks:

> *[In] descriptions of many biological phenomena . . . "understanding" means mastery of a sequence such as "A then B then C then D . . . ." If, for example, the letters represent stages of growth there is an obvious increase in complexity inherent in the process. Thus either omission or interchange of events signals a lack of understanding. Subsequent discussion with students [who gave incorrect answers on essay questions] showed that they really thought that the entire process was essentially equal to the sum of its parts, independent of order. It was as though in reading or hearing "then" the student was understanding "and." Now in a sense "then" does include "and," . . . but the sequential relationship is more restrictive, hence more precise, and it is this distinction that many students apparently fail to grasp.*

One might add that essentially the same problem frequently arises in connection with "if . . . then" statements of reasoning.

Shahn (1988) also goes on to illustrations with "because":

> *Six true/false questions were devised which were all of the form "A because B," and which were all unrelated to biology, for example, "Japanese cars are small because they use less gasoline." In each case the answer was false because either "A" and "B" were unrelated or the true statement should have been of the inverted form, "B because A." Too many students answer some of them incorrectly, indicating that there is indeed a problem.*
>
> *Generalizing from these two examples, it seems that students often misread conjunctions so that they mean "and." Often "and" is part of the meaning of "because" but not the entire meaning . . . .*

The problem here is not simply one of formal logic, and it is not eliminated by remedial exercises in formal logic. Although there indeed are similarities between formal logical operations on the one hand and scientific inference and explanation on the other, the processes are not identical. It is necessary to confront the problem directly in subject matter context and to allow the students to make errors and profit from the experience.

Many teachers find it difficult to believe that college students, in particular, have difficulties such as those Shahn describes. All I can do is emphatically confirm Shahn's report with my own experience, which even applies to a significant percentage of students in highly selected groups. To convince oneself, one must try such questions with one's own students. The results are almost invariably chastening.

There are, of course, many other linguistic problems relevant to our teaching, but it is impossible to give an exhaustive discussion here. The examples discussed above have been selected from among other possibilities because of their crucial relationship to the literacy we hope to convey in science teaching.

1.17 WHY BOTHER WITH THE UNDERPINNINGS?

It is easy, and even tempting, to brush off the problems of cognitive development posed in this chapter by adopting the view that students who have not broken through to mastery of such basic and simple reasoning modes do not deserve additional effort on the part of faculty and staff and do not belong in introductory physics courses. The problem should be taken care of in the schools and should not be allowed to deflect and dilute the process of higher education.

Enlightened self-interest, however, if not a sense of broader societal responsibility, dictates a less callous view: A large fraction of engineering-physics students have these difficulties. They would develop a far better grasp of physics, and would develop and mature far more rapidly as professionals, if they received appropriate guidance and help at the earliest stages. Among the students who fail or who simply disappear from our courses (or who never enroll in the first place because of deep fear and insecurity in the face of quantitative reasoning) are many potentially promising minority students as well as most of our future elementary school teachers, not to speak of many others in whom improved scientific literacy would lead to the capacity for wiser leadership, wiser executive decision making, or just wiser citizenship.

The problem should indeed be taken care of in the schools, but it has not been, and will not be taken care of in the near future, because the teachers, except for a very small minority, have not developed the necessary knowledge and skills. It must be strongly emphasized that this is *not* the fault of the *teachers*. The plight of the future teachers was blindly ignored when they were in college, and they were not helped to develop the abstract thinking and reasoning skills they need in their own classrooms. The vast majority of working teachers are individuals of dedication and good will, but they will not develop the necessary reasoning skills spontaneously. They need help, and this help must be forthcoming from the college–university level in both pre-service and in-service training.

Yet some university faculty, apparently without awareness of the damage being caused, pride themselves on attracting large student enrollments by offering science courses which avoid "math." Avoiding "math" almost invariably means avoiding any and all arithmetical reasoning with ratios and division, not just avoiding use of algebra or calculus. Future teachers, if they take any physical science at all, seek out courses of just this variety. Other courses simply let them sink (or get through by memorizing without understanding), and the inevitable result is the continuing graduation of teachers who are in need of remediation the instant they graduate.

If we wish to remove from the college domain the reasoning problems described in this chapter, we—college and university faculty—must, for the time being, accept the necessity of helping students (and in-service teachers) develop underpinnings such as those described. Until this obligation begins to be discharged, we shall simply continue putting the same degenerative signal into what amounts to a feedback loop and will not be relieved of the problem at the college level.

APPENDIX 1A

Sample Homework and Test Questions

1. Suppose we make a saline solution by dissolving 176 g of salt in 5.00 L of water. (The resulting total volume of the solution is very nearly 5.00 L.)
   (a) Calculate the concentration of the solution, explaining your reasoning briefly.
   (b) Using the result obtained in part (a), calculate how many cubic centimeters of solution must be taken in order to supply 10.0 g of salt. Explain your reasoning briefly.
   (c) Make up a problem which involves the density concept and in which the steps of reasoning are exactly parallel to the steps in (a) and (b) above. Be sure to select reasonable numerical values for the physical situation you describe. Present the solution of the problem, explaining the steps briefly.
2. We have a cylindrical container C as illustrated in the figure below. A second container D has the same shape as C, but the length scale, in all three dimensions, is larger by a factor of 1.80.

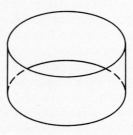

   Answer the following questions by using appropriate scaling *ratios* only. There should be no appeal to formulas for areas or volumes of special shapes. Evaluate final results in decimal form. Explain reasoning briefly in each instance.
   (a) How will the circumference of container D compare with that of container C, that is, what is the numerical value of the ratio C_D/C_C?
   (b) How many times larger is the cross-sectional area (i.e., the area of the base of D, denoted by S_D) than the cross-sectional area S_C?
   (c) If C contains 25.0 L of water when filled to the brim, how many liters of water will D hold when similarly filled?
3. A replica is made of the statue of a man on horseback. The total volume of the replica is 0.51 the volume of the original.

(a) How does the length of the man's arm in the replica compare with the length of the arm in the original?
(b) How does the total surface area of the replica compare with the total surface area of the original?

4. The earth has an equatorial radius of 3963 mi. (There are 5280 ft in one mile.)

   Imagine a string wrapped around the equator of a perfectly smooth earth. Suppose we now add 15.0 ft to the length of the string and shape the longer string into a smooth circle with its center still at the center of the earth.

   How far will the string now stand away from the surface of the earth? (Be sure to make the calculation in the simplest and most economical way; avoid doing irrelevant calculations and using irrelevant data. The sketch of an appropriate straight line graph can be more helpful than a stream of words in explaining your line of reasoning.)

5. Consider a bob on a string in uniform circular motion in a horizontal plane. Suppose that the tangential velocity v_t of the bob is increased by a factor of 2.35 while the radius of the circle is increased by a factor of 1.76. The mass of the bob remains unchanged at 145 g.

   How does the final centripetal force F_{cf} acting on the bob compare with the initial centripetal force F_{ci}?

   In showing your line of reasoning, use the language of *functional variation*: for example, in geometrical situations we argued that the area factor "varies as the square of the length factor"). It is *not* appropriate to substitute the given numbers directly into a formula since the numbers are *ratios* and are not themselves velocities or radii. Avoid using any data that might be irrelevant.

6. It is an empirical fact that the power output required of the engines of a boat or ship varies roughly as the cube of the velocity, that is, if you wish to double the velocity of the boat, you must increase the power output by a factor of eight.

   Consider a twin-screw boat with a mass of 2.0 metric tons (one metric ton is equivalent to 1000 kg or 2200 lb). The boat is moving at an initial velocity v_i. The captain increases the power output of the engines by a factor of 2.6.

   By what factor does he increase the kinetic energy of the boat, that is, how does the final kinetic energy K.E.$_f$ compare with the initial kinetic energy K.E.$_i$? (Explain reasoning briefly; use the language of functional variation, not formulas; avoid using irrelevant data; evaluate the final numerical answer in decimal form, do not leave an unevaluated expression.)

CHAPTER 2

Rectilinear Kinematics

2.1 INTRODUCTION

In *The Origins of Modern Science* the historian Herbert Butterfield remarks:

> *Of all the intellectual hurdles which the human mind has confronted and has overcome in the last fifteen hundred years the one which seems to me to have been the most amazing in character and the most stupendous in the scope of its consequences is the one relating to the problem of motion.*

The Greeks, with all their intellectual sophistication and mathematical skill, failed to invent the concepts of velocity and acceleration, failed to grasp the notion of an instantaneous quantity, and hence failed to penetrate to the law of inertia. Ideas of motion were continually belabored in the intervening years, but the breakthrough to formation and control of the concepts did not take place until the seventeenth century. This is a measure of the subtlety of the concepts and the justification for Butterfield's dramatic assertion; yet we expect our students to assimilate the whole sequence from two or three pages of cryptic text and a rapid lecture presentation. It should not be surprising that research indicates that very few students do master the basic kinematical ideas in the first years of introductory physics [Trowbridge and McDermott (1980, 1981)].

This chapter explores some of the reasons behind the existing failure and offers a few instructional strategies that might help students.

2.2 MISLEADING EQUATIONS AND TERMINOLOGY

A very common introduction to kinematics runs as follows: Suppose an object travels a distance d in an arbitrary time interval t. We define the average speed (or velocity) v by

$$v = \frac{d}{t} \qquad (2.2.1)$$

Subsequently acceleration a is introduced in a similar way as

$$a = \frac{v}{t} \tag{2.2.2}$$

and eventually equations such as

$$d = \frac{1}{2}at^2 \tag{2.2.3}$$

or

$$d = v_0 t + \frac{1}{2}at^2 \tag{2.2.4}$$

make their appearance. Equation 2.2.4 is then casually applied to cases of positive initial velocity and negative acceleration (e.g., throwing a ball vertically upward) in which the motion is not monotonic.

Such presentations are very misleading and essentially incorrect in certain very fundamental ways. In Eqs. 2.2.3 and 2.2.4, the symbol t does not denote an arbitrary time *interval* as it does in Eq. 2.2.1; it denotes clock readings (instants) measured from a zero setting. The symbol d in Eqs. 2.2.3 and 2.2.4 no longer denotes a distance traveled by the body; it denotes position numbers located as distances from some arbitrary origin, a point through which the body may never have passed. The students are not informed that the meaning of the symbols was changed in the derivations that followed Eq. 2.2.1, and many emerge with little understanding of either the physical concepts of velocity and acceleration or of the algebraic equations. They are hence forced to take refuge in memorizing calculational procedures that lead to "correct" numerical answers in end-of-chapter problems.

The presentation outlined above *must* be altered if students are to start kinematics with some hope of understanding the scheme. The shortcuts, omissions, and "simplifications," which are meant to reduce "complexity," are *not* in fact conducive to understanding; they are specious, and they make genuine understanding extremely difficult.

The concept of acceleration is inextricably connected to instantaneous velocity. It is impossible to deal clearly and correctly with instantaneous quantities without discriminating between instants (or "clock readings") and time intervals. It is impossible to deal with back-and-forth motion without discriminating between positions, changes in position, and distances traveled by the body (three *different* concepts to which the term "distance" is frequently indiscriminately applied.) These are indeed sophisticated ideas; that is why it took the human mind so long to penetrate them. It is unrealistic to expect students to make the penetration in the short time and through the shortcuts that are so frequently imposed.

2.3 EVENTS: POSITIONS AND CLOCK READINGS

The simplest and most realistic way to lead students into the kinematical concepts is to start with the concepts of "position" and "clock reading." (This, incidentally, paves the way from the very beginning for the notion of "event," which is so useful in introductory relativity.)

One can, for example, start with a rolling ball or moving cart on the laboratory table; make (or imagine) a "flash picture" that shows the object at uniform time intervals; place a scale behind the object; lead the students to see that the numbers on the scale do *not* represent distances traveled by the *object*; that, as distances, they are distances from an arbitrary origin at which the object may never have been located; that it takes two such numbers to give information about a change of position within a specified time interval; that we give such numbers the name "position numbers." (In my own classes, I usually have the students sketch hypothetical strobe pictures of their own as I lead them Socratically through the above sequence.)

Students must be led to see that a number on the position scale gives the location of a reference *point* on the moving object—the distance of the reference point from an arbitrary origin at which the object may never have been located—and that a position number itself does not refer to a length, that is, a position number has *zero* length.

We can now introduce a clock into our picture and associate each position of the moving object with a simultaneous clock reading. A clock reading is *not* a time interval. A clock reading is analogous to an object position (it literally is the position of the hand of the clock, if, for the precious moment, we abjure the digital era); it takes two clock readings to make a time interval; the zero clock reading may never have been involved in the sequence under consideration. A given object position and the corresponding clock reading are inextricably connected, and we call the combination an "event."

Now it becomes appropriate to couple the concept of "clock reading" with that of "instant." This must be done carefully and explicitly because the word "instant" is being taken out of everyday speech and given an unfamiliar meaning. To most students the word "instant" means a short time *interval:* for example, "I shall be there in an instant." They should be led to see that, just as positions have zero length, by definition, so clock readings or instants have zero duration, by definition.

If we use, say, the symbol s for position and (unfortunately, but conventionally) the symbol t for clock reading, we should not allow ourselves or the students to refer to s values as "distances" or to t values as "times" (to the student this invariably implies time *interval)*; it is wiser and more effective to use terms such as "position" and "clock reading" or "instant"; otherwise linguistic clarity is seriously compromised.

2.4 INSTANTANEOUS POSITION

If one carefully introduces the concepts of position and clock reading as outlined above, it is immediately possible to capitalize on this treatment by giving it deeper meaning and anticipating the more difficult notion of instantaneous velocity: If an object is moving continuously, how long does it stay at any one position number? This is not a trivial question, and most students have considerable difficulty with it. One must help them develop the following ideas:

The reference point we are using on our object is located at a particular position number *at* a corresponding clock reading (not *for* a clock reading; to the students the word "for" immediately implies a *finite* duration); how many seconds does the reference point spend at this position? (Many students will answer to the effect that the object spends a very short time interval at the given position.) How many seconds does a clock reading last? (Many students will again reveal their belief that the term represents a very short time *interval*, a very small number of seconds, despite having been through the development outlined in the preceding section.) We use the word "instant" as synonymous with "clock reading." How long does an instant last? And so forth. Students must be led to comprehend a clock reading, or instant, as lasting for *zero* seconds and the position as being occupied for zero seconds. It is important that they *say* these things themselves; for many students it is not enough to hear them said by someone else.

This sequence brings students their first exposure to the notion of an instantaneous quantity: instantaneous position. The notion is subtle and not easy to absorb, but it is considerably easier to absorb than "instantaneous velocity." Paving the way by introducing "instantaneous position" first makes the subsequent introduction of "instantaneous velocity" a recycling of the concepts "instant" and "instantaneous," and this significantly reduces some of the difficulties with "instantaneous velocity."

2.5 INTRODUCING THE CONCEPT OF "AVERAGE VELOCITY"

The most common way of introducing "average velocity" is by a statement to the effect that "average velocity over a given time interval is the change of position divided by the time interval over which the change occurred":

$$\bar{v} \equiv \frac{\Delta s}{\Delta t} \qquad (2.5.1)$$

There is nothing logically wrong with this, but starting the locution with the phrase "average velocity is . . ." leaves most students with the impression that the name "velocity" comes first as some kind of primitive they should "know," and the idea embodied in $\Delta s/\Delta t$ comes afterwards. Teaching is significantly strengthened, however, if one carefully abides by the precept "idea *first* and name *afterwards*," not just in this instance, but in the introduction of *every* new concept. The following approach is more effective than starting with the name:

Having first generated the position versus clock-reading description of the behavior of a moving object, an effective next step is to raise the question as to how one might now devise a calculation with s and t numbers, the result of which carries direct information concerning how fast the object was moving. This motivates examining the ratio $\Delta s/\Delta t$, without invoking a name, and interpreting its significance by using specific numerical examples of motion of the given body along a position scale: the number is large when the body moves rapidly; the number is small when the object moves slowly; the algebraic sign indicates the direction of motion, and so on. After the utility and meaning of the number are firmly established, it is convenient to give it a name, and the conventional name is "average velocity." Then one can stand back, explicitly indicate that the concept has been introduced in accordance with the precept "idea first and name afterwards," and explain why the precept is invoked.

A very effective contrast can then be provided by asking students to examine the ratio $\Delta t/\Delta s$. How does this number behave? Under what circumstances is it large? Under what circumstances is it small? What might be an appropriately descriptive name for this quantity? Allowing the students to invent a name impresses on them the fact that the initial idea is more significant than the name and that the idea comes first. (Geophysicists give this quantity the name "slowness"; it is useful in that science because the reciprocal of velocity arises automatically in connection with the use of Snell's Law in ray tracing of acoustic and seismic wave propagation.)

This approach immediately confronts students with the fact that scientific concepts are not objects "discovered" by an explorer but are abstractions deliberately created or invented by acts of human intelligence. (The same point is to be emphasized later in connection with the invention of the concept of "acceleration.")

This approach also allows a clear introduction to the notion of operational definition. Students should be led to articulate the entire "story" of the operations that go with the invention of "average velocity": creating the ideas of position and clock reading, observing two events with their corresponding values of s and t, calculating change of position Δs and the corresponding time interval Δt, dividing Δs by Δt, interpreting the physical significance of the result, and giving it a name. Very few students have ever encountered the idea of careful operational definition; to most of them "defining" a term means seeking out a synonym or memorizing a single pat phrase. They are initially very resistant to the idea of telling the entire story, describing every action that goes into the creation of a physical concept. Lecture presentation, however lucid, does not make the point. The concept of operational definition is registered only if students have the opportunity to write out such paragraphs of description in their own words and to have the writing evaluated for scientific precision and correctness of English usage.

The concept of velocity is usually introduced in connection with the simplest case, namely uniform motion. This is proper and desirable, but texts and teachers frequently overlook the fact that many students do not really know what the word "uniform" means in this context. It is a familiar English word, and students pass over it without thought as to the need of translation and interpretation. They should be asked what it means and should be led to descriptions such as "equal change in position in each succeeding second."

2.6 GRAPHS OF POSITION VERSUS CLOCK READING

Graphs of position versus clock reading are exploited to some degree in most texts. They offer a valuable alternative or supplement to verbal and algebraic treatments, offering students another way of manipulating the concepts being developed. Such graphs are most frequently (and very appropriately) used to provide a view of average velocity as the slope of a chord on the graph and to introduce instantaneous velocity as the slope of the tangent at a particular clock reading. They are also effectively used (along with velocity versus clock reading graphs) to assist the derivation of the kinematic equations for uniformly accelerated motion.

Unless they are explicitly led to do so, however, students do not consciously connect the graphs with actual, visualized motions; they treat them as uninterpreted abstractions. This is especially true of students who are still using concrete rather than formal patterns of reasoning (in the Piagetian sense of the terms). An effective way of reaching many students who have this difficulty is to give them problems in which they must translate from the graph to an actual motion and from an actual motion to its representation on a graph. [See McDermott, Rosenquist, and van Zee (1987).]

For example, students can be given a set of s versus t graphs and asked to execute the indicated motions along the edge of the table with their hand.

The direct, kinesthetic sensations attendant upon this exercise help reinforce the visualization and translation. Such exercises are best done qualitatively, sketching graphs and describing motions without use of numerical values. The reverse exercise is also important: motions of rolling balls or carts can be exhibited, and the student sketches the corresponding s versus t graphs. Another useful exercise is that of translating verbal descriptions into graphs. A powerful tool now available for such purposes is the microcomputer-based laboratory device of the sonic range finder, which exhibits graphs (directly on the computer screen) of motions executed by the student [Thornton (1987a) and (1987b).]

An especially important exercise with graphs is one in which students are asked to give verbal interpretations of various lengths in an s versus t diagram. For example, they should be able to identify a length parallel to the s-axis as representing a *change* in position. Similarly, they should be able to identify a length parallel to the t-axis as a time *interval*. Final contrast is provided by asking about the interpretation of a *diagonal* segment in an s versus t diagram. The majority of students do not initially have the courage of conviction to say that such a segment has no physical interpretation; they accord some spurious interpretation, most frequently a distance traversed by the body. Full understanding resides not only in knowing what something means, but in also knowing what it does *not* mean, and such exposure must be provided by the teacher (it is virtually never provided in the texts).

Another useful type of problem, rarely occurring in the texts, is that in which one examines the simultaneous behavior of two cars, say, moving at different uniform velocities and having different positions at some initial clock reading. In the light of given information, will one car pass the other? If so at what position and at what clock reading? Such problems should be solved *both* graphically and algebraically, not just in one mode; they provide a review of very basic ideas from ninth grade algebra and at the same time connect these ideas with a familiar physical situation. The great majority of students in introductory physics courses are very much in need of such review. Even many in calculus-based courses have severe difficulty setting up the simultaneous equations.

Still another question that initially offers great difficulty is the interpretation of a graph such as that in Fig. 2.6.1. Not having had such an opportunity before, few students have the courage to say that such a representation is meaningless; they need the opportunity to say that it is meaningless and to explain *why* it is

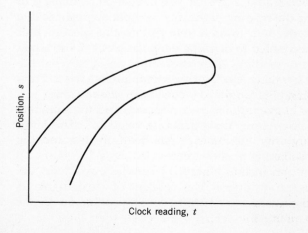

Figure 2.6.1 Meaningless s versus t graph.

meaningless. Such experience helps them acquire security for identifying nonsense or irrelevance on other occasions.

Sample problems of the type described in this section will be found in the appendix to this chapter.

2.7 INSTANTANEOUS VELOCITY

Acceleration cannot be carried beyond the level of being a protoconcept without engagement with the idea of instantaneous velocity. Many texts, particularly non-calculus ones, try to dodge this issue in the hope of making things "easier" for the students. The result is a specious treatment that cannot possibly lead to any genuine understanding of free fall or the law of inertia or the concept of force. Such treatments force students into memorizing calculational procedures and verbal routines that hold no meaning for them, and the result is an inevitable alienation from the subject.

I wish I knew some magic way of inculcating the concept of instantaneous velocity with no intellectual effort required from a passive student. That there is probably no such way is indicated by the long history of evolution of the motion concepts. Students must be given the chance to encounter these ideas *slowly* and with several episodes of cycling back to reencounter and reaffirm them as one proceeds through the study of kinematics and dynamics. Only a few students will absorb them on first encounter, but additional numbers break through in each subsequent episode. I wish to emphasize most strongly that I am *not* suggesting that one must stop and thrash around the concept of instantaneous velocity without moving on until every student has mastered it. This is both futile and impossible. Mastery develops slowly as the concept matures in the mind through use and application. The rate is very different with different learners. The cryptic presentations in the majority of texts, however, are almost useless as they stand. Some slowing up is essential, together with supplementation of the kind outlined in the preceding sections.

Starting with the uniform velocity case and the corresponding straight-line s versus t graphs, one can move to cases of speeding up and slowing down with corresponding curvature of the graphs, examine chords on the graphs and their connection to average velocities over arbitrary time intervals, and finally go to the tangents to the graphs *at* different clock readings. The slope of the tangent can be interpreted as that *uniform* velocity at which the object would continue moving if change ceased abruptly at that clock reading. The slope can also be connected in the minds of students with catching the reading of the moving speedometer needle in the car at the clock reading in question. (Merely referring to the speedometer needle is not enough. Students must be led to describe what the car is doing when the needle is stationary at a nonzero reading. Then they must describe what the car is doing when the needle is swinging clockwise or counterclockwise. Then they must be led to interpret the reading caught at a particular clock reading.)

I again strongly urge that the above inquiry be conducted, and the numbers examined *without* introduction of the name "instantaneous velocity." The latter term should be brought forth only after the concept has been created and the name becomes a response to the demands of convenience.

Locutions about velocity "for" an instant should be carefully avoided; "at" an instant is far more helpful and appropriate. The concept of "instantaneous position," developed earlier, can be invoked as a helpful comparison and a review of the notion of an instantaneous quantity.

Once the concept of instantaneous velocity is established, students should be led to precise articulation of an operational definition, describing all the actions and calculations that go into obtaining the number given this name. It should also be strongly emphasized that we have created a new concept, differing from "average velocity," even though the word "velocity" still appears. Students should be made explicitly aware of the process of redefinition that goes on continually in the creation and refinement of physical concepts. Such conscious awareness helps increase their security in the face of shifting meanings of technical terms. With "velocity," of course, another big shift occurs when we make the transition from rectilinear to two-dimensional motion.

If students are not led to give *verbal* interpretation of the velocity concepts, many of them continue to regard v as an abstraction to be manipulated in formulas and replaced by numbers rather than as something intuitively comprehensible. A first stage involves getting students (particularly those having trouble) to address a question such as "What does the term 'uniform velocity' mean? What information does it give us about the behavior of the moving object? Tell me in simple, nontechnical words of everyday experience." Some students will try to regurgitate the operational definition. Some will say something to the effect of "how fast it goes." Others will flounder around with various versions of the meaningless locutions about ratios discussed in Sections 1.8 and 1.9. Still others will talk about distance, or even position change, "over" time. (The latter locution is likely to be a trap for the teacher. The majority of students who use the word "over" are *not* thinking of the ratio, as one might like to believe. They are using it in the sense of "during" without conscious connection to the unit time interval.) One must persist until the student indicates that the number tells us how far the body goes in *one* second (or whatever time unit happens to be relevant). The "one" must be given firm emphasis; if it is hesitant or concealed, understanding is lacking.

Following this initial sequence, "average velocity" can be interpreted as that *uniform* velocity at which the object would have undergone the the same position change in the same time interval, and "instantaneous velocity" can be interpreted in the manner outlined earlier in this section. Each time it must be reemphasized that the number refers to what happens in *one* second.

Finally, the student must be led to see the distinction between the operational definition and the interpretation and must be helped to recognize that the interpretation does not constitute an adequate definition.

2.8 ALGEBRAIC SIGNS

If the course is one in which the full algebraic treatment of rectilinear motion is to be developed, it becomes important to lead students to see how the algebraic signs arise in the first place. However obvious it might be to us that the signs come from our uniting the number line with the position scale, this is not an insight that the students perceive or articulate spontaneously. They should be led to articulate in their own words that the algebraic sign that appears with the velocity is determined by Δs while Δt is intrinsically positive. They should then articulate the fact that the sign of Δs arises because of *our* introduction of the number line, and that *we* are therefore responsible for interpreting its meaning.

It is this personal responsibility for interpretation that most students do not discern. Without examination of the origin of the signs, they memorize the interpretation as an edict from text or teacher. This lack of insight subsequently almost completely blocks interpretation and comprehension of the algebraic signs of Δv, and the blind memorization continues.

2.9 ACCELERATION

There are still some authors who seem to think that life is made "easier" for the student by introducing acceleration as $a = v/t$, apparently failing to realize the confusion caused by using the same symbol v for either an instantaneous velocity or a *change* in instantaneous velocity. Fortunately, this treatment is now relatively rare, and most texts recognize the necessity of dealing with a change from one instantaneous velocity to another between corresponding clock readings. Thus, one now normally deals with some version of

$$\overline{a} \equiv \frac{\Delta v}{\Delta t} \qquad (2.9.1)$$

As with average and instantaneous velocity, I again urge adherence to the precept "idea first and name afterwards." Inquiry can first be directed toward devising a way of describing how fast velocity changes. The properties and behavior of $\Delta v/\Delta t$ can be examined and the name "acceleration" introduced after the meaning and usefulness of this ratio become apparent.

It takes many students, including ones in engineering-physics courses, a long time to begin to absorb some of the physical meaning. Understanding of the acceleration concept is *not* assured by the production of "correct answers" in the conventional end-of-chapter problems, and students having trouble with such problems are almost invariably unable to describe the meaning of "acceleration."

If asked to describe, in simple, everyday words, what "acceleration" means, many students respond "how fast it goes," with no very clear antecedent for the pronoun "it." If then asked to describe what "velocity" means, they give the same response. Some are surprised and a little troubled by the redundancy; others seem not to notice it. An effective approach is to go back to experience in an accelerating car and ask the student to invent a possible example with numerical values: select a velocity at a first clock reading; cite a possible velocity at a second clock reading. Do any of the numbers tabulated so far represent an acceleration? What must be done to obtain acceleration? Under what circumstances would the acceleration come out zero? How would you describe the meaning of the number in nontechnical, everyday language, that is, what does the number tell us about what is happening to the car? It usually takes substantial effort to lead students (especially those having trouble) to the point at which they say that the number tells us how much the velocity *changes* in *one* second.

One must be careful *not* to accept locutions such as "velocity per time" or "change of velocity over time." The majority of students using the word "over" are not thinking of the ratio but are using the word in the sense of "during," without explicit awareness of the connection to the unit interval. Some students interpret the statement "acceleration is the time rate of change of velocity" as "acceleration is the amount of time required to change the velocity." They fail to think about problems correctly until they can *say* things correctly.

Again, as in the case of "velocity," it is necessary to help students see the distinction between the operational definition and an interpretation.

Reversal of the preceding line of thought is also helpful, and even necessary, for many students: Suppose the acceleration of the car is 2.5 mi(hr)(s) and the velocity at this instant is 20 mi/hr. What will be the velocity at the end of the next second? At the end of the following second? And so on. Many students initially fail to make the simple translation of the numerical values without turning to a formula.

If the student has had some exposure to the phenomenon of free fall, it is useful to invoke the following: Have you worked with the number 10 m/s/s in connection with free fall? (Student answers: Yes.) What does this number mean, that is, what does it refer to or describe? (Student frequently answers: Gravity.) The word "gravity" refers to the whole phenomenon of attractive interaction between material objects. This number cannot possibly be "gravity"; what kind of *quantity* is it? Does it have any relation to kinematic concepts we have defined? (In this way, the student may finally be led to recognizing 10 m/s/s as an acceleration.) Now suppose we drop a ball from rest from a high position. What will be its velocity at the end of one second? At the end of the next second? The next? And so on. Suppose we throw a ball vertically upward and it leaves our hand with an instantaneous velocity of 30 m/s. What will be the velocity at the end of the first second? The next? The third? The fourth? The fifth?

Through sequences such as this, students make steps toward a grasp of meaning of the concept, steps not induced by the end-of-chapter problems. It should be clearly understood that I do not decry, or wish to eliminate, the problems. They are essential in the learning sequence, but they are *not* sufficient in themselves. They must be supplemented by the induction of phenomenological thinking of the variety being illustrated.

Again, if the course is one developing the full algebraic treatment of kinematics, it is essential to pause and help the students unravel the full meaning of the algebraic signs attending Δv. Unless this is done, few students ever come to understand the origin of the algebraic sign that goes with acceleration. They must be made to realize that the interpretation goes back to our introduction of the number line and is not an a priori dictum from above; that we must make the interpretation ourselves since we originated the scheme. This is best done by having them make up reasonable numbers for initial and final velocities of an object speeding up from an initial positive velocity, then slowing down from an initial positive velocity, then speeding up and slowing down from an initial negative velocity. The resulting Δv values should be listed to help reveal the pattern, and the algebraic signs should be explicitly interpreted.

I wish there were shortcuts for this exposure, but I do not know of any. The ideas are subtle and far from trivial. If ignored in the hope that penetration will occur spontaneously with passage of time, the chickens simply come home to roost later in dynamics. Most teachers are aware of the great difficulty students have with algebraic problems in dynamics: they ignore the signs; they avoid them; they treat them carelessly and incorrectly hoping to iron it all out in connection with the "right answer" at the back of the book. Seeds for this syndrome are usually planted when time is gained by avoiding confrontation with the algebraic signs of Δv. Settling the issue with respect to Δv does not remove all the subsequent difficulties with algebraic signs in dynamics, but it significantly helps to reduce them.

Developing the concept of acceleration provides another illustration of the fact that scientific concepts are created by acts of human imagination and intelligence—an illustration even more dramatic than that referred to in Section 2.5. Galileo's alter ego in the *Two New Sciences* puts forth two possible ways of describing change in velocity. We would recognize these as $\Delta v/\Delta s$ and $\Delta v/\Delta t$, respectively.

Galileo rejects the former on grounds that are not completely sound and adopts the latter, largely because he has the deeply rooted hunch that free fall, which is what he seeks to describe, is uniformly accelerated in the $\Delta v/\Delta t$ sense but not in the other. This episode vividly demonstrates the role of invention and shows that alternatives are sometimes possible. Furthermore, it demonstrates that the choice is sometimes dictated by criteria of elegance and simplicity, an idea that, at this stage of the game, is very startling to the students.

2.10 GRAPHS OF VELOCITY VERSUS CLOCK READING

The utility of s versus t graphs in providing opportunity to connect abstract concepts with concrete, kinesthetic experience has been discussed in Section 2.6. Much the same points can be made about v versus t graphs. Students should be led to translate such graphs into motion of their hand along the edge of the table and into verbal description. They should also translate verbal descriptions into graphs. The computer-connected sonic range finder is of great help in this context. [See Goldberg and Anderson (1989) for a description of learning diffculties observed among students who have been through conventional course treatments of kinematics.]

Just as in the case of s versus t graphs, students should be led to interpret the physical meaning of various segments on the v versus t graph: A segment parallel to the v-axis represents a *change* in velocity. A segment parallel to the t-axis represents a time interval. A diagonal segment has no physical interpretation. On this second go-around, quite a few students will have developed the courage of conviction to articulate the latter conclusion, and they derive considerable satisfaction and reinforcement from their ability to do so.

Some students, particularly disadvantaged students and many nonscience majors with scant experience in quantitative or graphical reasoning, have great difficulty interpreting v versus t graphs; they attempt to memorize rather than think through the problems provided. Many can be helped by alteration of the context: The ordinate can be changed to represent population growth rate; the rate of filling or emptying of a container; the rate of import of oil; and the like. The process of interpreting such graphs, especially when the rates are negative, seems to help students arrive at understanding more quickly than if confined to v versus t graphs alone. This illustrates the importance of looking at an abstraction in more than one way.

There is now the added dimension of going back and forth between position and velocity graphs. This is exploited to some degree in some texts, but rarely to the extent necessary to achieve grasp and understanding. Furthermore, these graphical operations are rarely tested for, and anything not tested for is disregarded by most students—especially those who need the exercises most. [A few sample problems are given in the appendix to this chapter. For investigations of student understanding of, and difficulties with, velocity graphs, see Brasell (1987) and Goldberg and Anderson (1989).]

2.11 AREAS

Difficulties that students have with area concepts have been discussed in Chapter 1. The study of kinematics provides a valuable opportunity to improve their understanding through application and use of the idea in a rich, substantive context.

Some texts provide a few limited exercises involving the evaluation of areas under graphs, but these are usually too limited by being restricted to rectangular and triangular cases in which students can use the simple geometrical formulas. Many students begin to appreciate the full force of the process and the meaning of the relations only when they have to evaluate the area of a figure for which no formula exists and for which they *must* count the squares. Again, problems of this kind are ignored in the homework unless they appear on the tests.

Dealing with areas from the earliest opportunity in kinematics opens a number of intellectual doors:

For students who have taken, or are taking, calculus, it provides experience with the interpretation of the concept of "integral" without the obscuring emphasis on an algorithm for evaluation of an integral. It is astonishing how many students come out of calculus courses with good grades and with complete blindness as to the interpretation of an integral in some related context such as kinematics. (This strongly suggests that mathematics instruction is as deficient in providing alternative ways of thinking about a concept and in providing pauses for interpretation and reflection as is much of physics instruction.)

For students who are not taking calculus, dealing with the areas becomes a way of dealing with, and comprehending, continuous change without the calculus formalism. If exploited at this juncture, it subsequently becomes a powerful tool in dealing with impulse-momentum and work-energy in an honest, rather than in a specious, way. It also paves the way, for example, for eventual understanding of what the household electric power meter is registering.

2.12 TOP OF THE FLIGHT

All teachers are familiar with the tremendous difficulty students have with situations in which instantaneous velocity is zero while acceleration is *not* zero: the ball at the top of its flight after being thrown vertically upward; the ball rolling up an inclined plane and back down; the pendulum at the end of its swing (although this is intrinsically a two-dimensional rather than a rectilinear problem). Students cannot bring themselves to believe that the acceleration is not zero when the velocity is zero. These situations require clear discrimination among ideas of acceleration, instantaneous velocity, and change in instantaneous velocity, and, at the time these situations are first encountered, the necessary concepts have not been firmly assimilated regardless of the lucidity of text and lecture presentations and regardless of the usual end-of-chapter exercises. There is also a fundamental linguistic obstacle that is inadvertently planted by texts, teachers, and the students themselves.

The latter obstacle arises from casual use of the word "stop," or the phrase "come to rest" in referring to the condition at the top of the flight. Describing the ball as "stopping" or as "coming to rest for an instant" is taken very literally by the students. To them the phrases mean "standing still for a while," and they literally think of the ball as coming to rest for a *finite* interval of time. Under these circumstances, the acceleration would certainly be zero.

A device that, in my experience, helps unsettle this misconception and redirects the student's thought is the following: Suppose you observe the ball, thrown vertically upward from the ground, from a platform or helicopter that rises at a *uniform* vertical velocity exactly equal to the initial velocity with which the ball leaves the hand of the thrower. Suppose you also release another ball (without throwing it) from the helicopter at the same instant the other ball is thrown upward. How will the two balls behave *relative to you*, as you are observing from the steadily rising helicopter?

When I first tried this sequence of questioning, I expected that many students, particularly slower thinkers, would have great difficulty with the change in frame of reference, and I was not very sanguine about its promise as an instructional device. To my surprise, I found that the majority of students respond correctly and perceptively when the questions are carefully and clearly phrased. They state that the two balls would appear to behave identically for each observer. They recognize that, from the point of view of the helicopter, both balls are falling (and accelerating) *all the time*. They recognize that nothing special is happening—no alteration of behavior—at the instant that the ground observer perceives them to be at the top of their flight. They begin to concede that the balls do *not* "stop" and that acceleration is taking place all the time, even at the instant the velocity is zero from the point of view of the observer on the ground.

All this reinforces the importance of talking about velocity *at* an instant rather than "for" an instant and continually emphasizing that any given velocity lasts for zero seconds. When the student begins to say, however tentatively and uncomfortably, that the velocity at the top of the flight is zero at an instant while acceleration at the same instant is not zero, he or she is approaching a major conceptual breakthrough—a step toward deeper grasp of the nature of instantaneous quantities and a step toward firmer distinction between velocity and acceleration.

The grasp can be strengthened by repeating the numerical exercise suggested in Section 2.9 with its rich connection to the algebraic signs: If we choose positive direction upward and the ball leaves our hand with a velocity of +30 m/s, what is the velocity at the end of the first second? [be sure to give the algebraic sign explicitly whenever you give a number] (Student: +20 m/s); at the end of the next second? (Student: +10 m/s); the next? (Student, tentatively: 0 m/s??). For how long does the ball have this velocity? (Student: 0 seconds??). What is the velocity at the end of the *next* second? (Student is likely to flounder, give a number without algebraic sign, and, if corrected, finally come forth uncertainly with −10 m/s???). What has been the acceleration *all* the time, throughout the *entire* history? (When the student finally comes forth with −10 m/s/s, a great many things begin to fall in place simultaneously.)

Finally, Mr. Brian Popp of our Physics Education Research Group has suggested a simple, compelling experiment: While driving a car up a gentle slope, put the car into neutral and coast. At the instant of zero velocity, abruptly put on the brake. The result is a heavy jolt, and the experiment should not be performed on too steep a slope. When the same experiment is performed in coasting to the instant of zero velocity on a level road, there is no jolt at all. This, of course, constitutes a preview of dynamics and the concept of force and can be exploited accordingly. One can also cycle back to this experiment when studying Newton's Second Law.

2.13 SOLVING KINEMATICS PROBLEMS

The usual numerical end-of-chapter problems on kinematics constitute valuable exercises for the students, and the concentration on less familiar aspects in this book implies no derogation of the problems. Quite a few texts now present the student with sensible, systematic schemes for approaching the solutions: draw a diagram of the physical situation; set up the position line, identifying positive and negative directions; translate the verbal statement into symbols so as to (1) tabulate the known quantities together with their symbols and (2) list the symbols of the unknown quantities; select the kinematic equation that gives the most efficient solution; make the necessary calculations; interpret the results.

When the text does not provide such help, the teacher should most certainly do so, together with posted or distributed solutions exemplifying the systematic approach.

What the teacher must be fully conscious of is the tremendous resistance many students bring to utilizing the systematic scheme despite its patent power and simplicity. In my experience, the great majority of students begin to take this process seriously *only* if its use is required on tests and only if substantial deductions are made when it is *not* used. The same resistance tends to manifest itself even more strongly later on in dynamics, and it can be reduced in marked degree if firm insistence on systematic procedure begins in kinematics.

There is another, less obvious and less frequently articulated, effect of firmly requiring use of the systematic problem-solving procedure. Most students at this early stage in their development refuse to put pencil to paper, or to analyze the verbal-to-symbol transitions that are essential, until they "see" the solution as a whole. Requiring that they institute the procedure propels them, willy nilly, into the problem, and the momentum thus acquired frequently carries them through to the solution. The increasing satisfaction gained from such experiences gradually makes them more willing to penetrate a new problem, with pencil and paper and inquiry, without waiting until the entire solution has been perceived. This is a very large step indeed in intellectual development and capacity for abstract logical reasoning.

2.14 USE OF COMPUTERS

Kinematics is, of course, a rich field for early experience with numerical calculation and the development of familiarity with elementary computer programming. The field is widely exploited accordingly, and published materials are available. [See, for example, Eisberg (1976).] Use of the computer in this context, however, has instructional feedback effects that are not always explicitly recognized. When a student has to program a numerical calculation, he or she is exposed in the most intimate possible way to the arithmetic in which an instantaneous acceleration, sustained for a short time interval, produces a small change in velocity; the new velocity, sustained for a short time interval, yields a new position; the new position gives a new acceleration, and so on. (The exercise is valuable even in the case of *uniform* acceleration.)

Very few students perceive or absorb this sequence of arithmetical connection among the kinematic concepts when they are exposed only to the closed algebraic equations for the case of uniform acceleration or for some of the special cases of varying acceleration. Programming (or even doing a few numerical calculations by hand) proves to be very revealing and helps register the full meaning of the concepts.

Although time is not available for every desirable activity in every course, anything that can be done to entice students into using their programmable hand calculators or home computers in this way pays dividends in improved understanding of the concepts of velocity and acceleration. There is more here than just enhancement of "computer literacy," although I have no intention of deprecating the latter.

2.15 RESEARCH ON FORMATION AND MASTERY OF THE CONCEPT OF VELOCITY

To most of us physics teachers the concept of "velocity" (or at least "speed") appears so simple and self-evident, so clearly connected with all our everyday experiences of motion, that it becomes hard to believe that students do not absorb its essentials from the usual text and lecture presentations. That thorough and effective intuitive grasp does not in fact develop so easily is clearly shown by the investigations conducted by Trowbridge and McDermott (1980).

In exploratory interviews, Trowbridge and McDermott found that students with no previous study of physics think of the word "speed" as a relation between distance traveled and the elapsed time but not necessarily as a ratio. Similarly, the word "acceleration" is used in a primitive sense of "speeding up" but not as a ratio. Trowbridge and McDermott describe the students at this stage as having "protoconcepts," rather than well formed concepts, connected with the standard technical terms. They then go on to show that the protoconcept stage persists to at least some degree in many students even after formal course development of the physical concepts.

Striking illustrations of what is transpiring in learners' minds are provided by student response to the following physical situation: The student being interviewed watches two balls rolling on parallel tracks (Fig. 2.15.1). Ball *A* travels with uniform motion from left to right while ball *B* travels in the same direction with an initial velocity greater than that of ball *A*. As ball *B* travels up a gentle incline, it slows down and eventually comes to rest. Ball *B* first passes ball *A*, but, a bit later, ball *A* passes ball *B*. The student observes the motions of the balls, first separately and then together, several times and has ample opportunity to absorb the whole picture visually. (The position vs. clock reading graph shown in Fig. 2.15.2 illustrates the motions just described, but this graph was not used in the interviews.)

During the course of the interview, students were asked, "Do these two balls ever have the same speed?" (The term "velocity" was used if the student had already been introduced to it.) Trowbridge and McDermott found that a substantial number of students (up to 30% in calculus-physics courses and larger percentages in less sophisticated courses) responded to this question by identifying the instants of *passing* rather than the instant near which the balls maintained an almost constant separation. The association of "same speed" with "passing" or "same position" was persistent and symptomatic and not idiosyncratic.

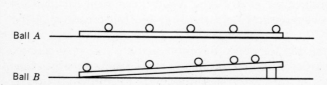

Figure 2.15.1 Speed comparison task: Motion of balls is from left to right. Ball A moves at uniform speed. Ball B starts off faster than A and slows down. There are two passing points. (See Fig. 2.15.2 for representative graphs.)

2.16 RESEARCH ON FORMATION AND MASTERY OF THE CONCEPT OF ACCELERATION

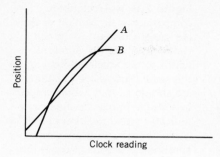

Figure 2.15.2 Position versus clock-reading graphs for motions described in Fig. 2.15.1. (These graphs were not used in interviews with students.)

When these students watched varying motions of two balls so arranged that they did *not* pass each other, they said the balls never had the same speed even though there was an instant at which the speeds were indeed the same. Many students view their own experience in cars passing each other in terms of having slower speed when one is behind, faster speed when ahead, and the same speed when "neck and neck" for a "while." (The reader interested in greater detail concerning the tasks and in direct quotations of student response should refer to the original paper.)

Trowbridge and McDermott summarize their investigation as follows:

In both pre- and postcourse interviews, failure on the speed comparison tasks was almost invariably due to to improper use of a position criterion to determine relative velocity. Although students who were unsuccessful could generally give an acceptable definition for velocity, they did not understand the concept well enough to be able to determine a procedure they could use in a real life situation for deciding if and when two objects have the same speed. Instead they fell back on the perceptually obvious phenomenon of passing. Some identified being ahead or being behind as being faster or slower. We refer to this use of position to determine relative velocity as the position–speed confusion. The use of the word "confusion" here should not be misconstrued to mean the mistaking of one fully developed concept for another. We are using the expression "confusion between speed (or velocity) and position" to refer to the indiscriminate use of nondifferentiated protoconcepts . . . .

Our research also has provided evidence that for some students certain preconceptions may be remarkably persistent . . . . Even on postcourse interviews, when difficulties occurred, they could be traced to the same confusion between speed and position that had been demonstrated during precourse interviews. The belief that a position criterion may be used to compare relative velocities seemed to remain intact in some students even after several weeks of instruction."

2.16 RESEARCH ON FORMATION AND MASTERY OF THE CONCEPT OF ACCELERATION

In addition to the investigation concerning the velocity concept, Trowbridge and McDermott (1981) also conducted a similar investigation with respect to acceleration.

In an exploratory sequence, students who had had some prior instruction in kinematics again viewed the motions described in Figs. 2.15.1 and 2.15.2 in the preceding section. When asked whether the two balls ever had the same accelera-

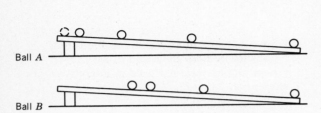

Figure 2.16.1 Acceleration comparison task: Motion of balls is from left to right. Balls roll in channels of slightly different width so the accelerations are not the same. Successive positions are shown as they would appear in a strobe photo. Dashed circle indicates initial position of ball A. Solid circles indicate corresponding positions at equal time intervals.

tion, some students said the accelerations were the same when the velocities were the same. When asked how they justified this conclusion, a typical response was "because your acceleration is that delta v over delta t. And at the point where you have the same velocity, you have the same delta t and the same delta v." These students were not discriminating between velocity and change of velocity. Further probing showed that the word "over" was being used in the sense of "during" and did not imply a ratio.

In a more sophisticated task, students viewed two balls rolling down inclined tracks with different accelerations. The motions they saw are described by Figs. 2.16.1 and 2.16.2. (The different accelerations are achieved by using as tracks two aluminum channels of slightly different width, making the accelerations different even though the slopes are the same.) Ball A is released first from a point several centimeters behind ball B. After rolling a few centimeters, ball A strikes the lever of a microswitch, which in turn releases ball B. As can be seen from the graph (which was not used in the interview), the balls have the same average velocity and the same final velocity. However, ball B, which rolls on the narrower channel, reaches that final velocity in a shorter time interval than ball A and has an acceleration about 15% greater. At the base of the incline, where they achieve the same final velocity, the balls roll side by side and then enter a tunnel. (The purpose of the tunnel is to deflect attention from any subsequent, irrelevant behavior.)

The balls were first rolled separately, and it was established that each one was accelerating. The students then viewed the two motions together so as to be able to compare them and were asked, "Do these two balls have the same or different accelerations?"

To encourage students to concentrate on the main conceptual issue rather than on subsidiary experimental details, specific guidance was provided. The interviewer explained that, to make the comparison, it is unnecessary to identify

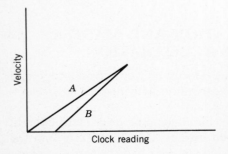

Figure 2.16.2 Velocity versus clock-reading graphs of motions shown in Fig. 2.16.1. Balls reach same velocity just as they enter a tunnel at the bottom of the incline. (Tunnel was used to deflect attention from events subsequent to balls reaching bottom of incline. These graphs were not used in the student interviews.)

2.16 RESEARCH ON FORMATION AND MASTERY OF THE CONCEPT OF ACCELERATION

the cause of the acceleration or to determine whether or not the balls, the channels, or the slopes are the same. The comparison of accelerations was to be made strictly on the basis of the motions observed. It was pointed out that ball B starts later than ball A. If students did not notice that the balls entered the tunnel at the same time and did not spontaneously compare final speeds, the interviewer asked questions that served to direct attention to these aspects. Thus students were assisted in concentrating on the observations necessary for comparing the accelerations.

Trowbridge and McDermott list a hierarchy of responses that emerged, running from the most naive to essentially correct as summarized in Table 2.16.1.

In precourse interviews, only 17% of students in a calculus-physics course were successful in this task, and other groups did even more poorly—down to zero percent success in a class of academically disadvantaged students.

In postcourse interviews, the success rate among the calculus-physics students rose to 38% while that among students in two noncalculus-physics courses averaged 25%. The academically disadvantaged group received specially careful instruction, not using this specific task, but addressed to encounter concrete phenomena and to improve capacity for ratio reasoning. The success rate in this group rose to 40%. (Greater detail, results with additional tasks, and information about scatter of the data will be found in the original paper.)

Table 2.16.1
Summary of Procedures Used by Students on Acceleration Comparison Task[a]

Procedure	Interpretation of Procedure
1. Balls have same acceleration because slopes of tracks are the same.	Nonkinematical approach
2. Balls have the same or different accelerations depending on their relative final positions.	Confusion between position and acceleration.
3. Balls have same acceleration because their final speeds are equal.	Confusion between velocity and acceleration.
4. Ball A has greater acceleration because it is overtaking ball B.	
5. Ball A has greater acceleration because it covers greater distance than ball B in the same time.	
6. Balls may have same acceleration because ball A covers greater distance than ball B in a longer time.	
7. Ball B has greater acceleration because its velocity changes by the same amount as the velocity of ball A but in a shorter distance.	Discrimination between velocity and change in velocity, but neglect of corresponding time interval.
8. Ball B has greater acceleration because its velocity catches up to that of A and thus changes by a greater amount.	
9. Ball B has greater acceleration because its velocity changes by a greater amount than velocity of ball A in the same time.	Qualitative understanding of acceleration as the ratio $\Delta v/\Delta t$.
10. Ball B has greater acceleration because its velocity changes by the same amount as the velocity of ball A in a shorter time.	

[a] From Trowbridge and McDermott (1981).

2.17 IMPLICATIONS OF THE RESEARCH RESULTS

These investigations dramatically illustrate the large gap that exists between the "protoconcepts" with which most students come to the study of kinematics and their grasp of the physical constructs put forth in text and lecture presentations. The investigations also show the high persistence of the gap in the face of conventional instruction.

Deficiencies in assimilation and understanding of the concepts remain concealed from us physics teachers partly because of our own wishful thinking regarding the lucidity of our presentations and partly because conventional homework problems and test questions do not reveal the true state of student thinking and comprehension. It is tempting to believe that adequate performance on conventional end-of-chapter problems indicates understanding, but, in fact, it does not.

Presentations can be refined and improved to some degree, and this is always worth doing, but it is illusory to expect that vividness and lucidity of exposition are sufficient in themselves. To help the learner assimilate abstract concepts, it is essential to engage the learner's mind in active use of the concepts in concrete situations. The concepts must be explicitly connected with immediate, visible, or kinesthetic experience. Furthermore, the learner should be led to confront and resolve the contradictions that result from his own misconceptions. [See Peters (1982) for additional examples with higher level students and for additional examples of useful questions.]

The gaps in understanding *cannot* be fully resolved for all students on the first passage through kinematics, even with better exercises and tests. Genuine learning of abstract ideas is a slow process and requires both time and repetition. Repetition without intervening time yields meager results. The most efficient approach is to move on through the subject matter but to keep returning and reinvoking the kinematical concepts in concrete, intuitive ways at every opportunity. As the ideas are reencountered in increasingly rich contexts, they are gradually assimilated—but at different rates by different individuals.

The necessary encounters must be generated through suggested observations, homework problems, and test questions that supplement the exercises prevalent in existing texts. The tasks used by Trowbridge and McDermott in their investigations are good examples; they have high instructional value. A few additional sample questions that provide such supplementation are illustrated in the appendix to this chapter. Teachers who explore and verify the learning problems described in this chapter will undoubtedly invent additional (and better) supplementary questions, as well as variations on the ones suggested. In doing so, they will be contributing to a pool that needs to be greatly expanded and made available in our journals and in textbooks. An instructional sequence designed to implement the insights gained in the researches described above is outlined by Rosenquist and McDermott (1987).

2.18 GALILEO AND THE BIRTH OF MODERN SCIENCE

The study of kinematics offers an excellent opportunity to bring out certain essential features and characteristics of scientific thought by examining the intellectual thrust of the *Discourses Concerning Two New Sciences*. What is important here is *not* priority of discovery or order of development; historical insight involves elements other than chronology. That Galileo had precursors in kinematics and theories of impetus is true but relatively insignificant in an introductory course. Fruitful insight at this juncture derives from looking at what Galileo himself emphasizes in his approach:

1. Galileo was explicitly conscious of the fact that he was *defining* new concepts and not "discovering" objects. He argues about the alternative definitions of acceleration discussed earlier in Section 2.9.

2. Galileo very consciously and explicitly restricted the *scope* of his inquiry in order to master and clarify one significant issue at a time. After some discussion (in the *Two New Sciences*) of the definition of acceleration and of instantaneous velocities of bodies in free fall, Sagredo, the impartial listener, suggests that,

> *From these considerations it appears to me that we may obtain a proper solution of the problem discussed by philosophers, namely, what* causes *the acceleration in the motion of heavy bodies?*

and Salviati (Galileo's alter ego) stops this line with,

> *The present does not seem to be the proper time to investigate the cause of acceleration of natural motion, concerning which various opinions have been expressed by various philosophers . . . . At present it is the purpose of our Author merely to investigate and to demonstrate some of the properties of accelerated motion, whatever the cause of this acceleration might be . . . .*

In other words Galileo firmly rejects an Aristotelian move to provide a complete explanation of all aspects of falling motion right from the beginning of the inquiry. Salviati's statement has a very modern stance: One of the most clearly notable characteristics of modern scientific investigation is the art of limiting the scope of inquiry in such a way as to ensure winning of one step of understanding at a time, avoiding the distraction and confusion introduced by premature or irrelevant questions. (But this procedure is, of course, not foolproof, and, in some cases, may serve to conceal important issues and inhibit solution of a problem. Deciding when and to what extent to restrict an inquiry is still the hallmark of individual genius.)

3. In "thinking away" the resistance of air to the motion of the falling body, Galileo explicitly introduces *idealization* into scientific thought. He recognizes that progress can be made in understanding nature without immediately dealing with natural phenomena in all their actual detail and complexity; that refinements can be developed subsequently through successive approximation. The bulk of our study of introductory physics is confined to such simplified and idealized situations, and students should be helped to remain explicitly aware of this strategy. One can hardly put the justification in more modern terms than did Galileo himself:

> *As to perturbations arising from the resistance of the medium, this is . . . considerable and does not, on account of its manifold forms, submit to fixed laws and exact description. Thus if we consider only the resistance which the air offers to motions studied by us, we shall see that it disturbs them all and disturbs them in an infinite variety of ways corresponding to the infinite variety in form, weight, and velocity of the projectiles . . . . Of these properties . . . infinite in number . . . it is not possible to give any exact description; hence in order to handle this matter in a scientific*

way, it is necessary to cut loose from these difficulties; and having discovered and demonstrated the theorems in the case of no resistance, to use them and apply them with such limitations as experience will teach.

4. Galileo's appeal to experimental evidence is frequently presented in a distorted and simplistic way by implying that the study of rolling down the inclined track was the "first experiment" and that observations and experiments were not made prior to this. Actually, there were many keen and skillful observers from classical times on down. The Greeks, for example, appealed to the resistance to compression of an inflated pig's bladder as direct evidence for the corporeality of air, and Aristotle's biological studies are still admired by modern biologists. The ancients, however, did not design experiments to test hypotheses. What was new in the *Two New Sciences* was the deliberate formation of a hypothesis (that $\Delta v/\Delta t$ is uniform in "naturally accelerated," that is, gravitationally accelerated, motion) and the design of an experiment to test the hypothesis.

5. Limited by a relatively crude method of measuring time intervals (weighing the amount of water that ran out of a large container), Galileo could make observations only over a few different inclinations of the track. To reach his most significant conclusions, he had to argue to the limiting (or asymptotic) cases. Since the acceleration proved to be uniform for all inclinations at which observations were possible, Galileo argues that one would expect this behavior to persist to the limit of an inclination of 90°, at which the object would be in free fall. He thus infers that free fall must also be uniformly accelerated.

He does not confine himself, however, to only the one limiting case; he also examines the other extreme, that of the level track or zero inclination:

. . . any velocity once imparted to a moving body will be rigidly maintained as long as the external causes of acceleration and retardation are removed, a condition which is found only on horizontal planes; for in the case of planes which slope downwards there is already present a cause of acceleration, while on planes sloping upward there is retardation; from this it follows that motion along a horizontal plane is perpetual; for if velocity be uniform, it cannot be diminished or slackened . . . .

Thus, by deep insight into one of the asymptotic cases, Galileo arrives at the first correct approach to the Law of Inertia: rather than ask what keeps a body moving, we should ask what causes it to stop.

Very few texts design situations in which students are led to think through limiting cases in order to draw insights or conclusions, or even simply to check the validity of results obtained in solving end-of-chapter problems. The situation just analyzed is one of the earliest in which students can confront such reasoning and sense its power; it is well worth exploiting for its intellectual content.

6. Neither Plato nor Aristotle believed mathematics relevant to description and understanding of the actual physical world. For Plato, uncertain physics was too far removed from the pure, abstract truth and reality of mathematical relationship: one can conceive of a line tangent to a circle, but the finest compass and straightedge will not construct a circle and a line with but one point in common. To Aristotle, the situation seemed inverted: to him reality lay in the forms, processes, and qualities of the physical world—aspects that could never be completely described in terms of the precise, abstract, unreal truths of mathematics. This dichotomy, deeply embedded in classical learning, was carried over into the

Renaissance with the classical revival. Galileo set out to overturn these views, and, in the process, he initiated the prodigiously fruitful line of mathematical physics that reached towering peaks in Newton, Laplace, Maxwell, Einstein, and Schrödinger and that plays its major role in our science today.

Galileo had previously argued that the Copernican system made Earth a heavenly body. Astronomy had always been a mathematical science. Since mathematics applied to the motion of the heavenly bodies, mathematics should apply to the earth. Westfall (1971) says, "If the immutable heavens alone offer a subject proper to mathematics, the earth had been promoted into that class . . . . To the mathematical science of bodies in equilibrium [Galileo] had added a mathematical science of bodies in motion."

In the *Two New Sciences* Galileo continually propagandizes the beauty and power of mathematics and illustrates its applicability to description of natural phenomena. After setting up what amount to the kinematics equations for uniformly accelerated motion, he asserts that he has discovered

> . . . *some properties of [naturally accelerated] motion which are worth knowing and which have not hitherto been either observed or demonstrated. Some superficial observations have been made, as, for instance, that the natural motion of a heavy falling body is continuously accelerated; but to just what extent this acceleration occurs has not yet been announced; for so far as I know, no one has yet pointed out that the distances traversed during equal intervals of time by a body falling from rest, stand to one another in the same ratio as the odd numbers beginning with unity.*

[To develop this result from the equation $\Delta s = (1/2)a(\Delta t)^2$ is a problem well worth assigning in homework. It makes the students think about the ratios instead of avoiding them by eternal substitution in formulas.]

To Galileo the occurrence of integer numbers in the description of a pervasive natural phenomenon had deep philosophical implications, showing that nature was in some sense "mathematical" and that mathematics could be successfully applied in natural philosophy. Such occurrences of integer numbers are fascinating to this day, whether it be in instances of resonance or standing waves, the Balmer formula, or quantum mechanics, as well as in the chemical Law of Multiple Proportions and in Mendel's evidence for discreteness somewhere in the genetic system.

In approaching formulation of the description of projectile motion, Galileo makes the first use of the principle of superposition:

> *In the preceding pages we have discussed the properties of uniform motion and of motion naturally accelerated . . . . I now propose to set forth those properties that belong to a body whose motion is* compounded *of two other motions, namely, one uniform and one accelerated . . . . This is the kind of motion seen in a moving projectile . . . .*

After setting up the description of projectile motion, he goes on to show that maximum range must be attained at an angle of elevation of 45° and then has Sagredo say:

> *The force of rigid demonstrations such as occur only in mathematics fills me with wonder and delight. From accounts given by gunners, I was already aware of the fact that, in the use of cannons and mortars, the maximum range . . . is obtained when the elevation is 45°; but to understand why this happens far outweighs the mere information obtained by testimony of others or even by repeated experiment.*

I hope that this section effectively illustrates the tremendous richness of the context. One can present the development of kinematics not only as a significant episode in intellectual history but also as an illustration of various facets of modern scientific thought and inquiry, and one can do this at an early stage with relatively simple subject matter. The development of such insights constitutes at least one part of the "general" or "liberal" education component of a science course, and, as such, it is at least as important for scientists and engineers as it is for nonscience majors. This is one component, albeit not the end-all, of scientific literacy. I contend that one of the most serious deficiencies of many introductory physics courses is the failure to incorporate an examination of such intellectual dimensions.

2.19 OBSERVATION AND INFERENCE

One aspect of abstract logical reasoning with which many students have great difficulty is that of discriminating between observation and inference. The principal reason is that they have been given virtually no practice in any of their schooling. Galileo's experiment with rolling balls on the inclined track offers an excellent opportunity for practice in a rich, nontrivial, context.

Given an account of the whole sequence (formation of the original hypothesis, design and execution of the experiment, interpretation of the experimental results), students should be asked to analyze the sequence and identify what was observed and what was inferred. Teachers who have not asked students for such performance will be astonished by the depth and extent of confusion and by the amount of guidance and help that must be provided. Furthermore, teachers should not expect the confusion to be remediated in this one exposure; the exercise must be repeated at every subsequent opportunity in other contexts.

In addition to discrimination between observation and inference, examination of the inclined track experiment affords one more valuable opportunity to deal with ratios. It should be analyzed, as Galileo analyzed it, to show that ratios of displacements from rest vary as the ratios of the squares of the corresponding time intervals and not simply by examining satisfaction of the algebraic formula. It might seem trivial to put so much emphasis on the ratios, but, unless we do this at every opportunity, we will not be helping the students overcome the grave difficulties and deficiencies described in Chapter 1.

APPENDIX 2A

Sample Homework and Test Questions

NOTE: Problems 1 and 2 lead the student to invoke kinesthetic experience in connection with forming the concepts of velocity and acceleration and in connection with interpretation of the conventional graphs describing rectilinear motion. The use of the acoustic range finder coupled to a microcomputer in the Microcomputer Based Laboratory (MBL) materials [Thornton (1987a) and (1987b)] greatly enhances the impact of such exercises by providing immediate visual display as well as immediate feedback, correction, and reinforcement.

1. Let the edge of the table be the straight line along which motion is to take place. Think of the $s = 0$ position as being near the center of the line with positive position numbers running toward the right and negative toward the left. Let your own hand be the moving object.

   Interpret each one of the position versus clock reading histories shown in the following diagrams by performing the indicated motion with your hand. Include all the details such as speeding up, slowing down, reversing direction, standing still, moving at uniform velocity, having your hand at the appropriate position at $t = 0$ and at the end of the history, and so on. Describe the motion in words as you execute it.

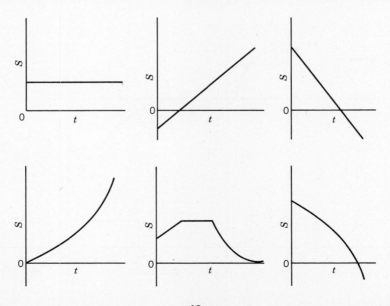

44 RECTILINEAR KINEMATICS

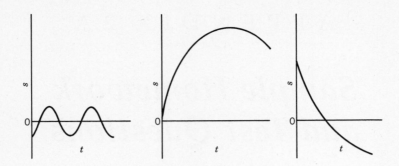

After having executed the motions with your hand, sketch the v versus t diagram corresponding to each of the s versus t diagrams. In your sketch, be sure to place the velocity diagram directly below the position diagram so that corresponding clock readings match up visually.

2. Let the edge of the table be the straight line along which motion is to take place. Think of the $s = 0$ position as being near the center of the line with positive position numbers running toward the right and negative toward the left. Let your own hand be the moving object.

Interpret each one of the following velocity versus clock reading histories by executing each motion with your hand. Does the diagram tell you where your hand should be at $t = 0$? Execute each motion more than once, each time placing your hand at a different initial position at $t = 0$. Describe the motion in words as you execute it.

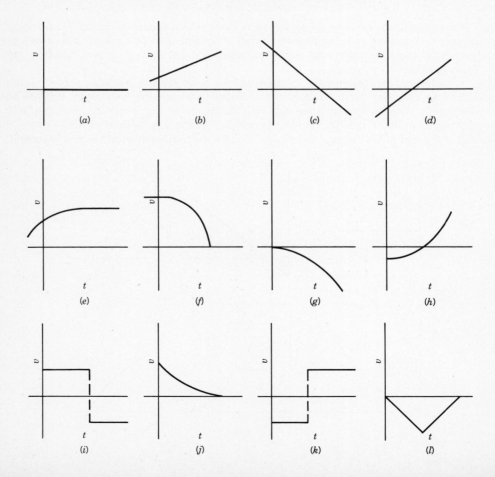

After having executed the motions and described them, sketch corresponding s versus t diagrams for each of the v versus t diagrams. Be sure to align the position diagrams directly above the velocity diagrams so that corresponding clock readings match up visually.

3. Cars A and B travel along the same straight road in the following manner: Car A is located at position $s = 2.4$ mi at clock reading $t = 0.00$ h and maintains a constant speed of 36.0 mi/h. Car B is located at $s = 0.0$ mi at clock reading $t = 0.50$ h and maintains a constant speed of 50.0 mi/h.

   At what clock reading will car B overtake and pass car A? At what position will the passing take place? How long a time after being at $s = 0.0$ will B overtake A? At the instant of being passed, how far will A have traveled from the position occupied at $t = 0.0$?

   Check yourself by solving this problem in *two* different ways: First solve it *graphically* by plotting the two s versus t histories on the same diagram and reading the required numbers off your graph. Then solve the problem *algebraically* by writing down *two* equations: one for the position s_A of car A as a function of clock reading t, and another for the position s_B of car B as a function of t. To do this, you must translate the verbal statement of the problem into symbols. You will now have two equations that you can solve simultaneously for the unknown quantities as in ninth grade algebra.

4. *Note to the instructor:* Most of the tasks used by Trowbridge and McDermott [(1980),(1981)] in their investigation of students' understanding of the concepts of velocity and acceleration can be adapted to instructional purposes, helping students master the concepts. The physical demonstrations can be set up in class or lecture and the questions asked, giving students opportunity to watch as many repetitions as they wish and to argue with each other about the answers. This is a very effective way of helping students confront the concepts intuitively, in concrete situations, and gain those insights that are not conveyed in the usual textbook problems. Trowbridge (1988) has prepared computer-based materials, under the title "Graphs and Tracks," that provide such exercises via the computer. He received a national award for these materials.

5. *Note to the instructor:* Peters (1982) describes an excellent demonstration, somewhat richer and more complex than those of Trowbridge and McDermott, and particularly suitable for engineering-physics courses. The apparatus is sketched in Fig. 2A-1. A glider slides down a slightly inclined air track, which has a bumper spring at its lower end. A standard horseshoe magnet is placed above the middle of the track so that the glider passes between the poles of the magnet without rubbing. The glider starts from rest at the upper end of the track, speeds up, moves at uniform speed (because of eddy current effects) between the poles of the magnet, speeds up along the lower portion of the track, bounces back from the bumper spring almost up to the magnet, then returns and bounces once more.

   Numerous repetitions of the motion were carried out in front of the class, and students were then asked to sketch, on a blank piece of paper, the s versus t and v versus t graphs. This exercise was given to an honors section of the calculus-physics course after rectilinear motion had been covered in class and s and v had been given precise meaning. Peters reports that only 30% of the students in the honors section represented the motion reasonably accurately on first experience with such a task. He also

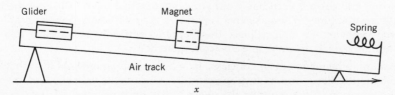

Figure 2A-1 Moving glider demonstration, [Peters (1982)]

describes and analyzes some of the more widely prevalent incorrect responses and types of confusion that were evident.

6. *Note to the instructor:* The following problem is an easy-to-grade, pencil-and-paper version of the Trowbridge–McDermott speed comparison task discussed in Section 2-15. In my own experience, statistics with respect to performance on this problem are surprisingly similar to those reported by Trowbridge and McDermott for performance on the concrete task. Encounter with this problem helps some students step beyond their protoconcepts and progress toward better discrimination between position and velocity.

The figure below shows position versus clock reading histories of rectilinear motion of two balls A and B rolling on parallel tracks.

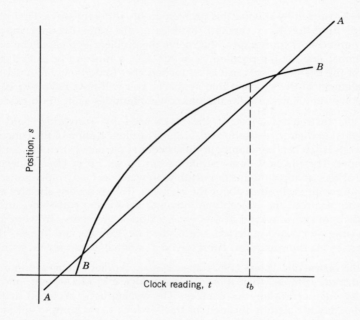

(a) Mark with the symbol t_a along the t-axis on the diagram any instant or instants at which one ball is passing the other.
(b) Which ball, A or B, is moving faster at clock reading t_b?
(c) Mark with the symbol t_c along the t-axis on the diagram any instant or instants at which the two balls have the same velocity.
(d) Over the period of time shown in the diagram, ball B is
 (1) speeding up all the time
 (2) slowing down all the time
 (3) speeding up part of the time and slowing down part of the time

(circle the correct statement)

7. Observation to be made outside of class: Take a ball (such as a tennis ball or any child's toy) and drop it vertically from your outstretched hand. Observe the bouncing carefully several times. Then sketch s versus t, v versus t, and a versus t graphs for the observed behavior. Be sure to place the diagrams vertically below each other so that corresponding clock readings line up appropriately.

8. Observation to be made outside of class: Take a sheet of paper, hold it so that the sheet is parallel to the floor, and let it drop vertically. Observe its behavior and the character of its acceleration. Now crumple the sheet into a tight ball and let it drop vertically. Compare the two cases in your own words, describing and interpreting the differences.

9. Observation to be made outside of class (or demonstration to be performed in class): Take a string about 3 m in length and attach weights (such as metal nuts or washers, or stones, or pieces of wood) at uniform intervals of 30 or 40 cm along the string. Standing on a chair, table, or ladder, as may be necessary, hold the string of weights vertically with the lowest weight at a distance above the floor equal to the spacing of weights along the string. Let the string fall and listen carefully to the clatter of the weights as they strike the floor. Describe the sound that you hear: does the clatter speed up, slow down, or remain uniform?

If the clatter does not remain uniform (i.e. uniform time intervals between weights striking the floor), how would you space the weights to make it uniform? Try the experiment.

10. *Note to the instructor:* The following type of exercise is helpful in leading students to perceive the difference between acceleration and velocity and to establish the connection between acceleration and change in velocity:

The lower diagram shows the acceleration versus clock reading history of a rectilinear motion. There are periods of uniform acceleration with very abrupt jumps from one acceleration value to another. This is quite possible physically: although the acceleration changes cannot actually take place instantaneously (i.e., in zero time interval), they can take place in time intervals very short compared to the scale employed on the graph. That is what is implied in this instance.

In the upper part of the diagram, plot a graph of the velocity versus clock reading history of this motion, assuming that the body starts from rest at $t = 0$. Describe what the graph would look like if the body had some nonzero initial velocity at $t = 0$.

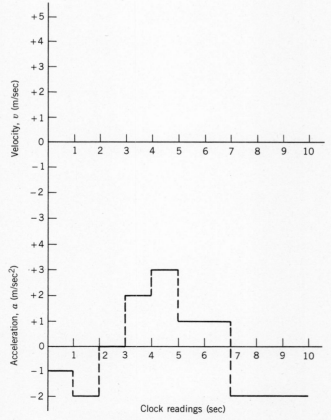

Acceleration (a) vs. clock reading (t)

11. *Note to the instructor:* Following is a type of problem that makes students confront a case of nonuniform acceleration and to recognize that the available kinematic equations are not applicable. Such encounter is important and illuminating, and yet it is very rarely generated in introductory courses.

    An object starts from rest at position $s = 0.0$ at clock reading $t = 0.0$. At clock reading $t = 5.0$ s it is observed to be at position $s = +40.0$ m and to have an instantaneous velocity $v = +11.0$ m/s.

    Examine the interconnections of the given data carefully. Was the acceleration of the object uniform or nonuniform? Explain your reasoning. Are the kinematic equations you have been using in class applicable to this case? Why or why not? Sketch the shape of the velocity versus clock reading graph that is implied by the data, that is, is the graph straight or curved? If it is curved, is it concave upward or downward?

12. *Note to the instructor:* The following question requires verbal interpretation of terms in an equation. Students almost never encounter such questions, yet the practice is an essential ingredient in learning and understanding. Similar question should be asked in connection with the equations derived subsequently for projectile motion.

    Consider the following familiar kinematical equation describing change of position with respect to clock reading in rectilinear, uniformly accelerated motion:

    $$\Delta s = v_0 t + \frac{1}{2} a t^2$$

    In your own words, give a physical interpretation of each of the two terms on the right-hand side of the equation.

CHAPTER 3

Elementary Dynamics

3.1 INTRODUCTION

In the study of physics, the Law of Inertia and the concept of force have, historically, been two of the most formidable stumbling blocks for students, and, as of the present time, more cognitive research has been done in this area than in any other. That the learning problem is formidable should not be surprising in view of how long it took the human mind to unravel these aspects of natural phenomena in the first place. Newcomers invariably have to relive at least some of the original hurdles and difficulties even though we shorten the time and smooth the way by providing guidance and instruction.

Most of our students come to us imbued with intuitive rules or notions that we are strongly tempted to call, pejoratively, "misconceptions." These intuitive notions are, however, neither perverse nor idiosyncratic; they are rooted in everyday experience, and they were initially held by all our predecessors. Our pedagogical orientation becomes sounder and more reasonable if we characterize these notions as understandable "*pre*conceptions," to be altered through concrete experience, rather than as ignorant "*mis*conceptions" to be removed instantaneously through verbal inculcation and a few demonstrations in which the student does not actively participate.

Researches (to be cited later in the body of this chapter) have repeatedly shown these preconceptions to be very deeply rooted and highly resistant to change. Furthermore, the views are not necessarily consistent and tend to shift from one physical situation to another, exhibiting contradictions that are not spontaneously perceived as such.

As with the kinematic concepts discussed in Chapter 2, one cannot expect the learner to acquire mastery of dynamics through verbal presentation alone, however lucid. Conventional end-of-chapter problems are also insufficient. This is not meant to disparage or advocate the elimination of such problems; they provide absolutely essential exercises in using the tools of the subject, and, without them, the student would never attain the capacity to apply and use the laws of motion. In existing texts, however, most end-of-chapter problems tend to concentrate on calculational procedures and on end results that rarely induce phenomenological, experiential thinking of the kind that research shows to be helpful in overcoming the conceptual barriers. It is shown repeatedly that ability to get correct, or partially correct, answers to the problems carries no assurance of genuine understanding of the basic concepts. End-of-chapter qualitative, phenomenological questions are also insufficient in themselves when not accompanied by concrete experience, Socratic guidance, and, eventually, testing.

Clear, vivid presentations, together with conventional quantitative problems, must be supplemented with questions and problems that engage the minds of learners in qualitative, phenomenological thinking. Learners must be confronted with direct experience, and with contradictions and inconsistencies, in such ways as to induce them to articulate lines of argument and reasoning in their own words and to lead them to abandon the deep-seated, plausible, intuitive preconceptions that impede development of the contra-intuitive but "correct" view. Most learners require several such encounters, distributed over time in increasingly rich context, and one must not expect to "rectify their disabilities" in one remedial session.

This chapter represents an effort to help the teacher become aware of some of the gaps that remain in many existing presentations and to give examples of supplemental treatments and exercises that seem to help the learner. Experience in using some of the hands-on approaches recommended is reported by Hake (1987) and by Tobias and Hake (1988) in a controlled experiment involving undergraduates as well as nonscience faculty colleagues at Indiana University.

3.2 LOGICAL STRUCTURE OF THE LAWS OF MOTION

The philosophical–epistemological basis of Newtonian Mechanics has been discussed at great length, over many years, in numerous treatises, and this is not an appropriate place to review this extensive literature. [An interested reader will find excellent summaries of modern views, relevant to physicists, in the papers by Eisenbud (1958) and Weinstock (1961) cited in the bibliography.] Before going on to description and analysis of student conceptual difficulties, however, it is appropriate to consider certain logical aspects of the Laws of Motion that are frequently ignored, or glossed over much too quickly, in many text presentations.

Many presentations start in by ignoring the fact that the words "force" and "mass," which, in everyday speech, are heavily loaded metaphors, are being taken out of everyday context and given very sophisticated technical meaning, completely unfamiliar to the learner. (It is even implied, in some presentations, that the student already knows the scientific meaning of the terms.) Students have, in general, not been made self-conscious about, or sensitive to, such semantic shifts, and they continue to endow the terms with the diffuse metaphorical meanings previously absorbed or encountered. It is helpful to make students explicitly conscious of the fact that the words remain the same but that the meanings are to be sharply revised.

This is a matter of operational definition, but many texts, unfortunately, either ignore operational definition entirely, proceeding as though the words have already been defined, or cryptically state a sequence that is essentially circular. The more elementary the text, the greater the tendency toward circularity and weakness of definition—apparently in the hope of making things "easier" for the learner. Given such presentations, there is no real hope of having students understand the concepts. How far one delves into operational definition of "force" and "mass"—with what degree of intensity, rigor, abstractness, and detail—is a matter of judgment for the teacher, but the matter should not go by default.

Widely different levels of sophistication are possible, and a teacher should make a choice reasonably matched to the students being addressed. Furthermore, the process of definition can be extended over time and need not be settled completely on the first encounter. One can start in some relatively unsophisticated way and help students refine the concepts by spiraling back to more rigorous definition as their grasp of the overall structure grows in later contexts.

There are two principal approaches to careful operational definition of "force" and "mass": one I shall call "Newtonian" for lack of a better term (Newton himself never actually propounded clear operational definitions of these terms); the other is associated with the name of Ernst Mach (1893).

In Mach's sequence, inertial mass is defined first. This is done by invoking the reaction car experiment, accepting as a law of nature the empirical observation that the ratio of the accelerations (and hence of the velocity changes) of the two bodies is a fixed property of the bodies, and defining the ratio of the masses as the inverse ratio of the accelerations. The net force acting on one body is then defined as the *ma* product for that body. [This is, of course, only a very cryptic summary of the more extensive line of argument. The reader interested in full detail will find an excellent presentation in Weinstock (1961)].

In scanning a number of widely used textbooks (I make no pretense of having carried out a full survey), I find that a significant minority use the Mach sequence. Since this sequence is basically sound and internally consistent, I shall not discuss the pedagogy in detail except to say that most of these presentations are so cryptic and so abstract that few students have any real chance of forming a sound operational grasp of the concepts from the text presentations. To induce such grasp, teachers would have to expand the development, give it far greater concreteness, and lead students to interpret, explain, and analyze in their own words.

Since the majority of the widely used texts adopt what I have called the "Newtonian" sequence (starting with force rather than inertial mass), and since I am myself partial to this approach because of its greater concreteness, I shall analyze this sequence in greater detail. I hasten to emphasize, however, that I do *not* put forth this sequence as the one-and-only correct presentation. There is no one "absolutely correct" or necessary road through this epistemological terrain. What counts eventually is the internal consistency of the network one elects to form. It is up to each teacher to select the variations he or she can help students articulate most clearly and compellingly, subject, of course, to the constraints of logical consistency and absence of circularity.

[A mathematically sophisticated version of the phenomenological sequence outlined in the following sections is to be found in Keller (1987).]

3.3 AN OPERATIONAL INTERPRETATION OF THE FIRST LAW

The Law of Inertia, or Newton's First Law as most of us call it, was not new with Newton. Galileo almost had it, and Descartes did have it, right. By the time of publication of the *Principia*, the First Law had become assimilated to the thinking of most active and productive natural philosophers even though, for some decades, the physics of motion continued to be taught out of scholastic texts. In the *Principia*, Newton does not arrogate the law to himself. He acknowledges the precedence of others and puts it forth as a declaration of independence from Aristotelian and impetus schools of thought.

Newcomers to dynamics, burdened with common sense ideas and rules about the behavior of moving bodies, have very great difficulty following this

52 ELEMENTARY DYNAMICS

breakthrough, and the learning problems this entails will be discussed in later sections. Here I wish to consider only one facet of the First Law: how to interpret it operationally in the sequence of definition of concepts.

Among the list of definitions at the beginning of the *Principia*, we find the following Definition IV:

> An impressed force is an action exerted upon a body in order to change its state, either of rest, or of uniform motion in a right line.

Then, as Law I of three Laws of Motion, we find:

> Every body continues in its state of rest, or of uniform motion in a right line, unless it is compelled to change that state by forces impressed upon it.

The circularity here is quite apparent, but it, in fact, does suggest how we might help a student interpret Law I in our modern sequence: Up to this point, we have generated only operational definitions of the concepts of kinematics, and "force" and "mass" remain undefined. Once we begin to accept the view that rest or uniform rectilinear motion are natural states of objects and that interactions with other objects are necessary to produce *changes* in such motion, we can interpret Law I as giving us a *qualitative* operational definition of "force," namely that action, by an agent external to the moving body, that imparts a *change* in velocity, and "change" includes both magnitude and direction.

This becomes a first step toward an operational definition of "force." The next steps come from construction of Law II.

3.4 OPERATIONAL DEFINITION OF A NUMERICAL SCALE OF FORCE

As indicated in the preceding section, intrinsically associated with enunciation of the Law of Inertia we discern a *qualitative* conception of force: any action, impressed externally, changing the velocity of a body. The next step is to refine the concept by making it quantitative.[1] At this point, more than one approach is possible. Newton, in fact, elected to associate "motive force," as he called it, with impulsive changes in momentum [for more detail on this aspect see Arons and Bork (1964)]. Our modern conception of force is different from Newton's and it is best to carry out the discussion in modern terms.

We start by visualizing operations we could perform with frictionless pucks on a level glass table top or on an air table (the *PSSC Physics* films on "Inertia" and "Inertial Mass," with Edward Purcell as narrator, in fact carry out something very close to the gedanken experiments to be described). Selecting a particular puck A, which becomes the standard body in our experiments, we impart rectilinear

---

[1] Here, incidentally, is an opportunity to make students explicitly aware of the fact that definitions of new concepts are rarely, if ever, generated completely, in full rigor, on the first encounter. One usually starts with an initial, tentative, even crude, definition and extends and refines it as insight deepens with use and application. This is precisely what happens with the concept of "velocity," where we start with a notion of average speed in rectilinear motion, refine the concept by infusing algebraic directions along the number line, refine it further into the concept of "instantaneous velocity," and finally generalize the vector properties in two and three dimensions. In each step of redefinition, the concept is altered significantly; it becomes, to all effect, a new concept even though the original name is retained. Our modes of instruction tend to lead students to concentrate on the name while losing track of the ideas behind it. It is an intellectually significant experience for the student to stand back and become explicitly conscious of the processes of definition and re-definition at such junctures.

accelerations by pulling it with a light spring, the extensions of which can be observed and marked on an initially unmarked card (Fig. 3.4.1).

Intuition tells us, correctly, that different strengths of pull impart different accelerations to body A. With a particular action or pull we shall associate the numerical value of the acceleration imparted and construct what amounts to a "force meter." Thus we imagine conducting the following experiments: (1) make a multiple-exposure photograph of accelerating puck A by flashing a light at successive uniform intervals of time; (2) from the sequence of increasing displacements in the photograph, we can determine whether the acceleration is uniform and whether the extension of the spring is constant.

All measurements of this type, whether made directly in the manner shown in Fig. 3.4.1 or accomplished in some indirect fashion, indicate that a constant spring extension is associated with a constant acceleration. Furthermore, we can satisfy ourselves that the effect is reproducible: the same spring extension imparts the same acceleration on different occasions and in different directions (right or left, north or south).[2] Having established confidence in the uniqueness and reproducibility of each experiment, we complete the scale of our force meter by labeling each needle position with the numerical value of acceleration imparted to puck A.

Thus, the numbers 1.00, 2.00, 3.00, and so on would be placed at needle positions under which accelerations of 1.00, 2.00, 3.00 m/sec^2, etc., were measured on the photographs. Noninteger values would be established in a similar way: the number 1.50 would *not* be entered half way between 1.00 and 2.00 but at the needle position that imparted an acceleration of 1.50 m/sec^2; similarly for force readings such as 2.36 or 3.82. In other words, the force scale is calibrated without any assumptions whatsoever concerning uniformity or nonuniformity in the stretching of the spring, that is, the spring is *not* assumed to obey Hooke's Law.

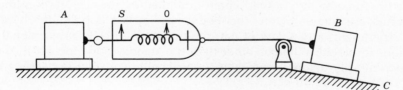

Figure 3.4.1 Frictionless puck B on incline C imparts uniform rectilinear acceleration to puck A. Acceleration can be changed by increasing or decreasing slope of incline. Needle attached to end of spring is at position O when acceleration is zero and spring is relaxed. Spring is extended, and needle is at position such as S when acceleration is imparted. (Note that *no* assumption is being made concerning spring linearity or the obeying of Hooke's Law.)

---

[2]Depending on the level of sophistication that is appropriate, one can take this opportunity to make additional, finer points: The spring must not be stretched so far that the needle fails to return to its initial, zero position at zero acceleration, but this behavior can always be checked between experiments. The care that must be exercised in calibrating the force meter is the same as that which must be exercised with clocks and meter sticks in measuring time intervals and lengths; precise measurements are to be made under conditions of controlled temperature and freedom from shock, bending, and other extraneous effects. In practice, knowledge of what effects are extraneous and how they must be controlled is rarely discerned a priori but is achieved through trial and error and successive approximations.

If puck A is constructed to match the international standard object called "one kilogram," we give the units marked on our force meter the name "Newtons." We now have a tentative definition of force on a numerical scale. The force numbers, which we shall denote by the symbol F, have *arbitrarily* been made identical with the numerical values of acceleration imparted to the standard body, puck A. Whether this arbitrary definition of a force scale is fruitful and useful can be determined only by appeal to nature through further experiments.

3.5 APPLICATION OF THE FORCE METER TO OTHER OBJECTS: INERTIAL MASS

If we now replace puck A by a different frictionless puck, denoted by D, we can impart accelerations to D using any reading we wish on the force meter. In such experiments we find that a fixed scale reading, such as 3.00 N, on the force meter imparts a constant and reproducible acceleration to D, but this acceleration is not, in general, 3.00 m/sec^2 as it is with puck A. Suppose the acceleration in this instance (force reading 3.00) turns out to be 1.50 m/sec^2. Note that it is *not* possible to tell what will be observed with still other force readings; one must proceed with the experiments. With other force readings, do we obtain results systematically and simply related to the one so far observed?

Table 3.5.1 illustrates results that would actually be obtained (column 3) and contrasts them with results that can be imagined but are not actually obtained (columns 4 and 5). Note the pedagogical importance of showing the student what is *not* the case as well as what is. Without such explicit contrast, the significance of the idea being presented is frequently unappreciated or incompletely understood. The best way for the student to grasp the idea contained in Table 3.5.1 is to sketch the F versus a graphs for the data in columns 3, 4, and 5.

Examining column 3 and the graph in Fig. 3.5.1, we see that it is possible to associate with puck D a *single* number, namely 2.00, which will, in each observation, give the force meter reading when multiplied by the acceleration imparted. Similar results are obtained with other bodies, as illustrated in Fig. 3.5.1. Thus we find, by experiment, a new law of nature: forces are directly proportional to the accelerations imparted to bodies other than the standard one for which the force scale was arbitrarily defined, and the proportionality constant is clearly a unique

Table 3.5.1
Accelerations a Imparted to Body D by Force Readings F Exerted by the Force Meter Defined in Section 3.4

(1) Applied force F (defined by acceleration imparted to A; units not named)	(2) Acceleration imparted to A, m/sec^2	(3) Observed acceleration imparted to D, m/sec^2	(4) Imagined possibilities of acceleration of D (not realized experimentally), m/sec^2	(5)
0.50	0.50	0.25	1.00	3.00
1.00	1.00	0.50	1.10	2.50
1.62	1.62	0.81	1.20	2.20
2.00	2.00	1.00	1.40	2.00
3.00	3.00	1.50	1.50	1.50
4.00	4.00	2.00	1.60	1.00

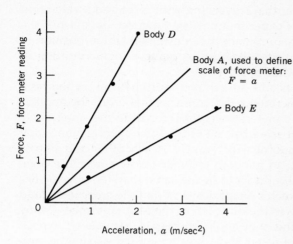

Figure 3.5.1 F versus a graph for bodies A, D, and E. Body D has larger inertia than A (given force meter reading imparts smaller acceleration). Body E exhibits smaller inertia than body A.

value, a *property* of each new body. (Note how this treatment can be directly connected with the straight-line ideas discussed in Section 1.11.)

Summarizing the argument: Once we have arbitrarily defined a force scale as in Section 3.4, it is found to be an experimental physical fact that F is proportional to a when different forces are applied to another body; that is, nature tells us that there exists a *single* number—a *property of the given body*—which is the proportionality constant. If we denote this proportionality constant by m, we write

$$F = ma \qquad (3.5.1)$$

where m, the property of the body being accelerated, is the slope of the corresponding straight line in Fig. 3.5.1. We give this property the name "inertial mass" or simply "mass," for short. The existence of this single, unique number for a given body is *not* just a matter of definition, as was the scale of force, nor is it deduced from theoretical principles; it is an experimental, physical *fact*—a law of nature—even though it was originally arrived at by conjecture rather than by direct experimental test.

Having arrived at this point, one can now lead students into discussion of the meaning of large and small values of m, comparing behavior of the bodies under action of the same force, and interpreting the significance of the fact that two entirely different bodies (different in size, shape, color, density, texture, and chemical composition) might have identical values of m, including the special value $m = 1.00$ kg.

Many students, teachers, and texts fall into the habit of using the term "mass" to denote an object, for example, speaking of "suspending a 10-kg mass." This linguistic carelessness is the source of certain kinds of confusion, especially later on, for example, when one wishes to distinguish between gravitational and inertial mass. It is best to avoid using the terms "object" and "mass" synonymously and to distinguish carefully between an object and its properties.

3.6 SUPERPOSITION OF MASSES AND FORCES

The preceding sections have shown how noncircular operational definitions of force and inertial mass can be constructed via the Second Law using what I have termed the Newtonian sequence, that is, starting with force and acceleration rather than with Mach's reaction car experiment. This, however, is still not the entire content of the Second Law. There remain the questions of superposition of forces and masses, and again one must appeal to experiment for verification of conjectures, however plausible the latter might be.

Experiment confirms that masses add (or subtract) arithmetically when bodies are combined (or separated.) Experiment also confirms that (with the application of two identically calibrated force meters) two equal forces in the same direction impart twice the acceleration imparted by one of the forces acting alone; that equal forces in opposite directions subtract (or "cancel")[3] each other and impart zero acceleration to any object; that, in general, colinear forces superpose algebraically; that forces at angles to each other add in the same manner as velocities and accelerations, thus behaving as vector quantities; and that the *acceleration* (and *not* the velocity of the body) is always in the direction of the resultant force. (Many students confuse the latter issue, and they must be helped to make it explicit through questions on homework and tests. Such questions must usually be supplied by the teacher; they are rarely given in texts.)

Finally, it is an additional empirical fact that orthogonal components of velocity, acceleration, and force are independent of each other in the realm of validity of Newtonian mechanics, while this is not the case when relativistic effects become significant.

3.7 TEXTBOOK PRESENTATIONS OF THE SECOND LAW

It is unfortunate that many textbooks, in their efforts to be "simple" or "easy" or concise, avoid careful operational definition and completely omit discussion of what aspects of the Second Law involve arbitrary definition and what aspects reflect a specific kind of order in nature. Such presentations leave the students with formulas:

$$\mathbf{F}_{net} = m\mathbf{a} \qquad (3.7.1)$$

or

$$F_{xnet} = ma_x \quad ; \quad F_{ynet} = ma_y \qquad (3.7.2)$$

but with virtually no understanding of the content and meaning of the Second Law.

It is important for students to realize that the algebraic statement is not self-contained and that it must be supplemented by a fairly extended text, giving a story of arbitrary definition and appeal to experiment along lines comparable to those illustrated in the preceding sections. Without the story, the formulas are sterile and unintelligible.

Ignoring these logical and conceptual aspects of the laws of motion, in order to make things seem "easier" or to achieve more extensive coverage, shows little

---

[3] One must be careful with the term "cancel" in this context. Many students tend to misunderstand and misuse it. Some have the notion that, when forces "cancel" each other, they cease to exist. Others confuse such "cancellation" with cancellation by *division* in algebra.

more than contempt for the minds of the students. Most students *can* understand these ideas if they are given time, opportunity, concrete experience, and suitable spiraling back from later context. Very few students can absorb or understand these ideas when subjected to the pace and brevity prevalent in most of our texts and courses, whether it be at high school level or at college calculus- or non-calculus-physics levels.

In most texts adopting the Mach sequence, the presentation is made so cryptically and so abstractly as to be quite meaningless to the majority of students, even though the conceptual development is sound and not circular. The questions of superposition are rarely made explicit or given any acknowledgment whatsoever. The verbal text, the qualifications and interpretations that accompany the Second Law, are entirely omitted. The more "elementary" the textbook, the more cryptic and less intelligible is likely to be the presentation.

The majority of widely used texts seem to adopt what purports to be the Newtonian sequence, but most of these start with "force" as though it were a primitive, already fully understood both qualitatively and numerically, and not requiring explicit operational definition. They then go on to "mass" as simply the proportionality constant between force and acceleration. The superposition questions are, for the most part, ignored.

Scanning some currently available texts for a few specific examples (with no pretense of complete coverage), I note that *PSSC Physics* (all editions) gives a simple, correct, and consistent presentation suitable for introductory levels. The treatment is (appropriately) less sophisticated than that outlined in the preceding sections, but it is quite reasonable for many introductory college contexts as well as for the high school level being addressed. Among college level calculus-physics texts, both Tipler (1982) and Resnick and Halliday (1977, 1985) give sound, albeit rather cryptic, presentations. The story outlined in Sections 3.3–3.7 is given in somewhat greater detail in my own (out of print) text [Arons (1965)]. I have yet to see a college level noncalculus-physics text that gives what I would regard as a sound, noncircular operational presentation of the Newtonian sequence.

3.8 WEIGHT AND MASS

In the development outlined above, it is to be noted that the term "weight" has never arisen at all, and this should be pointed out, repeatedly, to the students and extracted, in discussion, in their own words. In principle, all the procedures and experiments involved in the operational sequence could be performed in a space ship, away from gravitating bodies, or in a satellite in free fall. Making this explicit helps the students get started on forming the distinction between "weight" and "mass" and fixing the realization that the term "weight of an object" is the name given to a particular force: the *gravitational* force exerted *by* the earth *on* the object, imparting an acceleration of 9.80 m/sec^2.

In the initial stages, while students are still forming the distinction between the concepts, it is wise to maintain a rigid distinction of the units, speaking of mass in kilograms and weight only in newtons. Eventually, however, it is impossible to shield students from the looser usage that will be encountered in some technical literature as well as in everyday speech: they will certainly hear locutions such as "a weight of ten kilograms" or "a 3.00-kg weight. " It would be convenient to issue an edict forbidding such usage and wave a magic wand to have this edict enforced, but this will never be achieved (in spite of the most earnest efforts of some purists), and it is better to help the students interpret the inevitable usage as a shorthand reference to the force with which the earth attracts the given body: the phrase "3.00-kg weight" refers to an object on which the earth exerts a gravitational force of 3.00 × 9.80 = 29.4 newtons.

Parallel statements would be made, of course, in connection with the British Engineering (BE) system of units. Fortunately, while the country is still inching toward metrification, the majority of textbooks are leading the way by confining themselves to presentation of the SI system, leaving both the BE and cgs systems in abeyance (or placing them in such a way that the instructor can choose to leave them in abeyance.) This is the best way to handle the problem pedagogically, not only because SI is preempting the field, but also because throwing all the different systems of units at the students while they are still trying to unravel the concepts is gratuitous. If they need one of the other systems eventually, they can acquire it at a point where understanding of the basic concepts reduces the matter of units to triviality, and they can close the gap for themselves; it is only before understanding has been acquired that units form a major conceptual obstacle.

3.9 GRAVITATIONAL VERSUS INERTIAL MASS

We are confronted here with two operationally distinct concepts, yet students have very great difficulty forming the distinction. The difficulty arises partly from the fact that the operational definitions of force and inertial mass are rarely developed with sufficient clarity at the very beginning and partly from the purely linguistic confusion arising from use of the same name for two entirely different ideas.

It is true that one can argue the numerical equality of gravitational and inertial mass from the fact that all objects have the same acceleration in free fall (and this is essentially what Newton does), but this does not provide the student with an adequate *operational* distinction. Furthermore, the student is usually still struggling with the distinction between weight and mass, and invoking free fall at the beginning of the argument simply compounds the confusion. In my own experience, students can be helped to form the distinction by appeal to the following two clearly different gedanken experiments:

Experiment 1: Given the "force meter" operationally developed in Section 3.4 and Fig. 3.4.1, apply it to two different spherically shaped bodies, A and B, and determine their inertial masses through measurement of the accelerations imparted. Suppose we have selected A and B so that their inertial masses turn out to have a ratio of exactly two to one, i. e., $m_A/m_B = 2.00$.

Experiment 2: Now we take bodies A and B and bring them (one at a time) near one of the spheres (body C) at the end of a Cavendish balance. Body C is accelerated by the gravitational attraction, and the Cavendish balance begins to swing.[4] From the acceleration imparted to body

---

[4]The *PSSC* film "Forces" actually shows the execution of a similar experiment: The Cavendish balance consists of a meter stick suspended horizontally at its center from a high ceiling by means of a recording tape, which acts as the torsion suspension. Bottles of water hang at the ends of the meter stick. When the balance is stationary (a condition that was achieved only by taking refuge in an isolated, unused building), a box of sand is moved up close to one of the bottles. A spot of light reflected from a small mirror attached to the recording tape provides the optical lever, and the deflection of the spot of light is monitored in the film. This short (10 min) segment dramatically demonstrates the gravitational interaction between ordinary objects and is well worth showing in class if it is available. The only caveat is that the film is purely qualitative, and the objects, not being spherical, do not interact as point masses. With respect to our gedanken experiment, we should eventually be able to argue that our objects interacted as point masses, but this is a refinement that can come later.

C on the Cavendish balance, we determine the forces exerted on C by bodies A and B (separately) at a fixed distance between centers. We find *by experiment* that the force exerted by B on C is 2.00 times the force exerted by A on C.

Now we can emphasize the dramatic operational difference between the two experiments. Experiment 1, through the accelerations imparted to the two bodies by the same force, compares the property to which we have given the name "inertial mass." Experiment 2 has no a priori connection with experiment 1 at all; we are comparing an entirely different property and effect, namely the noncontact forces exerted by A and B, respectively, on a *third* body C. It is truly astonishing that the numerical ratio is exactly the same in both experiments and that this particular order in nature is confirmed experimentally in all circumstances, with all bodies, and, by sophisticated indirect measurements, to a fantastically high degree of precision.

How astonishing this is can be dramatized by pointing to the fact that an entirely different interaction between the spheres (say an electrostatic interaction if they are electrified by rubbing, or a magnetic interaction if they are ferromagnetic and are magnetized) exhibits a ratio of forces exerted on C that bears no relation whatsoever to the ratio of inertial masses of A and B. It is only in the *gravitational* interaction that the ratios are identical.

We now give the property defined operationally by the interaction observed in experiment 2 the name "gravitational mass." Using the same noun "mass" for the two entirely different properties constitutes a very unfortunate choice of terminology. It is responsible for much of the conceptual difficulty encountered by the students, but we are stuck with it and cannot change it by fiat. The best procedure is to keep using the adjectives together with the noun and to keep reemphasizing the operational distinction, giving students opportunity to describe it in their own words.

Students gain a clearer picture of the linguistic problem when they see that, regardless of the convention actually adopted, the language might have been quite different. Coulomb, in his great paper on the electrostatic interaction (before crystallization of the term "electrical charge"), refers to the "electrical masses" of his charged spheres. Inverting the analogy, we might just as well have talked about "gravitational charge." Some other term, neither "mass" nor "charge," would, of course, have been preferable, but we have no choice except to try to make the situation as clear as we can.

Once students have begun to acquire understanding of the preceding operational sequence and appreciate the complete independence of the two experiments, they can come back to the observation that all objects have the same acceleration in free fall and begin to discern the intimate connection among the various observations. Looking at the same idea in more than one way is a powerful aid to understanding the whole scheme, including the distinction between weight and mass.

3.10 UNDERSTANDING THE LAW OF INERTIA

Because of the obvious conceptual importance of the subject matter, the preconceptions students bring with them when starting the study of dynamics, and the difficulties they encounter with the Law of Inertia and the concept of force, have attracted extensive investigation and generated a substantial literature. A sampling of useful papers, giving far more extensive detail than can be incorporated here, is cited in the bibliography [Champagne, Klopfer, and Anderson (1980); Clement

(1982); di Sessa (1982); Gunstone, Champagne, and Klopfer (1981); Halloun and Hestenes (1985); McCloskey, Camarazza, and Green (1980); McCloskey (1983); McDermott (1984); Minstrell (1982); Viennot (1979); White (1983), (1984)].

Learners' difficulties in encompassing the Law of Inertia and the concept of force stem in large measure from the wealth of commonsense preconceptions and experiential "rules" that most of us assimilate to our view of the behavior of massive bodies before we are introduced to Newtonian physics. Some of these views are Aristotelian (e. g., the necessity of continued application of a push to keep a body moving—it is very difficult to abandon thinking of rest as a condition fundamentally different from that of motion, or to accept the view that, rather than asking what keeps a body moving, we should ask what causes it to stop), but many of these commonsense views are more closely related to the medieval notions of impetus associated with names such as Buridan and Oresme.

All investigations show these "naive" conceptions to be very deeply entrenched and very tenaciously held, and it is important for teachers to understand that student difficulties are not reflections of "stupidity" or recalcitrance. The difficulties are rooted in seemingly logical consequences of perceived order and experience and are vigorously reinforced by insistent use (or actually misuse) of words drawn from everyday speech (inertia, mass, force, momentum, energy, power, resistance) before these words have been given precise operational meaning in physics. Persistent misuse of the terms in thinking to oneself and in communicating with others is a major obstacle to breaking away from the naive preconceptions. (This is another reason for helping the students stand back and become very self-conscious about the process of operational definition—term by term.) Some teachers tend to minimize such problems by labeling them as "merely" a matter of language or semantics, apparently not realizing how formidable and significant the linguistic obstacles tend to be.

Investigations of understanding of the Law of Inertia further show that it is far from sufficient to inculcate the law verbally and supplement it with a few demonstrations of the behavior of frictionless pucks on a table or gliders on an air track. Many students will memorize and repeat the First Law quite correctly in words, but, when confronted with the necessity of making predictions and describing what happens in actual physical situations, concretely accessible to them, they revert repeatedly to the naive preconceptions and predictions, giving the disappointed teacher the sinking sensation of not having succeeded in teaching anything at all.

If one wishes to lead the majority, rather than a small minority, of students to understanding the Law of Inertia, one must accept the necessity of providing a wide array of experiences, both hands-on and hypothetical, in which students make their own errors, encounter the resulting contradictions, and, forced by these errors and contradictions, revise their preconceptions. Such experience cannot be provided and mastery developed, however, in one short remedial session. The ideas and initial experiences should be introduced, while development of the subject matter is continued without waiting for full mastery on first encounter. One then helps cultivate mastery and understanding through repeated spiraling back to qualitative application of the Law of Inertia in increasingly rich and sophisticated physical situations as the study of the science continues.

The most effective, albeit fairly expensive, physical situation I have been able to use to such purpose is one in which a full-size 50-lb block of dry ice, with its base smoothed to some degree, is placed on a large glass plate leveled up on a laboratory table. Students are then invited to perform literally "hands-on" experiments (using gloves, of course). A large array of very basic, vitally important, ideas can be developed Socratically in this context:[5]

1. How does the block behave once it is moving? What is the difference between this situation and the one in which ordinary objects slide on ordinary surfaces? (The way in which the block moves in ghostly splendor along the plate, especially at low velocity, without appreciable slowing down, makes a deep impression on most individuals who have never seen such effects.)

2. What action on our part is necessary to make the object move faster and faster, that is, accelerate continuously? (To many students it comes as a great surprise that they have to move faster and faster themselves to keep up with the block and to keep on exerting the accelerating force. Even though they previously saw the block move at uniform velocity in the absence of an external force, many of them have not translated this into the sensations that go with the exertion of a constant force on an accelerating object.)

3. What is the difference in behavior of the block when acted on by a steady push that keeps up with the block and when it is given a quick shove? (Many students have not had the opportunity to discriminate between a steady force and an impulse. In fact, to many students, the word "force" in the context of setting an object in motion means a quick shove rather than a steady action, and it is important to help them perceive the difference.)

4. How large a force is necessary to impart any acceleration at all to the block, that is, is there a threshold effect? (Everyday experience indicates that bodies are not set into motion until a certain minimum force is exerted; this is one of the eminently reasonable, naive rules that students bring with them initially.) The *PSSC* film called "A Million to One," in which a flea is hitched up and accelerates a massive dry ice puck, is well worth showing if it is available.

5. Suppose the block of dry ice is already moving: what must be done to make it slow down very slowly without changing the direction of its motion? Many students are inclined to apply an impulse rather than a gentle, continuous force. They must be guided into doing the latter, and they are usually astonished to find that they must allow their hand to retreat with the moving block. This experience helps reinforce the discrimination between impulse and steady force.

6. Suppose the block is moving to begin with, and we exert a steady force, either speeding the block up or slowing it down. How does the block behave? Now suppose we make our steady force smaller and smaller. How does the block behave? How will it behave when the force we are exerting reaches zero? (Note that what is deliberately constructed here is a *reversal* of the usual direction of presentation of the ideas: instead of using the zero force situation as the starting point, we are now starting with the nonzero net force and going toward the zero force condition. Many beginning students, at all levels, have very great trouble with the zero force case, despite all the

---

[5]Even though some situation other than the block of dry ice is invoked, the sequence of questions that follows is one through which most students should be led. The difficulties being intercepted are very widely prevalent among students in virtually all introductory physics courses.

preceding discussion and demonstration. Reversing the line of reasoning and experience, and seeing the situation both ways, helps in the acquisition of the desired insight.)

7. Suppose we exert two steady forces on the block in opposite directions, one with each hand. How does the block behave when the one force is larger than the other? When the forces are of equal magnitude?

8. Suppose the block is moving: what actions change the *direction* of its motion? (Here, once discrimination between the two has been developed, it is possible to explore the effects of both continuous actions and impulses.) What do you have to do to make the block move at right angles to its initial path? In some other specified direction? In an (approximate) circle? (The principal non-Newtonian expectation found among learners is that an initially moving object will move in the direction of the last impulsive push. It is important that they encounter the contra-intuitive phenomenon personally.)

9. What happens if you start the block spinning about a vertical axis? Without using any as yet undefined technical terminology, what are some implications of the observed behavior?

Some words of caution and advice about implementation of this experience: (1) Its essentially personal, hands-on nature tends to reinforce an idea, deeply embedded in many students, that accelerating effects (forces) are necessarily exerted only by animate beings. One should emphasize that contact interactions between inanimate objects (e. g., collisions, release of compressed springs, etc.) also impart acceleration. Noncontact interactions (electric or magnetic) can be introduced or referred to at the teacher's discretion. (2) Although a very small number of students may successfully explore the physical situation without Socratic guidance and emerge, on their own, with most of the insights listed above, the great majority do not carry out a genuine investigation or draw significant inferences under such circumstances. It is essential that the teacher provide guidance, but this is best done by asking questions and eliciting suggestions from the students rather than by giving a set of instructions to be followed. (3) The whole operation is at its best when, under minimum guidance from the teacher, the students suggest, try, argue, and interpret in their own words, carefully avoiding any, so far undefined, technical vocabulary.

There are, of course, other devices for providing some of the experiences outlined above. A massive dry ice puck does very well on the glass plate, although it is not as dramatic as the 50-lb block. Bricks (or other objects) can be piled on a slab of dry ice instead of using an entire block of the latter. A glass plate is not essential; any very smooth surface will do. A good bit can be done with pucks on an air table, although their rather small mass makes it difficult to perform some of the more delicate experiments, with small forces, using one's own hands. With both kinds of pucks, it is probably better to use some other device for application of a force—a weak rubber band or the stream of air from the hose of a vacuum cleaner operated in reverse, for example.

Another mode allowing for the development of individual experience is, of course, computer simulation, and many groups are developing instructional materials to this end [cf. di Sessa (1982); White (1984)]. Where the tactile, kinesthetic experience with real objects is impracticable, computer simulation is undoubtedly the next best mode. Computer simulation is also useful for providing more extended practice in thinking about a wide variety of examples. It is capable of supplying continual feedback regarding error and correctness and reinforcing the hands-on observations after the latter have been carried out. The weakest mode is that of lecture demonstration in which student participation is passive—limited to hearing assertions and to seeing effects produced by someone else.

Pencil-and-paper questions and exercises are also a useful component of instruction. They can be designed to help the student confront contradictions in his or her own thinking and to converge on genuine grasp and understanding. Such questions play an especially important role in homework and on tests and examinations; the appendix to this chapter contains selected examples.

3.11 WHAT WE SAY *CAN* HURT US: SOME LINGUISTIC PROBLEMS

There are natural tendencies in everyday speech that are inimical to development of understanding of the concept of force and the Law of Inertia. Teachers should become sensitive to these usages, learn to avoid them themselves, and divert students from their use. Some examples:

There is a very strong, almost universal, tendency to say that a force (or a net force) causes a body to "move." Students should be led to say "accelerate" instead of "move." The word "move" seriously obscures the issue and tends to sustain an Aristotelian view. Students who use it tend to fix on its connotation of "velocity" and lose sight of the primacy of "acceleration," particularly in the early stages when acceleration is still an unfamiliar concept and is incompletely distinguished from velocity.

A very common locution is that "force overcomes the inertia of a body." This encourages the student in thinking of inertia as a force to be "overcome" by other forces. (It is true that Newton himself listed "vis inertiae" as one of the forces to be discerned in nature, but he avoided confusing this with "motive forces" that impart changes in momentum.) In instruction, it is best to avoid any implication whatsoever that inertia is a kind of force.

"Force" is interpreted by many students as something *given* to, being a *property* of, or *resident* in a moving body or one being accelerated. (How much is this reinforced by our tendency to talk about forces "imparted" to a body? I myself find the latter locution difficult to avoid.) In any case, it is advisable to counter this notion and to emphasize external effect and interaction, as opposed to residence in the body.

The meaning of "net," "resultant," or "total" force (when forces are acting simultaneously on a given body) should be developed very carefully and explicitly. There is a strong tendency among students to think of some of the individual forces as having disappeared, or having been somehow obliterated, in the superposition, especially when some of the forces oppose each other and are "overcome" in the final effect. Some students, when one force "overcomes" an opposite force, see the dominant effect as acting alone, not as the algebraic or vector sum of the two.

Confusion between a continuous action and an impulsive shove in connection with "exerting a force" has been mentioned in the preceding section. The language requires explicit attention.

Many students proceed to talk about forces as "working" on objects when dynamic situations are being considered. It is advisable to intercept this locution and stick to the word "acting." Casual use of the word "working" invites confusion when one builds the energy concepts later.

3.12 THE THIRD LAW AND FREE BODY DIAGRAMS

The Third Law is, of course, part of the auxiliary "text" essential for full understanding of the concept of force. Without it there is no basis for separating two or more interacting objects and applying the Second Law to one object at a time, and, without it, students are seriously delayed in developing a comprehension of what object does what to which in familiar physical interactions. Those authors who develop the Second Law and then proceed to conservation of momentum as though that takes care of all the necessary physics, leave their students crippled through inadequate understanding of the force concept.

Once we are used to it, the idea articulated in the Third Law seems transparently simple, and teachers tend to become insensitive to the very great difficulty the majority of students encounter. Difficulties arise in a number of sources and become compounded for many students.

1. *Forces exerted by inanimate or "rigid" objects*. As pointed out in Section 3.10, many students have the preconception that forces can be exerted only by living beings, and they balk at the idea of a table, a floor, a block exerting a force on anything. As a college student once said to me in exasperation, "How can the table exert a force on the book? It has no p-p-power!" Thus, even though they see the table as a "barrier" to downward motion of the book, many students do not see it as exerting an upward force. Similarly, they do not see "resistance to movement" from a surrounding fluid medium, or from rubbing at surfaces, as a force.

This is *not* a trivial conceptual problem, and, since very few texts provide explicit help, it is up to the teacher to develop the insight. Most students are willing to accept the idea that deformed objects (e. g., springs) that return to their initial configuration are capable of exerting a force, and this provides an effective starting point. Because they are aware of the deformation, they can be led to admit that the bed, sofa, easy chair exert an upward force on the sitter, but they regard apparently "rigid" objects as being qualitatively different and do not readily visualize decreasing, but nonzero, deformation as rigidity increases. It is quite difficult to convey the realization that the table, floor, and block also deform—even when loaded with a sheet of paper. Minstrell (1982) describes how he finally convinced a group of students that the laboratory table deforms when loaded: He directed the beam of an overhead projector so that it was obliquely reflected from the surface of the table to an adjacent wall, thus making an optical lever. When the students saw the spot on the wall being displaced as a student walked on the table, they began to accept deformation of apparently rigid objects.[6]

---

[6]Students need explicit help and guidance in learning to visualize effects that elude direct sense perception. The deformation of apparently rigid objects in the context now under consideration is usually the first opportunity in a physics course, and its importance should not be underestimated. Later, such visualization is essential to understanding what happens in elastic and inelastic collisions, in deformations under tension and compression, in the breaking of a string, in the rupture of a container of water when the water freezes, in the propagation of longitudinal and shear waves in solids, in understanding that the far end of a long steel rod is not displaced at the same instant we push on the near end, and ultimately, to being prepared to accept finite time intervals for the transmission of electromagnetic effects (i. e., the invention of field theory). The sequence of visualization and concept building is best initiated at this, seemingly trivial, starting point.

2. *"Passive" versus "active" forces.* In light of the difficulties cited in paragraph (1), it turns out to be helpful for students to distinguish between two classes of forces, designated as "active" and "passive," respectively. Active forces are exemplified by animate pushes and pulls, the gravitational force, electric and magnetic forces. Passive forces are defined as those that arise, and adjust themselves, in response to active ones, for example, in compression of a spring, deformation of the table or floor under the load of a block, frictional forces, and so on. The increase (or "adjustment") of the passive force cannot take place indefinitely; it continues only to the point at which something breaks (table or floor or string) or gives way (as in sliding friction).

3. *Stating the Third Law.* The old, conventional jargon "for every action there is an equal and opposite reaction" has always been gibberish to the majority of students, and, fortunately, many authors are abandoning it. It is best to say "if one object exerts a force on a second, the second exerts an equal and opposite force on the first"—or some other, equally simple and straightforward, form. Even this simple a statement is not initially understood. Students, even when repeating the words correctly, do not do so with the clear realization that one is talking about *two* different forces, each acting on a *different* body. They need extended help in building this realization and making it explicit in diagrams and in their own words.

4. *Non-contact forces.* Confusion concerning the simultaneous presence of two different forces acting on different objects is enhanced by the fact that, at these early stages of development, we tend to concentrate almost exclusively on *contact* forces, and, in the case of contact forces, it is difficult to discern the two separate actions. Also, in the case of the only noncontact force usually considered (namely gravity), we postulate the interaction on the basis of the observed acceleration of free fall, and we are unable to demonstrate the force, equal and opposite to the weight of the object, that is exerted by the object on the earth. To most students this second force remains a source of mystery, confusion, and, in large measure, disbelief.

Without going into details about static electricity or magnetism, it is very helpful at this stage of the game to invoke these effects simply to the extent of demonstrating noncontact interactions made evident by the observed accelerations. Two charged pith balls visibly attract or repel each other without contact; thus we are forced to conclude that each experiences a separate force. Two bar magnets attract or repel each other without contact. Two air track gliders, with appropriately mounted magnets, undergo collisions without making contact. (This effect startles many students.)

A charged rod held in our hand attracts or repels a suspended pith ball or visibly accelerates bits of paper lying on the table. After discussion of the earlier demonstrations, it becomes plausible to the students that not only the pith ball and the bits of paper but also the rod experiences a force, even though the latter force eludes our physical sensation. The same applies to the case in which the magnet, held in our hand, accelerates small nails. With sufficiently strong magnets and more massive objects, the noncontact interaction can be sensed directly.

Given these demonstrations, the Third Law becomes much more plausible and intelligible to many students. Their force diagrams improve, and the gravitational force exerted by the book on the earth is accepted as reasonable and consistent, however undetectable it might be.

5. *Drawing free body diagrams*. It is a well-known phenomenon that many students, when they first start drawing free body force diagrams, produce pictures resembling a porcupine shot by an Indian hunting party—pointed entities stick out randomly in all directions. Practice in analyzing familiar, everyday situations is essential. As the randomness diminishes, many students still persist in showing the two equal and opposite forces of the Third Law acting on the same body. To at least some extent, these tendencies are fostered by many textbooks: A block is shown resting on the floor, and, to save space, the two interacting objects (block and floor) are not shown *separated*. The force exerted by the floor on the block and the force exerted by the block on the floor thus appear on the same picture instead of on well separated pictures, and the message about two different forces acting on different objects is completely obscured. Furthermore, the two forces are rarely described verbally right on the diagram itself.

Lecturing to students about these problems, telling them what should be done, and drawing diagrams *for* them produces very little effect. A more effective procedure is one that requires students to construct diagrams of their own (including redrawing the faulty diagrams in the textbook) under the following rules:

(a) *Both* objects in each relevant interaction should be shown: In the case of the book resting on the table, both the book and the surface of the table should be shown in well separated diagrams, even if the book is the principal focus of attention. The Third Law pair of forces between the book and the surface of the table should be shown, each on its appropriate diagram. In the early stages of such exercise, the earth should be shown as well, since it is the other object involved with the gravitational force acting on the book. (As time goes by, and the majority of students absorb the idea that, in the case of the weight of an object, the other member of the Third Law pair is visualized as acting at the center of the earth, one can begin to drop the requirement of including the earth.) When objects are connected by strings, there should always be a well separated force diagram of the connecting strings as well as of the other objects, even when the strings are regarded as "massless."

(b) Every force should be described in words right along with the diagram. A verbal description means indicating the nature of the force and stating what object exerts the given force on what, for example, gravitational force exerted by the earth on the book; gravitational force exerted by the book on the earth; normal contact force exerted by the book on the table; frictional force exerted by the table on the book; contact force exerted by the string on body A; contact force exerted by body B on the string; and so on.

(c) After the arrows are drawn and then described in words, each Third Law pair should be identified explicitly.

It is the *combination* of being aware of active, passive, contact, and noncontact forces, drawing arrows on well separated pictures, describing the forces in words, identifying Third Law pairs, and being corrected on their errors, that gradually leads students to understanding of the Third Law and the ability to set up problems and to apply the Second Law without guesswork and memorization. As in all other instances involving subtle concept formation, the practice must be spread out over time; attempts at quick remediation invariably fail.

3.13 LOGICAL STATUS OF THE THIRD LAW

In the *Principia* Newton felt it necessary to justify Law III , and he does this in the lengthy Scholium that follows the enunciation of the three Laws of Motion. First he cites papers that Wallis, Wren, and Huygens had (separately) contributed to the Royal Society in 1669 in which they each cited conservation of "quantity of motion" (momentum) in "impact" (collisions) as a fundamental law of motion [see Arons and Bork (1964)]. He then argues that such conservation follows from the Third Law and even implies that Wallis, Wren, and Huygens obtained their insights by having *used* the Third Law (something that is quite unlikely, since conservation of momentum in collisions had been recognized empirically for some time without clear articulation of a force concept.)

He cites colliding pendulum experiments of his own as providing corroborative evidence for momentum conservation and goes on to present the following argument appealing to "attractions," which at that time was the technical term for the (noncontact) electrostatic and magnetic interactions:

> *In attractions, I briefly demonstrate the thing after this manner. Suppose an obstacle is interposed to hinder the meeting of any two bodies A and B, attracting one the other: then if either body, as A, is more attracted towards the other body B, than the other body B is towards the first body A, the obstacle will be more strongly urged by the pressure of the body A than by the pressure of the body B, and therefore will not remain in equilibrium: but the stronger pressure will prevail, and will make the system of the two bodies, together with the obstacle, to move directly towards the parts on which B lies; and in free spaces, to go forwards in infinitum with a motion continually accelerated; which is absurd and contrary to the First Law . . . . I made the experiment on the loadstone and iron. If these, placed in proper vessels, are made to float by one another in standing water, neither of them will propel the other; but, by being equally attracted, they will sustain each other's pressure, and rest at last in equilibrium. [Note that Newton speaks of using a "loadstone and iron," not two loadstones, i. e., in his experiment one of the objects is passive.]*

It is very helpful to the students to invoke this example since it greatly expands and enriches the initial context in which the Third Law is usually presented. An analogous experiment is also easily performed with gliders on an air track.

Newtonian Theory is frequently referred to as an "action at a distance" theory, and the Third Law lies at the heart of this description. The Third Law says that *all* interacting objects exert equal and opposite forces on each other *instant by instant*, and this applies to widely separated gravitating bodies as well as to those exerting contact forces on each other: zero time elapses between a change occurring at one body and the effect of the change being felt at the other.

If we push on one end of a long rod, the other end of which is in contact with a block, the block does *not* exert an equal and opposite force on the rod at the same instant we push. A finite time interval elapses between our push and an effect at the block, the time interval being determined by the velocity of the elastic wave that passes down the rod. Thus, Newton's Third Law does *not* hold, instant by instant, for the forces at either end of the *rod*; it holds only layer by layer of material along the length of the rod, and momentum and energy are both conserved only by virtue of propagation of the elastic wave.

Throughout the later years of his life, Faraday was deeply concerned with analogous situations in electricity and magnetism: if two electrically charged particles are at rest, exerting equal and opposite forces on each other, and one of the particles is suddenly displaced, changing the force to which it is being subjected, does a time interval elapse before the the force on the other particle changes? Does the compass in the Oersted experiment begin its swing at the instant the current is initiated in the wire or does a finite time interval elapse? He constructed delicate mechanical equipment designed to detect such time intervals but, of course, never succeeded.

Maxwell appreciated the significance of these questions, and his invention of the first field theory provided an answer as well as a model for all subsequent field theories.

The point is that the Third Law does not always hold, and this is why modern physics has given primacy to Conservation of Momentum in the hierarchy of physical law. Although one would not discuss all these aspects with students at the time of first introduction of the Third Law, it is well to start laying the groundwork for eventual perception of where the law fails. The rod pushing the block makes a good starting point. The students are initially completely incredulous concerning the finite time interval, and the incredulity can be shaken by pulling on the block with a long slinky. One can spiral back to these questions, and fill in gaps, on arriving at discussions of mechanical wave phenomena and at the appropriate points in electricity and magnetism.

3.14 DISTRIBUTED FORCES

Very few textbooks lead the student to perceive that the single arrows representing the weight of an object, or the normal force on the object at an interface, or the frictional force at the interface, are a shorthand for the sum of distributed effects that must be added "chunk by chunk." This idea is usually left to implication, and only a very few students perceive the implication. Although naive students do not articulate the idea explicitly, they tend to hold the unexamined view that the arrows represent concentrated effects akin to actions such as pushing with a finger or pulling on a string. Later on, the lack of comprehension of distributed effects seriously impedes their understanding of the origin of buoyant forces acting on bodies in a fluid or hydrostatic pressure in general.

Summing the distributed effect does not seem to be an especially difficult idea for students to absorb once it is called to their attention. The point is that it *does* have to be called to their attention. If this is not done, a conceptual gap remains, and many students do not close this gap spontaneously until very much later in their development.

3.15 USE OF ARROWS TO REPRESENT FORCE, VELOCITY, AND ACCELERATION

While one is confined to a single context (forces alone, velocities alone, etc), the use of the arrow symbol to represent the given quantity causes no confusion. When we start dealing with situations in which forces are imparting acceleration to a body having nonzero velocity, however, use of arrows of identical form to represent all three different quantities does cause confusion in many students. They interpret velocity and acceleration arrows as forces acting on the body, and, in drawing their own force diagrams, they gratuitously insert velocity and acceleration arrows as additional forces. (Such confusion arises, for example, when one wishes to examine all the effects on an object in projectile motion; when one deals with objects in an accelerating car; or when one is concerned with forces applied to, and the velocity and acceleration of, a bob in circular motion.)

This confusion can be countered to some degree by slightly altering the notation. My own system is to use the ordinary arrow for force, a single-half-headed arrow for velocity, and a double-half-headed arrow for acceleration as in Fig. 3.15.1.

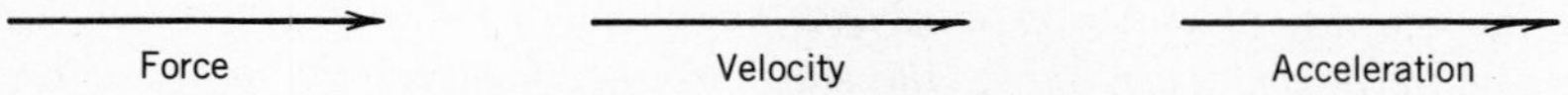

Force   Velocity   Acceleration

Figure 3.15.1 Using different arrows for different vector quantities

I ask the students to use this notation on tests and homework, and I use it myself in lecture presentations. The system is not onerous, and it helps reduce the inclusion of velocities and accelerations as forces on free body diagrams. There is, of course, nothing sacred about this particular notation, and any other form that distinguishes the quantities would serve equally well.

3.16 UNDERSTANDING TERRESTRIAL GRAVITATIONAL EFFECTS

Interviews with students show extensive misconceptions and confusion about "gravity" and gravitational effects—misconceptions that are rarely spontaneously articulated by the students, that frequently pass unnoticed by teachers, and that seriously impede understanding of the material being taught.

1. *Meaning of the word gravity.* One semantic problem, originating in early years and persisting to college level in many students, stems from an answer provided by many teachers and parents when the child asks, "Why do things fall?" A very common answer is, "*Because* of gravity. " (If you ask this question of a class of college students, you will get the indicated answer in the majority of cases. Only a few students are uneasy about such a facile answer and fewer still have the self-confidence to challenge it in the way it should be challenged.)

Children, as well as many adults, take this answer very literally: since the word "because" has been used, they uncritically jump to the conclusion that a *reason* has been given—that the "why" has been answered. They naively believe that a scientific name provides a reason; much of their experience with science in the schools has reinforced this acquiescence.

Students should be made aware of some of the history of term: the Greeks endowed bodies with the teleological properties of "gravity" and "levity," representing built-in desires or tendencies to seek the center of the earth or to rise toward the celestial domain; seventeenth-century science eliminated both the teleology and the term "levity" and applied the name "gravity" to the observed interaction between objects and the earth. With the Newtonian Synthesis, the meaning is expanded by the grand perception that the same effect that makes the apple fall also binds the moon to the earth and the earth and planets to the sun.

Finally, however, students must be made explicitly aware that the name does nothing more than conceal ignorance—that to this day, and despite the power of the Newtonian Synthesis and the beauty of the General Theory of Relativity, we have no mechanism for the interaction and no idea of how it "works."[7]

It is interesting to note what Galileo had to say about this matter. In the *Dialogue Concerning the Two Chief World Systems* one finds the following exchange:

> SIMPLICIO: *The cause of this effect [what it is that moves earthly things downward] is well known; everybody is aware that it is gravity.*
>
> SALVIATI: *You are wrong, Simplicio; what you ought to say is that everyone knows the it is called "gravity. " What I am asking you for is not the name of the thing, but its essence, of which essence you know not a bit more than you know about the essence of whatever moves the stars around. I accept the name which has been attached to it and which has been made a familiar household word by the continual experience that we have of it daily. But we do not really understand what principle or what force it is that moves stones downward . . . .*

It seems that the appropriate form of the dialogue has not changed very much over the interval of almost four hundred years.

Helping students see that names, as such, do not constitute knowledge or understanding, and coupling this with the emphasis on careful operational definition advocated throughout this book, does much to put students in the position of recognizing when they do not know the meaning of a technical term and to recognize when meaning has, or has not, been provided. My own observations show that many cease name-dropping of terms they have picked up but do not understand, and many report asking for meaning of technical terms in other (not necessarily science) courses.

2. *Meaning of "vertical" and "horizontal"*. Very few students possess clear operational definitions of "horizontal" and "vertical. " If asked how they might, as simply as possible, establish a precisely vertical direction right where they happen to be, many respond, "perpendicular to the ground. " If one suggests going over to the steep slope of a nearby hill and establishing the perpendicular to the ground, they back away from the initial suggestion, but few have anything with which to replace it. All told, very few students have established a clear connection between the direction of the force of gravity and the meaning of "horizontal" and "vertical"—either via the plumb bob or the carpenter's level.

3. *Air and gravity*. Many students, especially among the nonscience oriented, acquire the information that the air (or the atmosphere) "presses down on things" and translate this into an association with gravity. They thus tend to view gravity as imposing a downward push rather than a downward pull: Air presses down on the book on the table; gravity "disappears" when air is removed; many expect that objects would float around in an evacuated bell jar without the air to hold them down. Very large numbers of students expect an air-filled balloon, which is seen not to float in air, to float in an evacuated bell jar. One should allow these

---

[7]In using the words "mechanism" and "works" I am referring to processes that we visualize in terms of ordinary sense experience. We visualize such microscopic effects as gas pressure and diffusion, evaporation and condensation of liquids, crystallization and structure of solids, in terms of familiar behavior of macroscopic particles. We visualize invisible elastic waves in solids in terms of what we have seen happening on soft springs. We visualize the propagation of classical electromagnetic waves in terms of an analogy to mechanical shear waves. We have no corresponding forms of visualization for quantum mechanical effects or for gravitational interaction, "virtual" entities notwithstanding.

expectations to be openly articulated and brought to the surface and one should then counter them with suitable demonstration experiments.

4. *Meaning of "vacuum"*. A concomitant difficulty arises with the word "vacuum." Once in conducting a discussion of some observations of naked eye astronomy with a class of preservice elementary school teachers I casually referred to the "vacuum of outer space. " Noticing strange looks and sidelong glances among the students, I pursued the issue and finally discovered that, where I was thinking of space devoid of matter, most of the members of the class were thinking of the household appliance they used to clean rugs. They were left wondering what motivated me to talk about some mysterious cosmological vacuum cleaner. I forthwith brought out a pump, a hose, and a bell jar.

5. *Uses of the feather and coin tube*. The classical demonstration of the "feather and coin" tube (in which objects that clearly do not fall together in air do so in a vacuum) is well worth showing in virtually all classes. (The only students likely to have seen it are those who happen to have had an unusually good high school physics course.) Not only does this apparatus demonstrate Galileo's law of falling bodies, but it also offers the opportunity to raise the issues of "vertical" and "horizontal," discuss the meaning of the word "vacuum," and, for those who expect gravity to disappear in the absence of air, emphasize that this is not what happens.

6. *Meaning of g*. A very large number of students, including those in calculus-based physics courses, when asked what the symbol g stands for in kinematics and dynamics, respond "gravity. " They do *not* invoke the word "acceleration" at all. When the questioning is pursued further, it almost invariably emerges that students who respond this way have no understanding of any of the things they do with this symbol and are simply trying to memorize problem-solving procedures. They cannot clearly identify the kind of quantity the symbol represents, although many students seem to regard it as being more a force than anything else. It is necessary to get these students to the point at which they give a correct interpretation of g in their own words, with physical illustrations of its meaning in everyday experience.

7. *"Feeling" the weight of an object*. Teachers and textbooks frequently say that "we feel the weight of an object when we hold it" and imply that the same force acts on the table supporting the object. Granted that we can get away with this locution in everyday speech, it is very damaging, however, in a physics course in which we should be trying to ensure precise understanding of the scientific concepts and language we are creating.

The term "weight of an object" should be introduced, and then reserved *exclusively* for the gravitational force exerted by the earth on the object. Given this meaning, the force we feel when we hold an object is *not* the weight of the object but the contact force the object exerts on us. It is true that this contact force is sometimes numerically *equal* to the weight of the object, but the equality does not make it the same force. In fact, the two forces are not even numerically equal if something is pressing down or tugging upward on the object or if we are accelerating the object up or down. The distinction between the two forces is not trivial, and, if it is not maintained, a large measure of understanding of the scientific vocabulary is lost. Furthermore, understanding of the Third Law pair at the interface (the force exerted by the object on the table and the force exerted by the table on the object) is undermined.

8. *Weight and weightlessness*. Most teachers are aware of the unfortunate use (or misuse) of the word "weightlessness" in connection with satellites and space vehicles. There is not much we can do about the usage (any more than we shall be able to force people to say "mass" instead of "weight" when talking about a number of kilograms of potatoes in a grocery store). We can, however, give students an understanding of what is being described and why the terminology is unfortunate.

Some authors and teachers try to dodge the issue by suddenly switching the meaning of the word "weight" (usually without openly confessing that a switch is being made): after initially defining "weight" as the gravitational force exerted on the object by the earth, they switch to describing "weight" as the reading on the platform scale on which the object is supported, that is, they transfer the designation to apply to the normal force exerted by the platform on the object. As pointed out in the preceding paragraphs, not only is this *not* the gravitational force exerted by the earth on the object, but, in many circumstances, it is not even numerically *equal* to the gravitational force. Although this usage may *seem* to simplify matters for the learner, it is invariably disastrous and plants far more difficulty and confusion than it mitigates. As an illustration of such confusion: this usage reinforces the mistaken notion that the force of gravity indeed vanishes when an object is in free fall or when it is removed to appreciable distances from the earth.

The best procedure is to stick unswervingly to the initial definition of "weight" as the gravitational force acting on the object and help the student analyze the sensations he or she personally experiences under various circumstances:

First one must lead the student to realize that we do not sense or feel the gravitational force itself; we postulate its existence on the basis of the observation of acceleration in free fall and the definition of "force" as an action that imparts acceleration. When we jump from an elevated position, we do not feel something tugging on us as we are falling.

Next we lead the student to recognize that what we *do* sense or feel is the normal force exerted on us by the object we stand or sit on. This force is numerically equal to our own weight only if no one is sitting on our shoulders or trying to lift us, and only if we are not being accelerated either up or down. (Thus the student can be led to define the very special circumstances under which we "weigh ourselves.")

Now we proceed to explore what happens to the reading on the platform scale as we are accelerated up or down—say in an elevator. Most students have noted the sensations that go with such accelerations and are prepared for interpretation of the forces they experience: an upward force larger than the one normally felt when the acceleration is upward; an upward force smaller than the one normally felt when the acceleration is downward.

Finally one can argue to the limit: what happens to the upward force exerted on us by the platform as the downward acceleration gets closer and closer to that of free fall? Most students readily agree that the upward force on us, and the reading on the scale, go to zero.

One can now take up the matter of terminology: When we are in free fall, the gravitational force exerted on us by the earth has *not* become zero. What has become zero is the normal force at our feet—the force that we do sense directly. Under these circumstances we experience a strange sensation, one that might be called a "sensation of weightlessness." Hence arises the poor terminology in which the word "weightlessness" is used to describe the situation in a freely falling elevator or in a satellite. We must understand the confusing usage and not interpret the word as literally meaning that the gravitational forces have become zero.

9. Forces in free fall and in projectile motion. Many authors and teachers have become so accustomed to Galileo's law of free fall and to the usual idealizations ("thinking away" the ever-present frictional effects) that they are tempted to traverse this subject matter as quickly as possible in order to extend coverage to more "interesting" things. Unfortunately, the commonsense preconceptions pervading this area are very tenacious, and many students, if not given the necessary help, are left so far behind that they take refuge in memorizing and never really catch up.

(a) Many students, when they finally open up, tell me that they were "told," and that they can readily repeat the statement, that all objects fall together when dropped, but they have "never really believed it." They need to see and *discuss in their own words*: simple demonstrations such as the dropping of a sheet of paper side by side with a similar sheet crumpled up into a ball; the dropping of the sheet of paper placed on top of a falling book; stroboscopic pictures of large and small objects falling side by side; the feather and coin tube mentioned above, etc.

(b) After becoming convinced that all objects do indeed fall together in the absence of rubbing effects, many students will then switch to the view that, in order for this to happen, the forces acting on the different objects must all be the same. Countering this requires discussion and observation; a simple assertion on the part of the teacher produces little effect.

(c) Students should have the opportunity (in homework and on tests) to draw their own force diagrams (including both the object and the earth) for: an object dropped from rest; an object thrown vertically upward (on the way up, on the way down, and at the top of the flight); a frictionless puck sliding along an air table and then the same frictionless puck while flying through the air after having sailed off the table; a projectile at various points in its trajectory.

(d) The force diagrams in (c) should, in each case, be accompanied by a *separate* diagram showing the instantaneous velocity vector and by still another diagram showing the instantaneous acceleration vector. The juxtaposition of these various diagrams is significant in enhancing understanding since it makes the student view the same situation in entirely different ways.

10. Gunstone and White (1981) present a highly revealing set of student responses concerning the following situation: A bicycle wheel is mounted as a pulley with its axis 2 m above the laboratory bench. A cord, connecting a bucket of sand and a block of wood, equal in mass, is placed over the pulley, that is, the students see an Atwood machine with a bucket of sand at one end and a block of wood at the other. (The students participating in the investigation were first year students at Monash University in Melbourne, Australia—students who had not yet had university instruction in physics.) The students were then asked various questions, including ones that required making predictions as to what would happen when certain changes were made, and they were asked to write out the reasons for their answers.

(a) The participants were shown that the pulley rotated freely, and then the cord was placed over the pulley in such a way that the bucket was markedly higher than the block. The system remained stationary. The participants were asked "How does the weight of the bucket compare with the weight of the block?" Of the participants, 27% said the block was heavier, the largest proportion of these explaining their conclusion by pointing to the fact that the block was nearer to the floor and thus must be heavier. Another reason given by some students was to the effect that "Tension exists at both ends of the string. At the end towards the bucket the tension is less than at the end towards the block. This then causes the block to pull itself down and thereby raises the bucket."

(b) The students were then asked to predict what would happen if a large scoop of sand were added to the bucket. Now 30% predicted that the system would shift to a new equilibrium position with the bucket closer to the table and the block higher up.

(c) After it was shown that the system moved continuously after the scoop of sand was added to the bucket, the participants were asked to predict how the speeds of the bucket would compare at two marks—one High and one Low (near the table). Although 90% correctly predicted that the speed would be higher at the Low mark, some indicated that their prediction was based on knowledge that the gravitational force acting on the bucket increased as the bucket went down (or the force on block decreased as it rose.) Others stated that the acceleration of the bucket would be g. When the demonstration was made, 7% of the students reported observing the speeds to be equal at the two marks. The reconciliations of prediction and observation among these students included "no net force," "objects only accelerate in free fall," "friction," and "error in observation."

(d) The block and bucket (equal masses) were placed on the pulley so that they hung at the same level without motion. The block was then pulled down about 0.7 m and held. Students were asked to predict what would happen when the block was released. Only 54% predicted the system would remain stationary; 35% predicted return to the original position; 9% predicted the bucket would fall; 2% predicted the block would fall.

Gunstone and White give many more details in their informative paper; the preceding highlights have been selected for illustration. The moral of these illustrations is that we, as teachers, become so familiar with these basic concepts and phenomena that we regard them as too trivial to command any time in instruction. Only when questions of this variety are included in both homework and tests, however, do we begin to help the large number of students having such difficulties achieve understanding.

3.17 STRINGS AND TENSION

Many textbooks bring forth the word "tension" and start using it as though everyone must know what it means without operational definition. The student is confronted with the familiar problem in which one string is stretched by opposite forces of 50 N at each end while a second string, with one end attached to a wall, is pulled with a force of 50 N. The student wonders how it is possible for the tension in the string to be the same in each case and is unable to see why it is not 100 N in the first string.

There are two difficulties superposed here. One is that, when this situation is first encountered, many students have not fully assimilated the Third Law and, not drawing an adequate force diagram of the string, fail to see that the two situations are identical as far as the forces on the strings are concerned. The other difficulty, however, is that "tension" has not been defined.

One simple approach is to lead the student to imagine "cutting" a stretched string at some point along its length and drawing the forces acting on the two segments. (Not only is this a good exercise in using the Third Law, but it also introduces students to the examination of forces in the *interior* of objects. Up to this point all forces and force diagrams have usually been confined to external effects, and the realization has not been formed that one can, in imagination, "cut through" an object and show the forces at the selected cut.) Having drawn the equal and opposite forces acting on the two segments at the cut, one can give the name "tension at the cut or section" to the magnitude of the force acting on either segment. Tension and compression in rods or columns can then be defined in a similar way.

Having defined tension in this way, it is now a relatively simple matter, inviting valuable phenomenological thinking and visualization, to examine the tension in a massive rope (or chain or rod) as the object is accelerated by a force at one end. It is not necessary to solve quantitative problems! As one examines the tension "chunk-by-chunk" through the length of the object, it becomes apparent, through application of the Second Law, that it must decrease continuously from a value equal to that of the applied force at one end to zero at the other. One can then leave the further problem of how the tension varies when a rope is accelerated with two opposing forces, unequal in magnitude, at each end.

3.18 "MASSLESS" STRINGS

It is well known that "massless" strings are a source of significant conceptual trouble for many students. They have no intelligible operational definition of "massless"; they fail to see why the forces of tension should have equal magnitude at either end; they proceed to memorize problem-solving procedures without understanding what they are doing. The principal problem here is that, when massless strings come along in the text, many students have not yet fully assimilated the idea that the *difference* between the magnitudes of oppositely directed forces acting on an accelerated body depends on the mass of the body (when acceleration is fixed.) A careful and clear development of the "massless string" concept therefore not only helps students in this immediate kind of problem solving; it helps students register a vital aspect of the Second Law that has so far eluded them.

Understanding the meaning of "massless" is greatly facilitated by leading students through an operational definition of tension (as in the preceding section) and then proceeding with something like the following sequence: Suppose a rope of mass m_R is attached to a massive block, and we accelerate the system horizontally, with an acceleration a_R, by pulling on the end of the rope with a force T_1. Separate free body diagrams of the rope and the block should then be drawn, and students can be led to acknowledge that the block exerts a force on the rope at the opposite end; denote this force by T_2. Further discussion is usually required to make sure that students understand that T_2 must be smaller in magnitude than T_1 and that these two forces are also equal to the (different) tensions at the two ends of the rope.

Now they apply the Second Law to the rope, obtaining the expression:

$$T_1 - T_2 = m_R a_R \qquad (3.18.1)$$

Students must be led to *interpret* this expression. At this stage of development, very few students understand what it means to interpret an algebraic expression, and there is massive resistance to doing so. They should be led to say that the equation indicates that the two forces are equal in magnitude when the acceleration is zero and that the equation confirms the earlier, qualitative, conclusion that T_1 is larger than T_2 when the acceleration is not zero.

Now it is possible to get at the real point at issue: What happens to T_2 as the mass of the rope m_R is made smaller and smaller while the acceleration a_R is kept fixed? Having reached this point, most students are able to discern that T_2 becomes more and more nearly equal to T_1, that, in the limit, the two tensions are equal, and that this is the real meaning of the concept of "massless string" in the context of the textbook problems.

A somewhat more rigorous development, highly desirable for more sophisticated students, is to set up the algebra for the entire system (including the block, with mass m_B, ending with the expression

$$\frac{T_1}{T_2} = 1 + \frac{m_R}{m_B} \qquad (3.18.2)$$

Interpreting this equation shows that T_2 becomes very nearly equal to T_1 when m_R is very small compared to m_B, and shows the students that "masslessness" is, in the final analysis, a relative and not an absolute matter.

Such an analysis gives students in engineering-physics courses, for example, a very rudimentary exposure to theoretical formalism—an exposure that is, unfortunately, denied them in most instances through neglect of available opportunity. Teachers then wonder why the students seem to be so naive on such matters in more advanced courses.

3.19 THE "NORMAL" FORCE AT AN INTERFACE

The normal force N is usually first encountered in situations in which an object of mass m rests on a horizontal surface: the book on the table, the student's own body on the ground. In this special case the normal forces exerted by the book on the table and the table on the book happen to be equal in magnitude to mg, the weight of the book. Many students, not yet having formed a clear understanding of the force concept and of the Third Law, simply memorize the statement $N = mg$ more or less in self-defense and continue to stick to this equation in circumstances in which it is not applicable. [Locutions about "feeling the weight of the object when we hold it up," discussed in Section 3.16 (7), also feed this misconception.]

To forestall this difficulty, students should be led to visualize how the normal force varies when they exert an upward tug on the book and when they press down on it vertically, and this should be done as soon as possible after they have begun to accept the idea that the inanimate table is indeed capable of exerting such a force. They should be led to articulate the insight that, in fact, N is almost never equal to mg, and that the equality obtains only in the very special case in which there are no other vertical forces acting besides the pull of the earth.

Another exercise that is very helpful at this point, repeating some ideas but altering and enriching the context, is to press the book against the wall. Now the wall, another inanimate object, must be conceded as capable of exerting a normal force, and this normal force has nothing at all to do with mg; its magnitude is determined exclusively by the horizontal force we exert with our hand. (This situation is also useful for showing students that frictional forces do not necessarily depend on *mg*, a misconception they also pick up from the first encounter with friction on horizontal surfaces. See Section 3.21 for further discussion.) Inquiry into the behavior of the normal force when we press the book against the ceiling becomes a valuable homework exercise at this point.

Since the normal force is usually first encountered at horizontal surfaces, still other subtleties behind the concept go unnoticed and unarticulated. Many students, in fact, interpret the word "normal" in its sense of "usual" or "ordinary" rather than its geometrical sense of "perpendicular." The full meaning of the term does not become apparent until the confrontation with inclined surfaces, and, by this time, teachers frequently lose sight of the fact that the concept has not been convincingly explored, while many texts seem to take the attitude that it is too obvious to require discussion.

In my own experience, the physical situation shown in Fig. 3.19.1 is very helpful in raising and settling a good number of the issues involved. This apparatus is very widely used in showing composition and decomposition of forces (with actual numerical data being taken), and it is found in most preparation rooms. I have rarely seen it used, however, for explicitly generating the "normal force" concept by showing that the inclined plane exerts a force perpendicular to itself in the absence of friction.

After one balances the cart in the direction parallel to the plank, one proceeds to "replace," by loading the second string, the force exerted on the cart by the plank. Many students do not notice the direction of this string unless the direction is explicitly called to their attention. They must also be led to state the relationship between the force now being exerted by the string and the force previously being exerted by the plank.

Figure 3.19.1 Demonstrating that the so-called "normal" force at an interface is indeed normal to the interface.

A powerful impression is then made by shifting the cart up the plank (and then down the plank) so that the string is visibly inclined from the perpendicular, and watching the cart oscillate while returning to the position at which the string is again normal to the plank. A gasp is frequently heard when this demonstration is performed, clearly indicating that the observed effect was unexpected. In most cases, in order to get all the relevant ideas fully registered, it is necessary to continue the discussion as far as examining the components of force, and the accelerating effect on the cart, when the cart is displaced from the equilibrium position.

This demonstration is also valuable in helping students develop a better understanding of *orthogonal components* of forces, a matter that will be discussed in more detail in Section 4.3.

3.20 OBJECTS ARE NOT "THROWN BACKWARDS" WHEN ACCELERATED

Consider the following situations: (1) a ball is placed on a cart, and the cart is accelerated from rest; (2) a pendulum bob hangs from the roof of an accelerated car; (3) a person is sitting in a car that begins to accelerate.

If asked about any one of these cases, a great many students contend that the person, the bob, the ball are "thrown backwards" when the vehicle accelerates, and, if asked to draw force diagrams, they show a force acting in that direction. The source of the difficulty is, of course, a very natural and commonsense one: there is a strong inclination to put oneself into the accelerating frame of reference. These situations are far from trivial, and it is a mistake to consign them entirely to homework. At least one such situation should be discussed, with demonstration, in class.

No amount of previous discussion and definition of inertial frames of reference makes much impression on the majority of students until they encounter a noninertial frame and start confronting contradictions. In order to understand what an inertial frame *is*, one must begin to understand what it is *not*, and situations such as those proposed above are a first opportunity to make this point in rectilinear dynamics.

In the case of the ball on the cart (which can be assigned as a home experiment), most students are surprised to see that, although the ball rolls backward with respect to the cart, it moves forward with respect to the ground.

In the case of the pendulum bob, an excellent and very simple demonstration can be made by accelerating, in one's own hand, the top end of the string on which the bob hangs. Students can see the suspension point move forward while the bob retains its position relative to the floor. They can begin to discern that the bob is *not* thrown backwards relative to the floor and that acceleration of the bob begins only when the force exerted by the string acquires a nonzero horizontal component. (At this stage of development, many students are still very shaky about components of force and their accelerating effects, and this demonstration is particularly valuable because it invokes the concept of components in addition to frames of reference.)

Having examined cases 1 and 2 from the point of view of a bystander, students can now take up case 3 in which they are participants, as individuals in the accelerating car. They should be led to recognize explicitly that they are *not* thrown backwards but feel the force exerted on them by the back of the seat as the seat is accelerated—just as the pendulum bob experienced neither horizontal force nor acceleration until the inclined string began to pull it horizontally.

Not only does qualitative examination of these cases give students the opportunity for some valuable phenomenological thinking, helping them absorb the frame of reference concepts, but it also paves the way to better eventual comprehension of centripetal force and circular motion. The fact that time elapses between the two encounters is of vital importance, being conducive to learning.

If the teacher desires to do so, and if it is appropriate for the level of the students, the concept of "fictitious forces" can be introduced at this juncture.

3.21 FRICTION

Friction is a "passive" force in the sense defined in part 2 of Section 3.12; it adjusts itself in response to active effects. In fluids, the frictional resistance varies with the velocity of the moving object. At an interface between solids prior to slipping, the frictional force starts at zero, and, as the force tending to produce slipping increases, the frictional force increases until the interface "breaks," and slipping begins. I use the word "breaks" not in a literal sense but to emphasize the analogy between this situation and that in which bodies literally do break under loading as the normal force increases to a critical value—as in the case of piling weights on a table until it breaks. This is an analogy that students do not perceive unless it is made explicit, yet, when it is established, they acquire a better understanding of the nature of the effect.

That such understanding is initially lacking in many students becomes evident if one observes some of the things they do in attacking end-of-chapter problems. They tend to use the formula

$$f = \mu N \qquad (3.21.1)$$

for any and every frictional force whether slipping is about to occur or not. In other words, they do not explicitly realize that the frictional force might have any value between zero and the maximum referred to in the formula. It is not that the textbook has failed to present the formula properly; this is competently handled in most texts. The trouble is that the student has not been led to confront cases in which the value of the frictional force lies between zero and the maximum and thus fixes only on the formula. As in many other instances (e.g., the kinematic equations for uniformly accelerated motion), the student must be helped to see when an equation *does* apply and when it does *not* by dealing explicitly with cases in which it is inapplicable.

Another situation in which a force of static friction builds up from zero to its maximum value is that in which a frictional force acts to accelerate a body, as in the case of a block on an accelerating cart. The frictional force exerted on the block by the floor of the cart increases as the acceleration of the cart increases. Since there are no other horizontal forces acting on the block, this situation is fundamentally very different from the one in which a block is acted on by an external horizontal force while resting on a stationary platform, and many students have serious difficulty drawing a correct force diagram. Such situations are frequently encountered in end-of-chapter problems, but many students never acquire an understanding of the physics; they either never solve the problems correctly or they memorize procedures in which they plant μNs around without understanding what they are doing. It is most effective to develop and contrast the two situations (block on the floor and block on the accelerating cart) when the concept of the static coefficient is first being developed. Enlarging the context for the same concept is conducive to learning and understanding.

As pointed out in Section 3.19, many students pick up the misconception that a normal force N is always equal to mg because they first encounter the normal force in cases such as that of objects resting on a floor or table with no vertical forces acting other than the weight mg. This subsequently leads to their treating every frictional force as being equal to μmg regardless of what the normal force actually is.

An effective way of displacing this misconception is to examine the situation of the book pressed against the wall, where the normal force has no connection whatsoever with the weight of the book (see Section 3.19.) The problem should be posed as one requiring investigation of changes (not just as a single calculation with one set of given numbers): (1) Draw force diagrams of both the book and the wall. (2) Suppose the horizontal force we exert on the book is very *large*: what are the magnitudes of the frictional force and of the normal force? How is the frictional force related to the normal force under these circumstances? How is it related to the weight of the book? (3) Suppose we start decreasing the horizontal force we are exerting: What happens to both the frictional force and the normal force as the decrease proceeds? Under what circumstances does the book begin to slip downward along the wall? How is the frictional force related to the normal force once sliding begins? How is it related to the weight of the book?

Texts and teachers frequently tell students that "frictional forces always oppose motion" without examining this phraseology critically. Students interpret the word "motion" in this context as referring to motion of the *body* on which the frictional force acts, and, in this sense, the statement is not always true. It is true that frictional forces at solid interfaces always oppose *slipping of the surfaces*, but in many instances of everyday experience the frictional force is the one that accelerates the body under consideration: the frictional force exerted on our shoe by the ground accelerates us when we walk; the frictional force exerted by the road on the tires accelerates the car; the frictional force exerted on the block by the floor of the accelerating cart (in the illustration discussed above) accelerates the block.

Many students initially have quite a bit of trouble in visualizing the direction of frictional force on each of two objects at an interface. When this is the case, I find the following approach helpful: I suggest that they put their two hands together, palm to palm, and imagine one hand to be one of the two objects and the other hand the other. Then I suggest that, concentrating on each hand in turn, they slide one hand over the other in the direction in which the objects would tend to slide, feel the force exerted on the hand, and put that force on the corresponding object in the force diagram. The extent to which students find this device helpful is evident when one sees how many are rubbing their hands over each other during tests.

3.22 TWO WIDELY USED DEMONSTRATIONS OF "INERTIA"

Two excellent demonstrations are widely used to demonstrate what is frequently (much too casually) described, as "inertia":

1. The table cloth is yanked out from under a set of dishes, leaving the dishes on the table.
2. A massive block is suspended by a string from a rigid support, and an identical string hangs from the bottom of the block. When the lower string is pulled slowly downward, the upper string breaks; when the lower string is jerked downward, the lower string breaks.

There is much more involved here than just "inertia." Both of these situations are rich in physical phenomena, and students should be led to think about them in some detail in order to understand what is involved. Probably the best way to induce this thinking is to perform the demonstrations and ask enough leading questions (assigned as homework) to make it possible for the majority of students to fill in the gaps without getting bogged down.

In demonstration 1, if the dishes were glued to the table cloth, they would be yanked off the table. The demonstration depends on the fact that the coefficient of friction is sufficiently low to allow the interface to "break" (in the sense defined in Section 3.21 above) at a value of maximum frictional force sufficiently small to impart sufficiently small acceleration to the dishes. Even with a relatively small frictional force, however, the dishes would still be yanked off the table if the table cloth were very long, extending down the table well beyond the dishes. In other words, there is a time element involved, and the demonstration works because the time during which acceleration is imparted is short enough to make the displacement negligible.

The inertia of the dishes is indeed an important factor, but so are the others. Viewers of this demonstration are rarely given the opportunity to think it through and understand it fully. Part of the understanding depends on awareness of what *might* happen, of what is *not* the case—in addition to an awareness of what *is* the case and what *does* happen.

In demonstration 2, the crucial physical effect is the *stretching* of the strings to their breaking point. The stretching eludes direct sense perception and therefore has to be discerned in the imagination. Few students perform this act of imagination spontaneously, but it is not difficult to guide them into it. The key is again the element of time (as in demonstration 1, but in a somewhat different fashion): when the lower string is jerked, the low acceleration of the block allows the lower string to be stretched to breaking point before displacement of the block produces comparable stretching of the upper string; when the lower string is pulled slowly, both strings stretch without appreciable time delay, and the upper string is stretched to breaking point first because of the higher loading. [At a still higher level of sophistication, students could be encouraged to visualize the elastic waves that must propagate up and down through the components of the system preceding the displacements leading to breaking. See part 1 of Section 3.12 above and the accompanying footnote.]

Without visualization of the stretching of the strings, students acquire no understanding of the demonstration; they simply memorize, and repeat, that it had something to do with "inertia."

3.23 DIFFERENT KINDS OF "EQUALITIES"

A hidden source of confusion for many students, one rarely recognized and eliminated in course work, is the fact that the "equals" sign (=) means very different things in different contexts. Following are some examples:

Statements such as

$$\rho = \frac{M}{V} \quad \text{and} \quad \bar{v} = \frac{\Delta s}{\Delta t}$$

are actually definitions (or identities), rather than ordinary equalities and one should use the three-line symbol ($\equiv$) for "defined as" or "identical with" rather than the ordinary equals sign. (Some texts are now doing this, but the reason must still be discussed and emphasized to the students.)

The kinematic equations, however, are statements of functional equality (subject to the restriction to rectilinear motion and uniform acceleration) *derived* from the *definitions* of s, t, v, and a; they are like the equations the students have become familiar with in elementary algebra. The ordinary equals sign ($=$) is appropriate.

The equals sign in

$$F_{net} = ma$$

is not just an ordinary functional equality. It conceals the combination of arbitrary definition and laws of nature lying behind either the Machian or Newtonian approach to the Second Law (see Sections 3.2–3.6).

The statements

$$f_{max} = \mu N \quad \text{and} \quad F = kx \quad \text{(Hooke's Law)}$$

are of only limited validity and applicability. Hence the equals sign applies only under certain restrictions, the text of which must accompany the symbols.

The impulse-momentum and work-kinetic energy theorems, when derived from $F_{net} = ma$, reveal the remarkable and unanticipated equality of numbers calculated in entirely different ways: one way involves the necessity of knowing the history of variation of F_{net} as a function of either clock reading or position over the entire interval in question and finding the area under the appropriate graph; the other way involves only the initial and final velocity state of the body over the given time or position interval. The equations are therefore not merely functional relations; they have physical content and meaning not articulated in the equals sign alone.[8]

A common, but nevertheless confusing, use of the equals sign is in statements such as 2.00 m = 6.56 ft or 1 kg = 1000 g or, even worse, 1 kg = 2.20 lb.

These, of course, are not equalities at all but describe various kinds of *equivalence*, requiring an accompanying text of interpretation. Students recognize that the numbers shown in the statements are clearly not equal, but many are afraid to ask about the contradiction, sensing that the question is likely to be regarded as "stupid." Furthermore, it is difficult and confusing for many students to see

---

[8] From a still more advanced point of view, these theorems begin to exemplify the difference between a line integral (or inexact differential) on the one hand and a state function (or exact differential) on the other, and pave the way for more sophisticated utilization of these ideas, for example, in thermodynamics. It is not appropriate to belabor this mathematical aspect with students who are not ready for it and who are not going on to more advanced physics courses. It is something well worth returning to, however, from the perspective of a more advanced course, and, if the foundation has been started in an elementary way in the introductory course, understanding of the more advanced ideas is greatly enhanced.

that the statement $6P = S$, where P and S are *variables*, is profoundly different from the statement 2.00 m = 6.56 ft, where m and ft are *not* variables. Once they have mastered the meaning of something like $6P = S$, they are easily misled into thinking that the other relation implies that 2.00 times the number of meters is equal to 6.56 times the number of feet. (See Section 1.15 for additional detail.) Some equivalence symbol, distinctly different from the equals sign, would be much more appropriate in this context.

Since the same symbol (=) is used throughout these different contexts, students are not impelled to discern the profound differences in meaning unless the differences are discussed explicitly. One of the consequences of omission of such discussion is, for example, the notion that "*ma* is a force since $F = ma$." The result is the appearance of a force, labeled *ma* on the force diagram of an accelerated car or on the force diagram of a person seated in the car. Another consequence is the confusion attending translation of verbal statements into symbols that was discussed in Section 1.15.

These differences in the meanings of equals signs are, of course, not confined to physics; they permeate arithmetic and mathematics as well, and, if made clear in both areas, help develop student understanding that much more rapidly.

3.24 SOLVING PROBLEMS

The importance of helping students acquire the habit of a systematic approach to solving end-of-chapter problems has been discussed in Section 2.13, and, if such habit has been cultivated in kinematics, it provides a strong foundation for progress in the more subtle area of dynamics. Many textbooks give specific examples of a systematic approach, but many students resist using it unless they are required to do so as part of graded performance.

Guided drill sequences can be very helpful to students who find difficulty implementing the scheme outlined in their text. Such drill is provided, for example, in the computer-based Plato materials [see *Plato Physics I* (1983)]: the student is led to draw force diagrams first, set up a coordinate system, apply the Second Law, and carry out solutions. Not only does the systematic approach facilitate solving the problem; it helps break down the very high resistance, among many students, to putting pencil to paper, and thus beginning to analyze the problem, before the entire solution or "answer" has been "seen."

One systematic aspect of problem solving that is omitted by many textbooks and programs is the explicit writing out of the equation for continuity of acceleration when a problem deals with two or more interconnected bodies. In the case of the Atwood machine, for example, the symbol a is casually written down for both bodies without explicit recognition that the equality of the accelerations of the two bodies is one of the essential mathematical conditions describing the physical system (i. e., the nonstretching of the string).

This omission then causes students great difficulty if they confront more complicated problems (e. g., an Atwood machine suspended on one side of another Atwood machine) in which an equation relating the various accelerations is an essential part of the solution. Omission of the acceleration equation also makes students lose sight of the fact that, in the Atwood machine for example, the two accelerations are *not* the same if the string is stretching. They are also kept unaware of the fact that the derived results do not apply to the short but finite time interval during which the waves bounce back and forth along the string after the system is released. Greater care in making the acceleration condition explicit prepares students going on to more advanced physics for the very careful equation writing they must do in more sophisticated theoretical analyses.

Many textbooks fail to include certain valuable aspects of problem solving that can quite readily be added by interested teachers:

1. Problems should occasionally include information irrelevant to the solution. Part of understanding a physical situation, and of solving a problem concerning it, is being able to discriminate relevance and irrelevance. Such discrimination requires practice (it does *not* develop spontaneously); but students are very rarely given such practice. When suddenly confronted with irrelevant information, many students force it into their problem solution to make sure they have "used all the data. " Then, when made to realize the irrelevance, they complain about "difficulty" and "unfairness" with no realization that, in real problems, they will have to winnow relevance and irrelevance on their own. The point and purpose of injecting irrelevant information should be discussed explicitly; students are perfectly willing to accept such practice once they understand what it has to do with their own intellectual development.

2. Problems, where appropriate, should require development of a *complete* algebraic solution for the unknown quantity before substitution of any of the given numerical values. Many students initially do not understand what is meant by "complete algebraic solution," even if this is demonstrated in text or lecture. When called upon to set up such a solution themselves in a new problem, they believe that any algebraic relation they write down as a start satisfies the requirement. There is tremendous resistance to continuing through several steps of combining algebraic equations and solving for unknowns prior to making numerical substitutions. This resistance can be reduced only by fostering practice and, at least in some instances, forbidding numerical substitutions at intermediate points in a solution.

3. Another skill that requires practice, and which is inadequately cultivated in textbooks, is that of interpreting both numerical and algebraic problem solutions in words. If this is to be done, it must be elicited by the teacher in both tests and homework since it is not fostered elsewhere. This is a rather sophisticated intellectual mode, and very few students develop it spontaneously, but they show gratifying progress with practice—gratifying to themselves as well as to the teacher.

   In the initial stages and without the practice, a correct numerical or algebraic result does not necessarily indicate understanding on the part of the student. When one calls for, and examines, the interpretation, one frequently finds very serious errors, misapprehensions, gaps in understanding, and failure to perceive some of the content and implications of the solution.

   Interpretation of numerical results should include an assessment of whether or not the order of magnitude makes sense, with supporting argument. Interpretation of algebraic solutions should include examination of extreme, limiting, or asymptotic cases, especially when these limits show whether or not the result makes sense and thus constitutes a check on the correctness of the solution.

   Problem solving as an aspect of student cognitive development and performance is a subject of active research in many fields and on various fronts. There is a large and rapidly growing literature. No attempt will be made to discuss this area of investigation in this book, since the concentration here is on the formation of underlying concepts that *precede* application in problem solving rather than on problem solving, itself. Interested readers, however, can gain entry to problem-solving literature relevant to physics teaching through references such as Lapp (1940), Larkin (1981) and (1983), and Reif, et al (1981), (1976), (1982), and (1984).

APPENDIX 3A

Sample Homework and Test Questions

1. a Suppose you are sitting on a chair that stands on the ground. Draw well-separated force diagrams of your body, the chair, and the whole earth. Describe each force in words [describing in words means: indicating the nature of the force (gravitational, contact, frictional) and stating what object exerts that force on what]. Show the relative sizes of the forces by using a longer arrow for a larger force and equal-length arrows for forces equal in magnitude. Identify the Third Law pairs.
   b Suppose you are standing on the ground in a shed and pulling vertically downward on a string that is attached to the bottom of a block that hangs from the ceiling on a rope. Draw well-separated force diagrams for your body, the string, the block, the rope, the shed, and the whole earth. Then do all the various things called for in (a). Now repeat the exercise for the case in which you pull the string at about 45° from the vertical.

2. a Suppose you are in the act of jumping vertically upward: Your legs are flexed and pushing on the floor so that your body is being accelerated upward. Draw well-separated force diagrams of your body and of the earth. Show the relative magnitude of various forces; describe each force in words; identify the Third Law pairs.
   b Draw the force diagrams for the situation that obtains just after your body leaves contact with the floor on your way up in the jump, and do the other things called for in (a).
   c Repeat for the situation at the top of the jump, for some point on the way down, and for the situation just after you hit the ground and your bent legs are slowing you down.

3. Suppose you throw a ball vertically upward and catch it when it returns. Draw force diagrams for your body, the ball, and the earth: (a) while you are accelerating the ball upward; (b) while the ball is rising after having left your hand; (c) while the ball is falling; (d) while you are in the act of catching it. Describe forces in words; show relative magnitudes; identify Third Law pairs.

4. a Suppose you start walking or running. In either case, you start with an initial velocity of zero and accelerate to a nonzero velocity in the horizontal direction. This means that there must have been an unbalanced horizontal force acting on you during the acceleration. Proceed to analyze and sense what is going on by actually performing the actions. As you do so, sense the direction of the

horizontal force acting on *you*; then pretend you are the ground and visualize the direction of force you would feel. Now draw well-separated force diagrams of both you and the earth. Describe each force in words and identify Third Law pairs.

b How do the diagrams you have drawn in (a) differ from the ones when you are standing still? Label, on the appropriate diagram, the force that imparts acceleration to your body. What is the role of friction in this system? Could you walk or run in the absence of a frictional force between the soles of your shoes and the ground? Why or why not? Discuss, in terms of the force diagrams, what happens when you try to walk on an icy surface.

c Discuss the following statement (i. e., is it correct and accurate or is it incorrect? Explain your answer, and, if you believe the statement to be incorrect, alter it in such a way as to make it correct.): "When we walk or run in the forward direction, we push on the ground with a horizontal frictional force directed toward the rear. The ground, in turn, pushes on us with a horizontal frictional force in the forward direction. We are accelerated by the force exerted by the *ground*, not by the force that we exert."

5. Following the same sequence as in question 4, analyze the forces that act when a car accelerates along a road. (If you have available a spring-wound or electrically driven toy car, place it on your hand and sense the direction of the force exerted on you by the wheels as the car accelerates from rest along your hand.) Give a careful verbal description of the force that accelerates the car: how does this force originate and what object exerts it on what? Is it correct to say that "the car is accelerated by the force exerted by the engine"? Explain your answer.

6. A railroad car is in *uniform* rectilinear motion along its track. Observer A performs experiments inside the car while Observer B, outside the car at a fixed location along the track, watches these experiments as A comes by. In each of the following experiments with various objects, describe what each observer will see the object doing relative to his *own* frame of reference (i. e., how will A see the object behaving relative to the inside of the car, and how will B see it behaving relative to the ground?). As part of your description be sure to sketch a diagram of what each observer sees to be the trajectory of the moving object.

a Observer A puts a ball down on the perfectly level floor of the car. (Will the ball stay wherever he places it?)

b Observer A makes the ball roll on the floor in a straight line at uniform velocity relative to himself: (i) toward the front of the car; (ii) toward the back of the car; (iii) directly across the car.

c Observer A lets the ball fall out of his hand from a point several feet above the floor of the car.

d Observer A throws the ball vertically upward and catches it as it comes down.

e Observer A suspends a pendulum bob on a string from the roof of the car. How will the pendulum appear to hang?

f Observer A throws the ball with an initial horizontal velocity directed toward the front of the car.

g Observer A throws the ball with an initial horizontal velocity directed toward the rear of the car. (As part of this problem, consider the important special case in which the *magnitude* of this horizontal velocity is equal to the magnitude of the horizontal velocity of the car relative to the ground.)

h Observer A has an aquarium tank of water resting on the floor of the car. (How does the surface of the water behave—is it level or sloping?)

7. Consider once more all of the situations in question 6, but the car now has a uniform *acceleration* in the forward direction. Describe, with appropriate diagrams, what each observer sees relative to his frame of reference. What would be the difference between the personal sensations experienced by A and B?

8. Suppose you are sitting on a platform that is subject to very small frictional forces—for example, a large frictionless puck on a huge air table or a boat floating on water. You are initially at rest, and you have no paddles and no way of pushing yourself along the surface, but you wish to accelerate yourself and your platform to a finite velocity in some particular direction. You have available a basket of balls or stones.

   a What might you do to accelerate yourself in the direction you wish? What might you do if you wish to slow down or stop? Analyze this situation carefully, drawing well-separated force diagrams for a ball, yourself, and the platform. Explain, in terms of Newton's Laws, what is happening to make your transportation possible.

   b In the light of the thinking you have done in part (a), analyze, with appropriate force diagrams, what must be happening (i) when you paddle a boat through the water; (ii) when a propeller accelerates a boat or an airplane. Why is the action of the propeller still necessary once the boat or plane has been accelerated to desired velocity? Why is a propeller useless on a space ship?

   c In the light of the thinking you have done in part (b), describe what must be happening in rocket propulsion of a space ship: What happens to the fuel that is burned in a rocket engine? Draw well-separated force diagrams for the ship and for a parcel of gas (resulting from the burning of the fuel) being ejected from the ship. What must be done in order to slow down the space ship? What must be meant by the technical term "retro–rockets"?

9. We have given the name "weight" to the gravitational force exerted by the earth on all objects, including our own bodies. Note, however, that we do not have any direct feeling or sensation of the force of gravity—the pull of the earth on us. As we stand on the ground or sit on a chair, what we feel is not the pull of the earth but rather the upward force exerted on us by the ground or by the chair. When we jump from a height and are falling freely, our principal sensation (apart from the rushing air) is the *absence* of the upward force to which we are so accustomed, not a downward pull.

   a Verify the preceding statements by consciously examining your own sensations in these various circumstances: standing, sitting, jumping vertically upward, or jumping off a chair. (The interval during which you are out of contact with a support when you jump is very short, but you can still check on the fact that you do not feel anything pulling on you.)

   b Now suppose you are standing on a bathroom scale in an elevator that is standing still. Draw force diagrams of yourself, of the scale, and of the floor of the elevator where it interacts with the scale. How does the upward force exerted by the scale platform *on you* compare in magnitude with your weight? How do you interpret the reading on the scale? (Explain your answers in your own words.)

   c Suppose that, instead of standing still, the elevator is moving up or down with *uniform* velocity. Draw the same diagrams called for in (b), and answer the same questions. When the elevator is moving up or down at uniform velocity, how does the total upward force exerted by the cables on the elevator compare with the total weight of the elevator and its contents? (Explain your answer.)

   d Suppose that you are still standing on the scale in the elevator, and the elevator is *accelerating upward*. Stop and recall the sensations you have felt during the short interval when an elevator accelerates upward before attaining a uniform velocity. Draw the same force diagrams called for in (b). How must the magnitude of the upward force exerted by the scale platform on you compare with your own weight? How must the reading on the scale compare with the reading you make when the elevator is standing still or moving up or down at uniform velocity? Is your conclusion about the size of the upward force exerted on you by the scale consistent with the sensation you experience during the interval of upward acceleration? How must the total upward pull of the cables on the elevator compare with the weight of the elevator and its contents? What happens to the magnitude of the reading on the scale and the total upward pull of the cables as the acceleration is made larger and larger? (Explain your answers.)

e Suppose you are standing on the scale in the elevator, and the elevator is accelerating *downward* with an acceleration smaller than 9.80 m/sec^2. Go through exactly the same sequence of diagrams and questions as you did in part (d), including visualizing what happens as the downward acceleration is made larger and larger.

f Suppose the upward pull of the cables on the downward accelerating elevator is made smaller and smaller until the elevator is falling freely. Under these circumstances, what forces are acting on *you*? What is the magnitude of the upward force exerted on you by the scale platform? What is the numerical reading on the scale? Has your weight (the gravitational force exerted on you by the earth) changed?

g The elevator is completely enclosed, and you cannot see out; your frame of reference is the interior of the elevator. Describe what you would observe happening in the freely falling elevator as you performed various simple experiments such as: (i) holding a ball in your outstretched hand and letting it go; (ii) pushing the ball off gently in various directions (up, down, sideways, etc.); (iii) suspending an object on a spring balance; (iv) attaching a pendulum bob on a string to the ceiling of the elevator and trying to swing the pendulum. (v) What would your own sensations be like in the freely falling elevator? Could you "stand" on the floor as you do in ordinary circumstances? What would happen if you pushed yourself away from the floor or a wall? (vi) Invent some additional questions and thought experiments of your own.

h Suppose you are inside a capsule or elevator that is projected upward by a powerful catapult. (i) Describe the sensations you would have and the reading you would observe on the platform scale during the interval in which the catapult is projecting you upward. (ii) Describe the conditions you would observe in the capsule as it is rising vertically after having left the catapult. How do these conditions compare with those you described in the freely falling elevator?

i Is it, in principle, possible to accelerate an object downward with an acceleration larger than that of free fall, that is, larger than 9. 80 m/sec^2? What would have to be done to impart such an acceleration to a stone or to an elevator? What would conditions be like in an elevator accelerating downward with an acceleration larger than g? Repeat some of the thought experiments suggested in part (g) and describe what you would observe. Where would you be likely to find yourself standing during the interval in which such an acceleration is maintained?

10.a Suppose you are sitting in a car that is speeding up. Draw well-separated force diagrams of the following objects: your own body; the seat in which you are sitting (apart from the car); the car (apart from the seat); the road surface where the tires and road interact. Assume the car has rear-wheel drive. Describe each force in words; show larger forces with longer arrows; identify the Third Law pairs. Explain carefully in your own words how the force imparting acceleration to the car originates.

b Suppose a pendulum bob hangs from the ceiling of the car. Draw well-separated force diagrams for: the bob; the string on which it hangs; the car ceiling where it interacts with the string. Describe each force in words; identify the Third Law pairs.

11. Suppose you are sitting in a moving car that is slowing down under the influence of four-wheel brakes. Do for this case all the things called for in question 10. Explain carefully in your own words how the force slowing down the car originates.

NOTE: The preceding questions are designed to start with situations in which the student is called upon to analyze his own bodily sensations in terms of the dynamical concepts being developed. Insights so formed are then extended to situations that can only be visualized in the abstract or to unfamiliar situations that are "transformations" of the familiar ones.

12. Consider a pendulum bob on a string attached to the ceiling. The bob is pulled off to the left and let go. Sketch the trajectory (the path) that would be followed by the bob in each of the following instances: (i) the string is cut at the instant the bob is about halfway down to the lowest point in its swing; (ii) the string is cut at the instant the bob has reached the lowest point in the swing; (iii) the string is cut at the instant the bob is about halfway up to the highest point it would reach on the right; (iv) the string is cut at the instant the bob is *at* the highest point of its swing on the right.

13. In connection with his development of the concept of universal gravitation and the role of mass in the inverse square law, Newton did a highly significant experiment: He constructed a hollow pendulum bob into which he could put objects made of different materials, such as wood, iron, gold, copper, salt, and cloth. (One can measure the period of a pendulum very precisely by counting swings over a sufficiently long time.) Newton found that the period of the pendulum did not change observably with very different materials inside the bob. Why did Newton feel it important to do this experiment? What ideas does the null result support?

NOTE: Although many textbooks include qualitative questions intended to evoke some of the kinds of thinking described in this chapter, most fail to achieve their desired end for several reasons: the questions are fragmented and isolated in such a way that they do not help the student form a synthesis of the ideas under consideration; the questions are too sophisticated and too abrupt and do not provide sufficient Socratic guidance to the student who is having difficulty; there is insufficient spiraling back and repetition in subsequent, richer context. Apart from weaknesses in design, the questions also have very little effect because they are rarely, if ever, invoked in testing.

In addition to questions of the type illustrated above, most of the questions used in research on student learning, and described in the body of this chapter, are useful in helping students make the errors and encounter the contradictions that lead them to refine their conceptions and master the insights we wish to instill. These questions supplement and make more effective the conventional end-of-chapter problems provided in textbooks. The questions have very little effect, however, if they are not assigned, discussed, and included in testing and grading.

CHAPTER 4

Motion in Two Dimensions

4.1 VECTORS AND VECTOR ARITHMETIC

Most presentations of the concept of a vector start with the representation of displacements in two dimensions and develop the process of addition of such quantities in an intuitive way. This is, without question, the most reasonable and effective starting point, and students have relatively little difficulty with the ideas in the early stages. Trouble begins to set in, however, when abrupt jumps are made to other operations and to other vector quantities.

To generate the process of subtraction, for example, some texts simply assert that, to obtain the negative of a vector, one reverses the direction of the original arrow (or one simply multiplies by the scalar factor -1). Although the reason is quite obvious to *us*, it turns out to be far from obvious to many students; they hesitate to ask for a reason, and they memorize the assertion without understanding. By the time one wishes to find the vector change in velocity over a small angular displacement in circular motion, for example, the process has been forgotten, and the derivation of centripetal acceleration is not understood.

A more effective way of introducing the operation of subtraction is to adopt the systematic procedure of mathematics and ask what must be *added* to a given vector to obtain a zero vector. (In the technical terminology of mathematics, one is generating the "additive inverse.") This serves to define subtraction by tying it to the original starting point, namely addition, and gives the student logical continuity rather than abrupt change and unsupported assertion.

Many texts go on to assert that, since they also have "magnitude and direction," velocity, acceleration, and force are also vector quantities in the sense of obeying the same arithmetic as displacement. This, again, is far from obvious to the students; to them, adding velocity arrows appears very different from adding displacement arrows, and acceleration arrows are totally incomprehensible. The transition from one kind of quantity to another requires some discussion if vectors and their arithmetic are going to be used. It is not, in fact, "easier" for the learner if the logical questions are concealed or ignored.

To do this honestly requires at least some discussion of multiplying displacement vectors by constants larger and smaller than unity, of what happens when one divides the displacement by the time interval in which the displacement occurred, of units and dimensions of such quantities, of acceleration as a vector, and of the connection between force and acceleration. (It should be noted that the path is much smoother if attention was given to the algebraic signs of acceleration in rectilinear motion, as advocated in Section 2.8, and to careful operational definition of force as suggested in Sections 3.4–3.6.)

Another property of vectors, frequently taken for granted in instruction without being made explicit, is that of "movability." Many students tenaciously hold an initial view that vectors are "attached to points." One can see how this notion gets planted: displacements, the first vectors encountered, begin at a fixed position, and a sequence of displacements proceeds from point to point with the arrows head to tail, each tail rooted at the initial fixed position; velocity vectors appear to be attached to particles; forces act on objects at a point. When it comes to recognizing that forces can be considered as acting anywhere along their *line* of action or to transferring velocities from the diagram of circular motion in order to draw the vector diagram for change in velocity, one encounters resistance and disbelief. One is doing things that "don't make sense," are not understood, and are therefore avoided, or handled incorrectly, in the solving of problems on tests and homework. The fact that vectors are *not* attached to points requires explicit discussion if it is to be understood—and used in attacking problems.

Many students would benefit from more exercises and drill in graphical handling of vector arithmetic than are usually available in texts. Such drill, with immediate feedback and reinforcement, could well be supplied through computer-based materials exploiting currently available graphic capability. Unfortunately, at the time of writing, very few well-designed materials seem to be available commercially.

4.2 DEFINING A "VECTOR"

How rigorously one should pursue the definition of "vector" is a matter of judgment, and I myself do not feel that it should be pressed very hard in lower level courses or with students who are not going on to more advanced mathematics and physics. For the few front runners, however, and for students likely to go on as physics majors and engineers, this question should not be ignored or glossed over.

A first step is to show students that just saying "magnitude and direction" is *not* sufficient to define a vector. Finite angular displacements, for example, do have both magnitude and direction but are not vector quantities because they do not commute on addition. This fact can be readily demonstrated for students by having them rotate a book through two successive 90° displacements about two different axes and showing that the book winds up in two entirely different final orientations if the order of the rotations is reversed. This can be directly contrasted with the fact that displacements, velocities, and the like do commute when being added (or subtracted) and that this property is an essential part of the definition. (In order to understand what something *is*, one must also understand what it is *not*.)

Introducing the students to finite angular displacements has a very significant payoff later on when one wishes to show that angular velocity can be represented as a vector quantity. (This is usually asserted without any attempt at justification, but such unsupported assertions encourage memorization without understanding.) One can examine infinitesimal angular rotations to show that they *do* commute in the limit of zero displacement, and one is on the way to making the angular velocity vector a rational and plausible concept instead of a mystery to be memorized.

Another aspect worth bringing out is the mathematical view of vectors as "ordered *pairs*" of numbers. This ties in especially well, in due course, with the notion of rectangular components.

Commutation in addition and subtraction does not, of course, complete the definition of a vector. The final part of the definition resides in behavior with respect to transformation under rotation of coordinate axes, and this behavior is crucial to the final distinction between Cartesian and pseudo vectors. I find that an understanding of these distinctions, and of the need of extension of the basic definition beyond the requirement of commutation in addition and subtraction, comes far more easily to students in more advanced courses if they have the advantage of having been gradually exposed, in introductory courses, to the simpler ideas outlined above, instead of suddenly encountering all of them de novo at the advanced level. Such preparation, in other words, operates to give students a very much better grasp of the nature of the vector cross-product.

4.3 COMPONENTS OF VECTORS

The concept of orthogonal (or Cartesian) components of vectors seems so simple and transparent to teachers, and manipulations, when the Cartesian axes are given in a problem, are so easily memorized by students, that many significant student difficulties in this area go unnoticed. Interviews with students, however, reveal very significant gaps in understanding.

Consider the two diagrams shown in Fig. 4.3.1. If one draws diagram (*a*) and asks the student to "show *graphically* how large an effect the vector represented by the arrow (perhaps a force or a velocity) has along the direction indicated by the line," many students find themselves at a loss and are unable to answer the question. If one draws diagram (*b*) and asks the same question, still more students are unable to answer. (In the latter case the difficulty has been enhanced by the fact that the line does not pass through the tail of the arrow. See comments in Section 4.1 concerning this aspect.) Without an angle marked with a familiar symbol, with no Cartesian axes shown in familiar orientations, and with the word "component" not used, nothing triggers the student to bring out the memorized formulas with sines or cosines. In other words, students exhibiting this difficulty have not formed an understanding of the concept.

There are quite a few ways of introducing concrete thinking and experience that help students assimilate the concept, and I do not see any particular one as superior to all others. It is necessary, however, to give the idea more attention

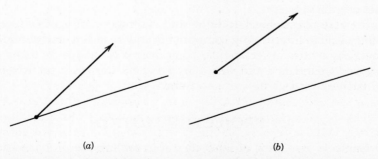

(*a*) (*b*)

Figure 4.3.1 What is the magnitude of "effect" of the vector in the direction indicated by the line?

than is accorded in most texts, and I have, over the years, narrowed down on the following mode in my own practice: Starting with an instance in which the arrow in Fig. 4.3.1b represents a displacement and the line represents a wall, we set up the corresponding situation in the classroom (or lecture room). The beam of an overhead projector is directed perpendicular to the wall so that the shadow of the displaced object (a student, a hand, a ball) is seen to be displaced along the wall while the object itself is displaced as indicated by the arrow. (We concentrate at this point only on the change of position and not on the motion.) We describe the displacement of the shadow as the "effect," along the plane (or direction) of the wall, of the actual displacement of the object in the room. A similar observation is then made of the displacement of the shadow along the orthogonal wall.

We then investigate a few changes in order to enrich the context: what happens to the two "effects" when the original displacement is parallel to one wall and perpendicular to the other? Under what circumstances are the two "effects" equal in magnitude? What happens to each "effect" as the magnitude of the original displacement is kept fixed while the angle relative to one wall is increased or decreased? Must the axes along which we desire to measure "effects" be oriented only horizontally and vertically?

After such concrete experience, the students can quite easily be led to describe the corresponding pencil-and-paper procedure of dropping perpendiculars to the orthogonal axes and to recognize how the relevant sine and cosine expressions arise. Now that the *idea* has been fully established operationally, it is appropriate to introduce the technical *name*: "rectangular components of the displacement."

In following stages, at a pace consistent with the readiness and sophistication of the students, one can examine (or visualize) the behavior of the two shadows of a continuously moving object and extend the concept to include components of velocity and acceleration. Logical extension to the notion of components of force follows from whatever has been done in constructing the concept of "force" to begin with. For many students, however, it is an important part of concept building actually to pull a block (or some other object) along the floor or table with a string oriented at various different angles to the horizontal and to sense that their pull has two simultaneous but separate effects, one vertical and one horizontal, and that the magnitudes of these two effects vary with the angle of the string in essentially the same manner as did the magnitudes of the shadow displacements and velocities along the wall.

After one has built such an underlying structure, going back to a question such as that asked in connection with Fig. 4.3.1 (but with the picture rotated into an entirely different orientation) becomes a first test of whether the student has begun to absorb the concept.

As in the case of graphical addition and subtraction of vectors (Section 4.1), supplementary drill in interpreting components and in adding and subtracting vectors arithmetically by use of rectangular components is needed by many students. Computer-based materials could very effectively fill the gap usually left by textbooks and teachers for lack of space and time.

4.4 PROJECTILE MOTION

Examining Galileo's own view of what he did in solving the problem of projectile motion and examining, at the same time, some of the logical and epistemological questions arising in the story provides an especially valuable opportunity to enhance the "scientific literacy" of the entire spectrum of students in introductory physics courses—from nonscience majors to future engineers and physicists. The degree of mathematical sophistication invoked can be very different for different groups, but the important ideas can still be brought out and discussed in an intellectually honest way with any group of students.

Galileo's solution of the problem of projectile motion in the idealized limit of zero air resistance represents one of the very first deliberate uses of the concept of superposition in science, and, in the *Two New Sciences*, he describes his approach, and its conceptual importance, with utmost clarity:

> *In the preceding pages we have discussed the properties of uniform motion and of motion naturally accelerated. . . . I now propose to set forth those properties which belong to a body whose motion is compounded of two other motions, namely, one uniform and one naturally accelerated. . . . This is the kind of motion seen in a moving projectile; its origin I conceive to be as follows: Imagine any particle projected along a horizontal plane without friction. . . . This particle will move along this plane with a motion that is uniform and perpetual, provided the plane has no limits. But if the plane is limited and elevated, then the moving particle, which we imagine to be a heavy one, will, on passing over the edge of the plane, acquire, in addition to its previous uniform and perpetual motion, a downward propensity due to its own weight; so that the resulting motion. . . is compounded of one which is uniform and horizontal and of another which is vertical and naturally accelerated.*

The key phrases in the preceding quotation are "resulting motion" and "compounded of." With these deceptively simple terms, Galileo reveals how far he has come from the scholastic point of view in which motion was seen only as a whole and never conceived as compounded. These phrases also articulate his inductive guess that the horizontal and vertical motions of a projectile do not influence each other, that is, that they behave as though each alone were present and that the net effect is a simple combination of the two independent motions calculated separately. This is a hypothesis about *physics* and is not just a matter of definition; verification is required—just as with the hypothesis that free fall is uniformly accelerated.

To justify the superposition, we must know the answers to *two* separate questions: (1) Does imparting a horizontal velocity to a particle in any way alter the vertical acceleration and velocities it normally acquires in free fall along a straight line? (2) Conversely, does the presence of a vertical acceleration and vertical velocity alter the horizontal velocity a particle might initially have? It must be emphasized that these are indeed two *separate* questions and that, if one is true, the converse does *not* automatically follow.

Galileo, of course, could not test these conditions experimentally with any great degree of precision (although he does claim that an object dropped from the top of the mast of a moving ship hits the deck at the base of the mast). He had to depend principally on internal consistency and overall agreement of the derived results with experiment. In this modern age of electronic devices and high-speed photography, it is possible to make direct tests that can be shown to the students. The widely reproduced stroboscopic photograph (all editions of *PSSC Physics* and many other texts) showing that two balls released simultaneously, one dropped vertically and the other having an initial horizontal velocity, occupy the same vertical levels at successive intervals of time, gives an affirmative answer to question 1. Films showing that an object dropped from a mast or vertical standard moving at uniform velocity falls directly along the mast and lands at its base, give the affirmative answer to question 2. (The logical necessity of examining *both* of these questions is, unfortunately, glossed over in many presentations.)

To help register an understanding of the physics of projectile motion, it is important to have students draw *separate* diagrams of (1) the force acting on the projectile at various different points of the trajectory (e. g., a point on the way up; the top of the flight; a point on the way down); (2) the acceleration at the same points; (3) the horizontal and vertical velocity components at each point; (4) the total vector velocity at each point. A similar array of diagrams should be drawn for the frictionless puck while it is moving along the table and for its trajectory after it flies off the edge of the table. Illustrations in texts sometimes have arrows for different quantities on the same diagram. This is invariably a source of serious confusion.

Still another valuable exercise is to sketch the trajectory that would be followed by the pendulum bob after its string is cut: (1) at some point in the downward swing; (2) at the lowest point of the swing; (3) at some point during the upward part of the swing; and (4) at the instant of the end of the swing. (The role and importance of such exercises was initially discussed in Section 3.16, and a statement of this problem is given in item 12 in the appendix to Chapter 3.)

In the *Two New Sciences* Galileo is doing much more than just presenting his mature insights into strength of materials and kinematics. He is also propagandizing the use of mathematics in the description and understanding of natural phenomena. After deriving the equation of the trajectory in projectile motion, he proceeds to show that the range must be a maximum at an angle of elevation of 45°.[1] His delight in this result is apparent in his having Sagredo say:

> *The force of rigid demonstrations such as occur only in mathematics fills me with wonder and delight. From accounts given by gunners, I was already aware of the fact that in the use of cannons and mortars the maximum range . . . is obtained when the angle of elevation is 45°. . . ; but to understand why this happens far outweighs the mere information obtained by the testimony of others or even by repeated experiment.*

If the above quotation is used, it should be accompanied by a discussion of what scientists mean when they use the phrase "understand why . . ." in this and similar contexts. Students take this phrase far too literally unless it is explicitly qualified as referring to explanation *in terms of* an array of simpler, plausible, preferably well established, ideas that underlie a very much wider range of phenomena but that themselves have no explanation and may have an endless regression of unanswered "why" questions behind them, as, for example, why all objects fall with the same acceleration regardless of their weight or why the orthogonal components of projectile motion are independent of each other.

4.5 PHENOMENOLOGICAL THINKING AND REASONING

The purpose of this section is to illustrate certain kinds of thinking and reasoning that few students enter into spontaneously but that can be cultivated and

---

[1] For students who have gotten far enough with the calculus, this is a valuable opportunity to find the extremum by differentiation and to ascertain whether it is a maximum or a minimum; they have rarely had the experience of using this mathematical technique in a physical problem. For students who have not had calculus, it is perfectly possible to find the maximum in an intuitive, non-rigorous way. This is well worth doing both for the sake of appreciating Galileo's argument concerning the role of mathematics and for the sake of connections with other physical situations in which similar maxima, associated with the product of the sine and cosine functions, occur.

enhanced through practice. The practice is best afforded through structured sequences of questions, preferably related to situations that are easily set up or that are commonly encountered in everyday experience. The specific examples given are not essential in themselves. They are merely illustrative, and many alternate versions and variations, of equal or greater effectiveness, can be generated by the individual teacher to best suit a particular group of students.

The algebraic results describing projectile motion, for example, should not simply be left for substitution of numerical values in end-of-chapter problems. Students should be led to interpret the algebraic results: What factors determine the range of the projectile? How does the total vector velocity change through the course of the flight? If a projectile is fired horizontally, what factors determine how far it has dropped below its initial level as the horizontal distance from the firing point increases? How can this vertical drop be decreased at a given range? What is the significance of the fact that the mass of the projectile does not appear in the equations?

Some years ago, on a Ph.D. qualifying examination, I asked why the cathode beam was not observed to be deflected by gravity when the electric and magnetic deflections were so pronounced. Upwards of 40% of the students taking the examination answered that this was due to the smallness of the electron mass, clearly indicating that they had not been led to engage in some of the thinking suggested above.

A situation that can be invoked to help students register a deeper understanding of Newton's Second Law is that of the ubiquitous problem, with zero friction and a massless string, illustrated in Fig. 4.5.1. This problem is usually posed for algebraic solution or numerical calculation (or both), but students are rarely asked to bring out the essential physics by interpreting their results. They should be asked to extract what their algebraic results say would happen to the acceleration of the system and the tension in the string if m_B were made very much larger than m_A or very much smaller than m_A. Do these predictions make physical sense? Why or why not?

Finally, and most importantly, they should be asked how the force exerted by the string on the cart compares in magnitude with the weight of m_B, "comparing" meaning indicating whether equal, larger, or smaller. (I have, over the years, been asking this question of graduate students, most of whom are teaching assistants and have been helping students with this homework problem. Initially, virtually all of them have said that the two forces are equal in magnitude, and only about 30% changed their minds when I asked whether they would like to reconsider their answer.) Giving the response "equal" is an indication of inadequate comprehension of the phenomena in question and of the phenomenological implications of the Second Law. Being able to get correct answers to the conventional problems is no assurance of understanding.

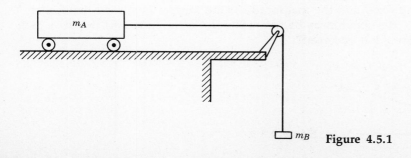

Figure 4.5.1

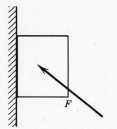

Figure 4.5.2

A simple situation that affords a good exercise in phenomenological thinking and that does not lend itself to blind substitution in formulas is shown in Fig. 4.5.2: A person holds a block against a wall by exerting an inclined force **F**. There is friction between the block and the wall. Does the block move up or down or stand still?

First, students should be led to draw the force diagrams for the block, the wall, and the hand holding the block, in accordance with the approach advocated in Section 3.16. The simplest version of the problem is then to give numerical values for the force, the mass of the block, and the coefficient of friction and to ask what happens in the given circumstances. The most sophisticated version is to ask for a discussion in algebraic terms alone. (The latter is an excellent exercise for students who are interested in theoretical work.) The point of this form of the question is that students are not told what to calculate; they must make some decisions as to what to do, what to look for, and how to interpret the results. This induces some phenomenological thinking and militates against reliance on memorized procedures.

The preceding problem might not be very interesting if it were an isolated situation unrelated to others to be encountered later, but this need not be the case. A valuable and closely related problem is that of the person "stuck" to the wall of the rotating cylinder in the amusement park. Another related problem is that of the electrically charged balloon sticking to the wall of the room. When students encounter these problems, weeks apart, as the course progresses, one begins to see the grins and glances that betoken recognition of an old friend in a new context. Were it not for such repetition or recycling, the ideas would be forgotten.

Finally, a two-dimensional situation rich in physical effects but rarely exploited as effectively as it might be is shown in Fig. 4.5.3. As usual, the force diagrams should be drawn first. Then the questions can begin: If the block is not sliding, is the frictional force necessarily zero? What range of values might the frictional force have? If the block is sliding, is the frictional force equal to $\mu m g$? Why or why not? What happens to the acceleration as the angle θ is increased, starting at zero, while the magnitude of the force **F** is kept fixed? What happens if the angle is kept fixed and the magnitude of the force is increased indefinitely? What are some of the differences between the given situation and that in which the direction of force **F** is reversed?

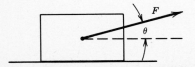

Figure 4.5.3

4.6 RADIAN MEASURE AND π

Very few students in introductory physics courses have an understanding of radian measure, even if they have been exposed to the concept somewhere along the line. A few may have memorized something about arc length and radius, but they are rarely able to recover the definition precisely or to use it for dealing with angular displacement or angular velocity. For a great many students the problem begins not with radian measure itself but with the meaning of π.

If asked what π means, they are likely to mutter a formula, either for circumference or area, without being sure of which one the formula is for, but they are unable to explain what the symbol means or where the formulas came from. What is needed here, of course, is not a development of π to huge numbers of significant figures by rigorous methods of the calculus but evidence from direct measurement of circles—the primitive kind of evidence that must have been noted in ancient times. The calculation by methods of analysis can come later for those who continue to that level.

My own procedure with college students who do not understand the meaning of π is to have them use string and rulers to measure the circumference of every cylindrical object they can find around the laboratory and to measure the diameters by placing the objects between well-squared wooden blocks. They keep a running graph of circumference versus diameter. (It should be noted that this exercise is *preceded* by some of the graphing and arithmetical reasoning described in Sections 1.10 and 1.11, and the students have acquired some sense for the meaning of the slope of a straight line graph.) Almost every time I have conducted such a session (with both pre- and in-service elementary teachers, for example) a voice has sounded through the room with something like "So *that's* what they meant by π!!" This "discovery" *must* be followed by exercises and interpretation such as that illustrated in Chapter 1; otherwise the incipient understanding is lost.

Once the meaning of π is understood at the verbal level defined above, students can be led to see that the ratio S/R (where S denotes the length of intercepted arc and R denotes the radius) must have a fixed value for any given angle regardless of the size of R; that the value runs from 0 to 6.28 for the complete circle; and that, because there is a unique value for each angle, the value of S/R can be used as a measure of the angle subtended. (Very few students in any introductory course I have taught, including calculus-physics, have known the meaning of the word "subtend. " If this term is used, one must be sure to define it explicitly.)

To enrich the context and add significance to S/R, it is worth pointing out that the ratios to which we give the names "sine," "cosine," and "tangent" also have fixed values for any given angle, regardless of the size of the circle on which the ratios are taken, and thus also measure the size of angles. These ratios, however, do not vary *linearly* with the size of angle (conceived of as a fraction of a complete circle) as does S/R, and they are therefore used in an entirely different way.

Finally, the ratio S/R gives us a "natural" way of measuring angles by virtue of its connection to π, the intrinsic property of *all* circles; it is not "artificial" as is measurement in degrees. The principal property of the number given by S/R is that it is *dimensionless*; it is a "pure" number—this being part of the meaning of "natural" in this context. It should always be emphasized in this connection, however, that, although this new angular measure is dimensionless, it is *not* unitless: the *unit* is called "one radian" and can be carried around accordingly in calculations, but it must not be confused with *dimensions* such as mass, length, and time.

It should be noted that the preceding discussion follows the precept "idea first and name afterwards" advocated in Chapter 2. We thoroughly examine the properties of the ratio S/R before writing the expression

$$\theta \equiv S/R \qquad (4.6.1)$$

for subsequent use. This sequence of development helps reduce the tendency to memorize the "formula" without understanding that it is a *definition*, invented because of its practical utility.

Students who go on to levels at which they will encounter substitution of the value of the angle itself for the sine or tangent at small angles are helped by being led to confront the fact that the substitution is valid only in radian measure and not in degrees. This is very easily done with the now ubiquitous hand calculator, and most students enjoy making use of their calculators in this way. One way of presenting the problem is to suggest that they tabulate angles in both degrees and radians along with the sines and tangents, starting at larger angles and going toward smaller ones, and discover the angle at which the three values agree to two significant figures, three significant figures, four, and so on.

Finally, for students taking the calculus, it should be emphasized that the formulas for the derivatives of the sine and cosine functions are valid only if the angle is measured in radians since the limit relations

$$\lim_{\Delta\theta \to 0} \frac{\sin\Delta\theta}{\Delta\theta} = 1 \quad \text{and} \quad \lim_{\Delta\theta \to 0} \frac{(1 - \cos\Delta\theta)}{\Delta\theta} = 1 \qquad (4.6.2)$$

used in the derivations are valid only if the angle is measured in radians.

The principal reason that students previously exposed to radian measure have no understanding of the concept is not that it was not "explained" correctly (the explanations are usually basically sound); it is that they have never been required to discuss and explain any of the reasoning in their own words. Without such requirement, the tendency is to memorize formulas and avoid the visceral effort that accompanies striving for understanding.

4.7 ROTATIONAL KINEMATICS

The kinematics of circular motion in a plane is usually glossed over very quickly because of the obvious parallelism to rectilinear motion. For students who have genuinely mastered the concepts and relations of rectilinear kinematics, this is appropriate since unnecessary repetition would waste their time. As pointed out in Chapter 2, however, many students do not master the concepts on the first go-around, and some form of spiraling back is essential. The altered context makes a somewhat more careful treatment very worthwhile for this group, and the pace can be a bit more rapid than previously.

At this stage, polar coordinates are new to many students. Even if they have seen them in mathematics, they have not associated such coordinates with representation of physical events, and they should be given a chance to absorb the physical connections. They must see that angular positions, like positions on the number line, do not, in general, represent displacements of the moving object, which may never have occupied the zero position. They must have a chance to see how the algebraic signs arise for positive and negative angular displacements, for positive and negative angular velocities, and for positive and negative angular accelerations. The difficulties in discriminating between the various concepts (described in Chapter 2) arise over again for many students.

Another layer of difficulty is that of the arithmetic reasoning connecting angular displacement with linear displacement along the arc, angular velocity with tangential velocity, angular acceleration with tangential acceleration. The reasoning involves not only an understanding of radian measure (Section 4.6) but also the perception that instantaneous tangential velocity corresponds in magnitude to the length of arc that would be swept out in one second if the angular velocity remained constant at the given instantaneous value. Many students must actually have the concrete experience of rolling a length of string off a cylinder in order to grasp the relations. While this is being done, it saves time to have them also examine the rolling of a wheel and to absorb the connection between angular velocity and distance rolled in absence of slipping. The two contexts reinforce each other, and better understanding ensues.

After the kinematic relations have been developed, it is effective to demonstrate rectilinear and circular motions that correspond to each other. An especially useful concrete example is furnished by a wheel with a weight hanging from a string wound around the axle. If the weight is initially wound up to the axle, the wheel accelerates uniformly as the weight falls, and one can compare the parameters and the equations representing the observed unidirectional angular and rectilinear motions respectively. If the weight starts near the floor and the wheel is given an initial spin, the wheel slows down as the weight rises, passes through zero instantaneous velocity, and reverses direction, speeding up as the weight falls. Many students do not perceive that the behavior of the wheel in these circumstances is identical with the behavior of a ball thrown vertically upward until they are led to articulate the connection.

As pointed out in Sections 1.6–1.10, many students in introductory physics courses start out with serious difficulties due to lack of mastery of arithmetical reasoning with ratios and division. Those who have not acquired such mastery by the time they arrive at rotational kinematics have great trouble with the simple connections among angular velocity, period of revolution, and frequency of revolution. If not required to explain the algebraic relations in their own words, they memorize the relations without understanding and are unable to use them in a sequence of reasoning. In solving problems, they mechanically substitute numbers in formulas, hoping to arrive at the "answer" given at the back of the book. Having no other frame of reference, they consider the task accomplished when the "answer" emerges, and they are unaware that they have no understanding of the analysis. For such students, encounter with the simple rotational quantities is a valuable opportunity to spiral back, in altered context, and strengthen their facility with ratio reasoning.

4.8 PRECONCEPTIONS REGARDING CIRCULAR MOTION

Many students come to the study of dynamics of a particle in circular motion with a number of very deeply rooted preconceptions—preconceptions that tenaciously persist beyond earlier science or physics courses where the subject might have been considered. One such preconception is the expectation that curvilinear motion imparted to an object by some constraint persists after the constraint is removed. For example, when shown the apparatus in Fig. 4.8.1 in which a ball is set in motion around the inside of a open hoop lying on a table, and asked what will happen when the ball reaches the open section, many students expect the

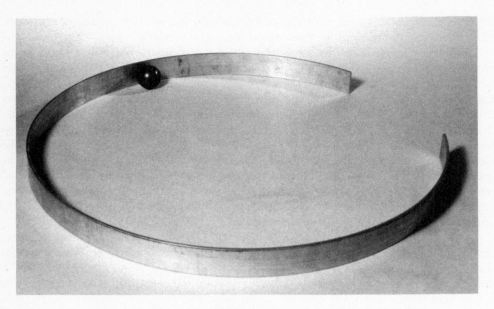

Figure 4.8.1 Ball rolling around on a table inside an incompletely closed hoop.

ball to follow a curved path at the opening and continue around the inside of the hoop. One hears gasps of astonishment when the demonstration is performed, and the ball leaves the hoop on the tangent line without continuing around.

McCloskey and his co-workers [McCloskey, et al. (1980); McCloskey (1983)] report similar results in interviews with students who are asked to predict the exit path of a bead fired around a spiral glass tube lying on the table. Many students predict the exit path to be curved in the same direction that the tube was curved; there is a strong expectation of persistence in a curved path after the constraint is removed.

There is much to be said for starting study of the dynamics of circular motion of a particle with at least some of the following demonstrations: (1) the hands-on imparting of approximately circular motion to the large block of dry ice on a glass plate as suggested in paragraph 8 of Section 3.10 (if at all possible); (2) the demonstration of Fig. 4.8.1; (3) cutting the string holding a puck in circular motion on an air table (or holding a dry ice puck in circular motion on a glass plate); (4) the curved tube experiments used by McCloskey.

Students should be invited to predict what will happen in each case—and to argue about their predictions—before the demonstration is performed. Such procedure is far more effective than prior assertion of the expected result by the teacher.

A second deeply rooted preconception is that of the presence of a force in the direction of motion: the push imparted to the object "stays with the object and keeps it moving"; the bob or puck lags behind the string, and the string is pulling it in the direction of motion.[2] One could define the "Compleat Optimist"

---

[2] Many students, when they whirl a bob on a string, believe they are pulling the bob in the tangential direction. Since the application of torque accelerating the bob in the first place, and then keeping it going in the face of frictional resistance, is actually a rather complicated process, this view should not be surprising. One should not jump to the idealization of negligible torque as though it were obvious to everyone. The notion that the string keeps the bob moving by pulling in the tangential direction provides an additional reason for doing experiments with an inward force supplied by a solid body, such as the hoop in Fig. 4.7.1; it is more difficult to rationalize the presence of a tangential force in this situation.

as the teacher who expects all his students to have shed this idea after lucid treatment of the Law of Inertia for rectilinear motion along lines such as those of Chapter 3. One should not be surprised, however, that the expectation is not realized; the yield, as in chemical reactions, is never 100 percent. The context has been altered from rectilinear to circular motion, and, although some students have indeed mastered the Newtonian view, there remain significant numbers who have not. The altered context should be seized upon as an opportunity to spiral back and bring additional recruits to the Newtonian level.

A third, and very widely prevalent, preconception concerning circular motion is that the object is being "thrown outward." A principal source of this (apart from hearing it said by poorly prepared teachers in poorly taught science courses) is one's own sensation in going around curves. It is for this reason that it is very advantageous to start laying the groundwork to counter this idea back in the discussion of rectilinear motion as suggested in Section 3.20. Once one begins to realize that he or she is *not* thrown backward in a car that is speeding up, or forward in a car that is slowing down, one is far better prepared to discern what is happening in going around a curve, and the way is then better prepared for formation of the concept of centripetal force.

4.9 CENTRIPETAL FORCE EXERTED BY COLINEAR FORCES

Since the concept of centripetal force and the relation

$$F_{\text{cent}} = \frac{mv^2}{R} \tag{4.9.1}$$

are most commonly developed in connection with a bob (or frictionless puck) attached to a string and moving in a horizontal circle, there is a strong tendency for many students henceforth to connect the term "centripetal force" with the pull of a string regardless of whatever other effects might be present.

Although it might seem trivial to the teacher, many students are helped if one emphasizes the absence of a string in the case of the ball rolling around the inside of the hoop (Fig. 4.8.1) and in the case of our own body in the car going around a curve on a level road. Students need to be made explicitly conscious of the fact that a centripetal effect might be a *push* toward the center rather than a pull.

A valuable problem at this point is that of the car on the unbanked road. It offers the opportunity to calculate centripetal forces that are not exerted by strings: the centripetal force exerted by the seat on one's own body and the frictional force exerted by the road surface on the car. (This problem, stopping at this point, paves the way for subsequent treatment of banking of the road, the intervening time helping in the assimilation of the ideas.)

An element of physical insight rarely made explicit for the student is the fact that in these various circumstances we are invoking a passive force (exerted by the string, or hoop, or seat, or road surface) that adjusts itself to the magnitude required by the given radius and angular velocity just as the normal force exerted by the table on a block adjusts itself to the magnitude required by the weight of the block, or the frictional force exerted by the floor on a block (before sliding) adjusts itself to the external push or pull on the block. Unless led to articulate this insight, many students fail to see that the string will break, that the hoop will bend, and that the car will slide when the demand for a still larger force can no longer be sustained. To see this connection among these disparate situations is highly conducive to learning.

Situations depending on passive forces must eventually be set in contrast with those involving active forces such as gravity or electrostatic attraction. In these latter cases, a change in tangential velocity is not compensated by a change in the passive force, and the radius must change in consequence. Many students do not penetrate to an understanding of the physics of these contrasting situations unless they have the opportunity to discuss the differences. Such discussion is frequently omitted because treatment of the active force cases comes quite some time after the initial development of circular motion, and advantage is not taken of the opportunity to spiral back. (No pun intended.)

Furthermore, because situations other than the one represented by the equality in Eq. 4.9.1—that is, with forces larger or smaller than the value demanded in the equation—are almost never examined, students acquire the impression that *any* inwardly directed force is to be called a "centripetal force." This impression needs to be countered by asking how the bob will behave, at a given radius and tangential velocity, if the inward force is made larger than mv^2/R or smaller than mv^2/R. It is not at all necessary to work out the consequences mathematically; this is far too formidable a task. All one needs is the qualitative insight that the particle will deviate either inward or outward from the prescribed circle. When the student has been led to see the broader perspective (i. e., what is *not* the case as well as what is), the operational definition of "centripetal force" begins to take on firmer meaning: the term applies *only* to the magnitude of the force that imparts the particular acceleration mv^2/R to the particle of mass m and keeps it in the given circular path.

A still more serious problem about the meaning of "centripetal force" emerges in circumstances where the centripetal force is imparted by two or more effects—as in case of the bob on the string revolving in a vertical circle or in the case of the car on the loop-the-loop. Although it should be emphasized from the very beginning that the term "centripetal force" refers always to the *net* force that imparts the acceleration v^2/R to the object in question, the word "net" means very little to students until they encounter the case of superposed forces. With the bob at the top or bottom of its circle, they first tend to regard the centripetal force as being just the pull of the string and do not recognize that, in this case, the term refers to the algebraic sum of the pull of the string and the weight of the bob. The analogous difficulty arises in the case of the loop-the-loop.

This misconception regarding the meaning of the technical term "centripetal force" then serves to impede understanding of the physical phenomenon—namely that the active force (the weight of the bob or car), contributing to the centripetal force, remains unchanged, while the passive force (pull of the string or push of the track) adjusts itself to precisely the value that makes the algebraic sum of the two forces impart the centripetal acceleration required.

Many students solve the end-of-chapter problems that ask for the smallest tangential velocity that allows the bob or car to negotiate the top of the circle (or the height at which the car must start on its initial track) by memorizing an established procedure and emerge with no understanding of the role of the two superposed forces or of the full meaning of the term "centripetal force."

The weakness of these conventional problems is that they concentrate all attention on the crossover condition and fail to lead the student to examine what must happen, in general, on either side of the crossover: What is the magnitude of the tension in the string at the top of the circle if the tangential velocity of the bob is greater than the critical value? What happens to the tension as the tangential velocity is decreased? Under what circumstances does the tension become zero? What happens if the tangential velocity is decreased further? The whole sequence must be analyzed and interpreted if understanding is to be cultivated.

The most effective way of doing this is to lead the students to set up the equation for tension in the string at the top of the circle:

$$T = \frac{mv^2}{R} - mg \qquad (4.9.2)$$

and examine how T changes as one causes v to decrease from some initially high value. They then encounter both the crossover condition and the necessity of interpreting the meaning in the change of sign of T at the critical value of v. This is a valuable exercise in interpreting mathematical results. A parallel analysis can then be invoked to contrast the variation of T at the bottom of the circle with its variation at the top.

Throughout these discussions it should be emphasized that T is *not* the centripetal force; that the centripetal force at the top of the circle is given by $T + mg$ and at the bottom of the circle by $T - mg$. This begins to eliminate the notion that T is always the centripetal force.

Although it might now appear that examination of the normal force exerted by the track on the car in the loop-the-loop has become a trivial repetition, this is actually not the case. The context is just sufficiently altered so that analysis of the variation of the normal force exerted by the track at the top and bottom of the loop makes a good test question. Students who have mastered the insights enjoy doing well, while those still struggling have a valuable reencounter. Furthermore, there is a connection between this situation and the elevator sequence developed in question 9 in the Appendix to Chapter 3: the sensation experienced by the loop-the-loop rider is similar to the sensation imparted by an elevator in upward acceleration.

Still another useful variation is the case of a car going over the crown of a hill, the crown being treated as essentially circular in shape. Here one finds oneself on the outside of a "track" instead of on the inside (as in the loop-the-loop). Again, the situation is slightly different from that of the preceding cases, and the change is just enough to be a sensitive probe for understanding. This situation ties in with a sensation many students have experienced: that of seeming about to "take off" when speeding over the top of a hill. It also connects with the sensation in an elevator accelerating downward. Tying all these cases together in homework provides a valuable learning experience.

Relatively few students acquire firm mastery of force diagrams during the initial exposure to rectilinear motion, and it is advisable to utilize all available force problems in circular motion as an opportunity to spiral back and strengthen grasp of the force concept. For this purpose students should be required, on both tests and homework, to draw the force diagrams, describe forces in words, and identify Third Law pairs in the manner recommended in Section 3.12. This means that the diagrams should include not only the moving particle but also the string and whatever is at the other end of the string; the track as well as the car in the loop-the-loop; one's own body in the car, the car, and the road surface when going around a curve on an unbanked road.

4.10 CENTRIPETAL FORCE EXERTED BY NONCOLINEAR FORCES

In the following illustrations, students should, of course, be required to draw force diagrams for *all* of the interacting objects in the manner outlined in Section 3.12.

1. *Banking of a road.* This problem is usually treated in a manner similar to the treatment of the car negotiating the loop-the-loop. Only the crossover condition is solved for, and many students fail to achieve understanding of the physics. In this instance, it is particularly important to start with an understanding of what is involved in negotiating a curve on an *unbanked* road: the location of the center of the circular motion, the role of friction, and the origin of skidding. This is why treatment of this case is so strongly recommended in the preceding section.

After the unbanked case is understood, one can lead students to pursue what happens as the angle of banking is increased from zero. First, however, they must be led to define an appropriate set of coordinates. Because virtually all inclined plane problems previously encountered have been most conveniently handled with Cartesian coordinates parallel and perpendicular to the plane, most students have developed the conditioned reflex that one always chooses this orientation of the axes. They must be brought to the realization that, in this instance, it is better to take one of the axes in the direction of the required centripetal force (i. e., horizontally), else one invites a great deal of misery.

Having established the most appropriate coordinate axes, students should then be led to the *qualitative* insight that, as the angle of banking increases and the normal force exerted by the road surface on the car develops an increasing horizontal component, the magnitude of the required frictional force decreases. The optimum condition then becomes clearly defined and can be solved for. The examination should not stop here, however; it should continue into what happens when the critical angle is exceeded, and students should articulate the renewed demand on the frictional force, now acting outward rather than inward.

2. *Amusement park ride in which one can take one's feet off the floor while remaining stuck to the outer wall of a rotating cylinder.* This problem affords a very useful spiraling back to connect with situations in which a box was pressed against a vertical wall by a horizontal force and did not slide down. The two problems should now be juxtaposed and examined for similarities and differences in the physics. The rotating case should be examined in full, setting up the appropriate equations, imagining continuous change in angular velocity (starting at either a high or a low value), passing through the critical value, and continuing on to the other side. This exercise is especially valuable for students going on to more advanced levels because it requires thinking in terms of inequalities more than in terms of equalities — a kind of thinking they very rarely have the opportunity to do in elementary physics.

3. *Direction of* **g** *(the local vertical) on a spherical (i.e., rigid), but rotating, earth.* (Although this is not an especially formidable problem, it offers considerable difficulty to most students, even honors students in a calculus-physics course, and it therefore absorbs a substantial amount of time. My own practice is to toss it out as an extra-credit challenge in ordinary classes but to assign and discuss it in honors sections of students going on to higher levels.)

This highly idealized problem can first be examined purely qualitatively with great conceptual profit; a more quantitative treatment can come later, if at all.

In my own experience, the most effective way is to establish a diagram of the rotating earth such as that in Fig. 4.10.1 and imagine suspending a plumb bob on a string. Students must first be led to recognize that the centers of circular motion at different latitudes lie on the intersection of the plane of the latitude circle and the axis of rotation rather than coinciding at the center of the sphere. For most students, this is a first encounter with the rotating sphere, and they are not yet familiar with this simple aspect; the encounter is useful since it prepares them for more complicated cases later.

The students are then led to struggle with the vector diagram of the forces acting on the plumb bob: they must recognize that the gravitational force is fixed in magnitude and directed toward the center of the sphere; they must then discern that the passive force exerted by the string must adjust itself in magnitude and direction in such a way that the resultant of the two forces provides the requisite centripetal force directed along the perpendicular to the axis of rotation. The force exerted by the string is, of course, equal and opposite to the "local weight" of the bob, and its direction defines the local vertical. It is then apparent that **g** is directed toward the center of the earth at both the equator and the poles and that, at other latitudes, there is a deviation in the direction shown in Fig. 4.10.1. From these results students can begin to see how the figure of the oblate, nonrigid earth would be formed, and one can visualize what would happen with increasing and decreasing angular velocity.

It should be noted that the introduction of the string is a very useful device. It is much easier to deal concretely with this situation than to try to visualize what happens with a freely falling body.

Once the vector diagram has been drawn, a quantitative solution is virtually at hand. Neglecting small terms, the result is

$$\alpha \cong \left(\frac{a\omega^2}{g}\right) \sin 2\theta \qquad (4.10.1)$$

where α denotes the angle of deviation of local vertical from the radial direction to the center of the sphere, a the radius and ω the angular velocity of the earth, and θ the angle of latitude. It is apparent that the angle of deviation is largest at a

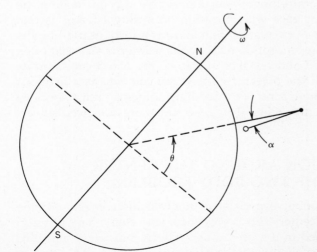

Figure 4.10.1 Plumb bob suspended above a spherical rotating earth.

latitude of 45°. (The magnitude of this maximum deviation is close to six minutes of arc.)

4.11 FRAMES OF REFERENCE AND FICTITIOUS FORCES

Most teachers are concerned about how to deal with the concept of "centrifugal force" since students come to physics courses with this vocabulary firmly embedded through frequent encounter in school science and in popular literature. The usage is also massively reinforced by the personal sensation of being "thrown outward" when going around a curve. It seems to me that the most appropriate treatment has been clear for a long time and is well handled in many texts. An especially fine treatment is available in the PSSC film "Frames of Reference," narrated by Hume and Ivey of the University of Toronto. (I have long felt that this film can stake a legitimate claim to being the best instructional physics film ever made. The physics is excellent. The film is not dated even though it is in black and white, and its humor stands up under repeated viewing.) The approach is to define what is meant by a "fictitious force" and identify "centrifugal force" as being in that category.

If one does not wish to get involved in the subtlety of fictitious forces (and such a decision is perfectly legitimate, especially in terminal courses), the best approach is to deny the validity of centrifugal force as a concept applicable in the inertial frames of reference in which we are applying the Newtonian Laws and to eliminate the use of the term. As indicated both in Chapter 3 and in preceding sections of this chapter, it can be compellingly argued that there is *no* force throwing an object outward in circular motion and that only a net *inward* force is necessary in order to impart centripetal acceleration.

For students going on to more advanced levels, however, it is worth developing the idea of the fictitious force since it will, of necessity, be encountered in noninertial frames of reference, especially the frame of the rotating earth itself in subjects such as meteorology, oceanography, or geophysics. Here the important fictitious force will be the Coriolis force rather than the centrifugal force, but it is wise to start with the simpler concept.

Some texts and teachers elect to accept the term "centrifugal force" and apply it to the force exerted by the bob on the string or by the string on the peg to which it is tied. This is a most unfortunate and undesirable way of treating the concept. The unwritten text that goes with the term "centrifugal force" is that this is a force acting *on the object* in circular motion and *not* a force exerted by the object on something else. The preconception held by the learner is also that of an outward force on the object itself and not that of a force the object exerts on its surroundings. A force acting on the object itself is legitimate *only* as a fictitious force in the noninertial frame. There is no point saddling students with a specious version that they will have to change when they get to the more advanced subject matter.

4.12 REVOLUTION AROUND THE CENTER OF MASS: THE TWO-BODY PROBLEM

Many treatments of circular motion, especially in noncalculus courses, confine themselves to the case of a fixed center. This is proper for very elementary physical situations such as those discussed in the preceding sections, but, since allusions are usually also made to the revolution of the moon around the earth and of the planets around the sun, it leaves students with the impression that the latter motions are identical with the former. Furthermore, any mention of lunar and solar tides is specious without some accompanying consideration of the two-body problem.

I have never had a student ask, for example, why we should view the earth as revolving around the sun when the gravitational interaction, as a centripetal force, might just as well make the sun revolve around the earth? There is valuable and important physical understanding and phenomenology buried here, and the question should be elicited since it does not arise spontaneously.

This need not be done with full dynamical analysis, including formation of the concept of reduced mass, although such treatment is well within the scope of the usual calculus-physics course. It is possible to handle the question entirely qualitatively with students unprepared for an analytical approach.

I have, for example, used the following approach with classes of general education students and elementary school teachers: Early in the course [Arons 1977], I pose the question as to why we believe the earth and planets revolve around the sun rather than vice versa, and we spend weeks working up toward an answer to the question. After having developed kinematics, Newton's Laws, the concept of terrestrial gravity, and the concepts of centripetal acceleration and centripetal force, we read about Newton's suggestion that gravity might extend to all the interactions of the celestial domain and that it might provide the necessary central force for the observed motions. The question remains as to what revolves around what.

Some days after having raised this question and allowed it to simmer, I go to the air table and tie two identical pucks together with a string. I tell the class that I am going to give the pucks a push such that they will tend to go into circular motion with as little translation as possible, and I ask for a prediction of the location of the center of the circular motion. The first time I did this I expected a wide range of response, especially because I hinted that I would hold one puck still and push the other. To my surprise there was a chorus of expectation of circular motion around the midpoint between the two pucks. (Our intuition for symmetries of this kind is apparently very deeply embedded and extends to individuals without much prior experience with the phenomena.)

Having established the behavior of the identical pucks, I proceed to load one of the pucks, and I ask again for a prediction of the center of circular motion. The chorus places the center closer to the more massive puck. When I ask what will happen as the mass increases further and becomes very much larger than that of the unaltered puck, the expectation of a center of rotation very close to the center of the massive puck emerges very clearly. Almost always someone in class exclaims, usually in the startled tone that accompanies discovery, "That must be the earth and the sun!" We are now on our way; there remains only the problem of developing evidence as to relative masses.

If one has a bit more time, with a somewhat more sophisticated class, one can set down and interpret the equation

$$mr\omega^2 = MR\omega^2 \qquad (4.12.1)$$

for the two-body case and bring out the role of the center of mass of the system. Although this is desirable, I submit that it is not essential as a first-level approach.

If one essays to discuss the lunar tides, however, other than by pure hand waving, it is essential to develop the significance of revolution of both the earth and the moon around the center of mass of the system. If the earth were fixed, with the moon revolving around it, there would be only a diurnal (rather than a semidiurnal) lunar tide, and its height would be devastating. The semidiurnal tide, and its modest height, stem from the fact that the earth is in free fall toward the center of mass around which the revolution is taking place.

One way of taking up this issue is to extend the ideas developed in the gedanken experiments in a freely falling elevator in question 9 of the Appendix to Chapter 3. (This is another valuable opportunity for spiraling back.) Suppose we make the height of the elevator very large—so large that it reveals the gradient of the earth's gravitational force, with the force being measurably smaller at the top of the elevator than at the bottom.

In free fall, the entire body of the elevator accelerates downward with the acceleration at the center of gravity; let us denote the magnitude of this acceleration by g_{cg}. (The elevator itself is under stress since the gravitational force at the bottom is greater than that at the top.) If we release a ball in the vicinity of the center of gravity, it will stay where we release it. If we release a ball at the top of the elevator, however, it will "fall" *upward* relative to the elevator with an acceleration of magnitude $g_{cg} - g_t$, where g_t denotes the free fall acceleration relative to the earth at that elevation. Similarly, if we release a ball near the bottom of the elevator, it will accelerate "downward" relative to the elevator with acceleration $g_b - g_{cg}$. These upward and downward accelerations would be noticeable only because the gravitational effect of the elevator itself would be negligible.

If the elevator had a powerful gravitational field of its own originating at its center, this gravitational force would completely overpower the effects described in the preceding paragraph, and the ball would fall toward the center of the elevator whether it were released at the top or the bottom.

The situation of the earth in the earth–moon system is to some degree analogous to that of the freely falling elevator: the earth is in free fall toward the center of mass of the system; the acceleration of the entire earth is that of its center of gravity; the earth is subjected to the gradient of the lunar gravitational force and is under stress. Relative to the *earth*, there is now a residual (fictitious) force *away* from the moon on the far side of the earth and *toward* the moon on the near side. This "splitting" accounts for the semidiurnal lunar tide.

At this point the analogy to the elevator begins to break down, and one must be careful. The stresses produced by the residual lunar forces result in earth tides in the solid earth; although measurable, these are very small. Our interest is in the oceanic tide. Here the sphericity of the earth becomes a critical factor. We resolve the residual lunar forces into radial and tangential components with respect to the earth. The radial component is completely overpowered by the earth's own gravity and has a negligible effect (as in the case of the freely falling elevator with a powerful gravitational field of its own.) It is the tangential component that is capable of displacing ocean water until the hydrostatic pressure gradient due to the slope of the surface balances the tangential effect of the residual lunar gravity on either side of the earth.

Many textbook discussions of the semidiurnal lunar tide tend to leave the impression that it is the residual force directly "underneath" the moon (and toward the moon) on the near side, and the residual fictitious force radially away from the moon on the far side, that pile up the waters and produce the semidiurnal tides. This is entirely incorrect, as indicated above; it is the tangential force that dominates the observed effects.

Lamb (1932) presents a rather neat alternative view of the semidiurnal lunar tide. He shows that, if there were *two* identical moons, each with half the mass of the existing moon, orbiting the earth 180° apart, the earth would be fixed rather than in free fall, but the tidal situation would be identical with the one that exists. [More detailed discussion of all these matters can be found in Tsantes (1974), Arons (1979), and Lamb (1932), as well as in additional references cited in these sources.]

A final word about the two-body problem: it arises again, in a highly significant way, in the quantum theory of the hydrogen atom. To obtain agreement between the theoretically calculated and empirically observed Rydberg constant, one must invoke the reduced mass of the electron–proton system rather than just the mass of the electron. The latter usage implies a fixed proton while the former acknowledges dynamics relative to the center of mass.

An interesting physical insight attends awareness of the difference between the earth–moon and electron–proton systems. If the earth were fixed, the period of revolution of the moon would be given by the (Kepler's Third Law) form

$$T_1^2 = \left(\frac{4\pi^2}{Gm_E}\right) H^3 \qquad (4.12.2)$$

while the period of revolution about the common center of mass is given by

$$T_2^2 = \left[\frac{4\pi^2}{G(m_E + m_M)}\right] H^3 \qquad (4.12.3)$$

where H denotes the distance between centers of the earth and moon. Thus the actual sidereal period of the system is slightly less than it would be if the earth were fixed. Equation 4.12.3 does not contain the full expression for the reduced mass because of the cancelation arising from the mass dependence of the Law of Gravitation. The full expression for reduced mass *does* appear, however, in the electron–proton case because the force law (Coulomb's Law) does not involve the masses.

Many students coming to the hydrogen atom problem with no previous inkling of the phenomena attending the two-body situation, are confused and astonished by the unfamiliar physics. Thus, students going on to more advanced subject matter can be significantly helped by at least a qualitative exposure to two-body physics in the introductory course.

4.13 TORQUE

Once the preceding conceptual structure is established, it may seem that the concept of "torque" is a simple extension that requires relatively little effort and attention. For many students, however, this is not the case. They are still struggling to grasp the large array of new concepts, the analytical formulations, and the connection to phenomena. In many instances, the expression for torque is asserted much too quickly, with far too little motivation or connection with experience. Furthermore, the effectiveness of development is enhanced if one adheres to the precept "idea first and name afterwards."

In addition to appealing to kinesthetic experience in the acceleration of rotation with cranks and various lengths of crank arms, it frequently helps to draw on students' previous experience with balancing (as in the see-saw). Lessons on balancing are quite common and widespread in school science, and invoking concrete experience from the past does much to reduce the feeling of fear and insecurity that attends new abstract formulations.

If students have had previous exposure to beam balancing along the lines of the prototype illustrated in Fig. 4.13.1, it is likely that they have described the condition for balance in terms of ratios of the form

$$\frac{W_1}{W_2} = \frac{L_2}{L_1} \quad \text{or} \quad \frac{L_1}{W_2} = \frac{L_2}{W_1} \qquad (4.13.1)$$

(They may not have written out the expressions algebraically, but their thinking has usually been in this mode, principally with integer ratios.)

One can draw on this background in the following manner: Set up a situation such as that illustrated in Fig. 4.13.1; elicit from the students a description of the balance condition; and also elicit both algebraic statements in Eqs. 4.13.1. Then point out that the balance condition in the form of ratios "scrambles" the forces and the lever arms, that is, the subscripts 1 and 2 both appear on each side of the equation. Now ask whether one can restate the balance condition so that all the subscript 1s are on one side of an equation and all the 2s on the other and elicit the form

$$W_1 L_1 = W_2 L_2 \qquad (4.13.2)$$

It can then be pointed out that Eq. 4.13.2 states the balance condition in such a way that the left side of the equation contains only terms from the left side of the balance and the right side only terms from the right side of the balance. Thus the quantities $W_1 L_1$ and $W_2 L_2$ can each be regarded (or interpreted) as "effects" intrinsic to each side, and that, even more specifically, they can be interpreted as measuring a "turning effect" applied to each side. This interpretation is strongly reinforced by showing that the beam is unbalanced and rotated counterclockwise if $W_1 L_1$ is made greater than $W_2 L_2$, and is unbalanced and rotated clockwise if the inequality is reversed. (It is important to invoke departure from the balance condition; otherwise many students fail to see what is *not* the case in addition to what *is*).

Next, it is useful to set up a situation with two weights on one side of the fulcrum and one on the other as shown in Fig. 4.13.2. Now students can be led to see that the condition for balance can no longer be expressed in a form such as that of Eqs. 4.13.1 but that the form of Eq. 4.13.2 still works. The equilibrium condition becomes

$$W_1 L_1 = W_2 L_2 + W_3 L_3 \qquad (4.13.3)$$

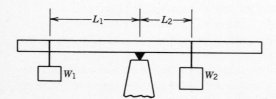

Figure 4.13.1

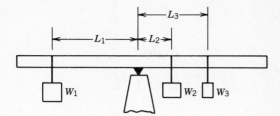

Figure 4.13.2

In fact, turning effects, as defined by the *WL* product, are simply additive! The term "lever arm of the applied force" might be introduced at this point.

Having established the significance and utility of the *WL* product, one can now examine a few situations in which pushes and pulls are applied to the beam by a force other than a hanging weight (pull on the string by hand, for example) and point out that we have been dealing, initially, with a very special situation— one in which the applied force has been perpendicular to the lever arm. How should the turning effect be calculated when the force and lever arm are not mutually perpendicular?

This is an opportunity to spiral back, in a new context, to the concept of force components. One can resolve the applied force into components along, and perpendicular to, the lever arm and examine the effect of each component. Since the component along the lever arm is directed through the turning point, it has zero turning effect. The turning effect is determined by the size of the component perpendicular to the lever arm, and so on. This is a valuable opportunity to reinforce awareness that vectors are not anchored to points but are to be shifted, in our imagination, along their lines of action. It is also very important to show the alternative calculation of the turning effect in which one finds the effective lever arm for the total applied force, without resolving the latter into components.

Many students have very great initial difficulty in making these calculations. The geometry is far from transparent to them, especially when constructions have to be added to an initial diagram. Much more practice is needed than is usually afforded in end-of-chapter problems. Practice should include a wide variety of configurations (rotated into different orientations when represented on a page) and the labeling of different angles in the different configurations. Problems should include zero length lever arms as well as lever arms that must be visualized and are not concretely represented in the given drawing, and problems should include algebraic formulations as well as numerical calculations. This is another case in which practice could be very effectively implemented through well-designed, computer-based, modules, but, as of the time of writing, good materials are not available.

After the concept of "turning effect" has been firmly registered along lines such as those illustrated above, it becomes appropriate to introduce the name "torque." Such concrete and relatively modest development paves the way for subsequent treatment of torque as a vector cross-product and as the agent of change of angular momentum and makes the latter abstract extensions far more intelligible to many students.

APPENDIX 4A

Sample Homework and Test Questions

NOTE: The following qualitative questions on force diagrams illustrate the spiraling back that is helpful to many students. They review ideas developed at the level discussed in Chapter 3 and combine them with ideas discussed in this chapter. Despite the previous exercises, many students will still have difficulty with these questions when they are first posed. They are helped by the practice. Other similar questions, connected to actual or possible personal experience, are needed. The interaction of permanent magnets might well be included, especially if magnets were used to show noncontact interaction between gliders on an air track.

1. A person sits on a box resting on the platform of a merry-go-round and holds a bob suspended on a string. The merry-go-round is turning. Draw well-separated force diagrams of the bob, the string, the person, the box, and the platform. Show forces of equal magnitude with arrows of equal length; show larger forces with longer arrows, etc. Describe each force in words, and identify the Third Law pairs.

2. A person stands at the edge of a rotating merry-go-round and leans in such a way as to feel most comfortable and well balanced during the rotation. Hanging from the railing near the person is a pendulum (a bob on a string.) It is an observed fact that, under these conditions, the body of the person is aligned parallel to the string of the pendulum. Draw force diagrams of the bob, the string, the railing, the person, and the platform. Explain how it comes about that the person is most comfortable under the conditions described. Describe the connection between this situation and the situation of a car going around a curve at the speed for which the curve is banked. A person sits inside the car, and a pendulum bob is suspended from the roof.

CHAPTER 5

Momentum and Energy

5.1 INTRODUCTION

There are several sources of difficulty for students entering study of the momentum and energy concepts: (1) the necessity of defining the *system* that will be under consideration; (2) distinguishing clearly between open and closed systems; (3) mastering the large number of new technical terms that are generated—many of them drawn from everyday speech but endowed with altered meaning—by connecting the new technical terms with the algebraic expressions and calculations with which they are associated; (4) mastering the *relations* among the technical terms and using them to describe and analyze familiar, everyday phenomena.

These difficulties can be significantly reduced if students arrive at this level with prior practice in: (1) stating operational definitions in their own words; (2) translating symbols into words and words into symbols; (3) describing familiar phenomena in the technical vocabulary previously developed.

If the material on kinematics and the laws of motion has been covered too rapidly, however, without opportunity for the kind of practice just referred to, the difficulties with the abstractions now encountered are seriously compounded, and many students fall by the wayside. This is not to say that the previous material must have been fully *mastered*; the exposure must have been such that the student is ready to progress into the new abstractions by using procedures that are becoming familiar through encounter in earlier, more concrete, contexts. Concepts not fully mastered previously will become more firmly assimilated through the spiraling back afforded in the new, richer context.

A serious impediment to understanding of the energy concepts, and of the concept of "work" in particular, resides in the way many texts introduce these ideas. The concept of "work" and the Work–Kinetic Energy Theorem are initially developed for the very special case of point masses or particles, that is, the case in which the displacement of the net external force is equal to the displacement of the center of mass of the object on which the force acts. The indicated relations are then extended, sometimes fallaciously, to systems (such as rotating or deformable systems) with internal degrees of freedom in which the displacement of the external forces is *not* equal to the displacement of the center of mass. The vocabulary becomes very confusing; many commonly made statements are basically incorrect. Many of the better students are very uneasy and feel that something is missing, but they fail to pursue the issue for fear of appearing "stupid;" weaker students simply memorize without comprehension.

This problem requires more careful attention in texts, lectures, and homework; it is discussed in Sections 5.6–5.14.

5.2 DEVELOPING THE VOCABULARY

Having become very used to the concepts over an extended period of time, teachers and textbook authors tend to lose sight of how bewildering the large array of new technical vocabulary is to many students. Confusion is also augmented by the fact that new meaning is being given to seemingly familiar words such as "momentum", "energy", "work", and "impulse." Besides the individual nouns themselves, we have phrases such as "impulse delivered (or work done) by the force **P**"; "impulse delivered (or work done) by the *net* force $\mathbf{F}_{net}$"; "*change* of momentum of body A"; "*change* in kinetic energy of body B"; "work done *against* force P or force f"; "change in potential energy of system S".

Furthermore, in rectilinear situations, the algebraic sign associated with change in momentum of a body has very different meaning from the algebraic sign arising in connection with change in its kinetic energy.

For most students, whether in algebra- or calculus-based courses, reading text descriptions, hearing the lecturer apply the vocabulary to a few sample cases, and doing the end-of-chapter problems do *not* provide sufficient exercise in use of the vocabulary. Understanding the meaning of the terms and phrases, and the ability to use them in describing physical phenomena, do not develop until students have had the opportunity to establish sequences of *interconnection* between words and numbers in an array of simple physical situations. Furthermore, there is a good bit of *memorizing* to be done, and students fail to recognize the necessity of memorizing vocabulary unless such necessity is explicitly pointed out and acquisition is tested for.

A few sample problems that help develop this capacity are given in the appendix to this chapter. The reader will note that the *individual segments* of these problems are to be found in most textbooks. The gap left by the textbooks resides in the fact that the segments are presented separately and are rarely linked into one integrated sequence, applying the vocabulary to a rich, extended context. Any teacher who has never given, on a test, a simple sequential problem such as one of those illustrated in the appendix will be shocked at how poor the entire class performance is the first time this is done. The performance is extremely poor even in the calculus-physics course and even among students who have done the homework.

Firm grasp of the connection between the names and phrases on the one hand and the numerical calculations on the other begins to develop only after *several* homework and test exercises of the type illustrated in the appendix.

5.3 DESCRIBING EVERYDAY PHENOMENA

In addition to establishing the connection between numerical calculations and the names and phrases discussed in the preceding section, assimilation of the significance of the energy and momentum concepts is strongly enhanced by leading students to describe familiar phenomena, purely qualitatively, in the new vocabulary.

Many common experiences, confined to essentially rectilinear motions, lend themselves to such examination:

Throwing a ball vertically upward and catching it on its return (or allowing it to strike the ground).

Throwing a ball horizontally and having it caught by another person.

Throwing a ball horizontally so that it bounces back from a wall.

Throwing a ball of putty horizontally so that it sticks to a wall.

Jumping vertically upward and landing back on the ground.

Jumping vertically up and down on a trampoline.

Climbing a rope.

Walking and running (including starting, stopping, and maintaining uniform speed).

Pushing oneself off from a wall while standing on roller skates and coasting to a stop.

Approaching a wall on roller skates and stopping by cushioning the collision through flexing of one's outstretched arms.

Accelerating a car, maintaining uniform speed, and then slowing down and stopping.

Pushing a box along the floor in the presence of friction (including cases of speeding up, maintaining uniform velocity, and slowing down and stopping, and including the momentum and energy changes of one's own body as changes are imposed on the box).

Rowing a boat or driving it by means of engine and propeller.

Accelerating or slowing down a space vehicle by means of rocket propulsion in free space.

Stretching (horizontally) a spring fastened to a wall.

Stretching a spring horizontally by pulling its ends in opposite directions.

Sitting down on a soft chair.

Hanging an object on a spring fastened to the ceiling.

Students should be led to describe the energy and momentum changes taking place for all the interacting bodies, including the earth. It should be noted that some of the forces playing major roles in these phenomena are zero-work forces, and one must be very careful in the description of the energy transformations (cf. sections beginning with Section 5.6.)

Performance in such verbal descriptions is initially extremely poor among virtually all students, even among those of higher ability. The performance improves only with practice that is accompanied by evaluation and correction of the verbal statements. As the verbal performance improves, visible improvement in understanding is exhibited in problem solving. Practice, as in other instances that have been discussed, must recur and must be spread out over time. Quick "remediation" is ineffective, especially with students below the highest ability levels; "mastery" is never attained on first encounter.

Fortunately, one can keep spiraling back in increasingly richer context if one simply remembers to take advantage of the opportunities for verbal description that arise in connection with circular motion, thermodynamics, electric and magnetic phenomena, waves and light, and with phenomena on molecular, atomic, and subatomic scales.

5.4 FORCE AND RATE OF CHANGE OF LINEAR MOMENTUM

Tempted by what seems (in hindsight) to be prescience on Newton's part on the one hand and by the primacy of momentum in special relativity on the other, some textbook authors have been turning to emphasizing the Second Law form

$$\mathbf{F}_{net} = \frac{d\mathbf{p}}{dt} \qquad (5.4.1)$$

where

$$\mathbf{p} = m\mathbf{v} \qquad (5.4.2)$$

in connection with development of the impulse and momentum concepts and in connection with conservation of momentum.

There is nothing wrong with this[1] except in those instances in which it is implied that this form is more general than

$$\mathbf{F}_{net} = m\mathbf{a} \qquad (5.4.3)$$

because it allows for variation in mass as well as velocity.

Equation 5.4.1 is valid, in general, only for *closed* systems. It happens to apply to one very special open system, namely that in which the frame of reference is such that the velocity **v** is not only the instantaneous velocity of the object undergoing change in momentum but is simultaneously the *relative* velocity of the incoming or outgoing material. It is *not* applicable to other open systems, especially, for example, the rocket.

These questions are now properly treated in many texts [e.g., Resnick & Halliday (1977), Section 9.7], but incorrect or misleading statements are still sufficiently frequent to necessitate wariness and care on the part of the teacher.

5.5 HEAT AND TEMPERATURE

It is well known to most teachers that many students in introductory courses do not discriminate between the terms "heat" and "temperature" and tend to use the words synonymously when referring to thermal phenomena. This confusion will not go away by being ignored. It arises because, in the bulk of earlier student experience, the words have been used as though they were both primitives, with meaning obvious to everyone. The students have not been led to articulate simple

---

[1] What is being referred to here is the logical soundness of the concept. A serious question can be raised as to the pedagogical wisdom of adopting this approach too early in an introductory course in which the majority of students (even at the engineering-physics level) still find *dv/dt* mysterious and momentum bordering on the arcane.

operational definitions even though this can readily be done in elementary school science. Since this has not been taken care of earlier, it is necessary to give it some attention in introductory physics.

Fortunately, the task is not difficult because one can appeal to a large array of everyday experiences with thermal phenomena. A simple, basic strategy is to accept initial intuitive knowledge of thermometers and thermometer readings as primitives, not requiring elaborate definition to begin with.[2] One can then lead students to articulate descriptions of everyday experiences of which they are intuitively aware but which they have rarely made explicit:

> When a bucket of hot water is put out in a room, the reading of the thermometer in the bucket always decreases, approaching the reading of the thermometer on the wall of the room.
>
> When a bucket of cold water is put out in a room, the reading of the thermometer in the bucket always increases, approaching the reading of the thermometer on the wall of the room.
>
> When two objects at different temperatures are brought in contact with each other, the temperature of the higher one always drops while that of the lower one rises until the two temperatures are equal.
>
> When two objects at the same temperature are brought together, no changes take place.

These descriptions, extended by others elicited from the students, can be used to generate the concept of "thermal interaction" between bodies initially at different temperatures and the concept of "thermal equilibrium". Thus, one can cultivate explicit awareness of the prevalence of interaction and of the general trend toward thermal equilibrium in familiar phenomena.

The next step in the strategy is to lead students to recognize that, despite the powerful insight afforded by recognition of interaction and of the trend toward thermal equilibrium, the thermometer readings do not tell the whole story. Something else must be happening. Recognition of this fact can be elicited by drawing attention to the following observations and experiences in which thermometer readings are carefully *controlled* while other conditions are varied:

> If we "insulate" the buckets of hot and cold water put out in the room by covering them with layers of cotton or of material such as that inserted in the walls of houses (or if we put the water in Thermos bottles), we markedly alter the *time* it takes for the thermometer readings to reach equality with the reading on the wall. The initial and final thermometer readings are all still the same as they were without the insulation, but something has changed that the thermometers do not tell us about.
>
> Similar time differences arise in the cooling down to outdoor temperature of two houses, with and without wall insulation, when heating is stopped inside. (It is surprising to find how many students are unaware of this and cannot really articulate the point and purpose of wall insulation.)

---

[2] It can be pointed out that, ultimately, there is more to the *concept* of temperature than this, but that, as with many other concepts, we go through successive steps of definition and redefinition in progressing to deeper insights from our initial crude starting point.

If we start with two containers of water, one large and one small, at the same temperature and place them to heat over identical burners, it takes a longer time and more fuel to elevate the temperature of the larger amount of water to the same final value as that of the smaller.

If we freeze or melt pure materials, the freezing or melting does not occur without interaction with a surrounding "bath" at either lower or higher temperature than that of the material undergoing the change. Despite the ongoing interaction during freezing or melting, however, the material undergoes *no* change in temperature at all until it is either all frozen or all melted (if students are not aware of this fact, it should be demonstrated).

All these illustrations, the last one especially dramatically, testify to the fact that thermometer readings do not tell the entire story of thermal interaction, that something else must be happening, and that an additional concept (or concepts) must be invented.[3] We give the name "transfer of heat from the higher to the lower temperature body" to the *process* of interaction implied by the preceding list of observations, and a recounting of this story constitutes a *qualitative* operational definition of the new term—which we all too soon, and all too cryptically, abbreviate to the one word "heat".

Quantifying "transfer of heat" then follows in the usual way: Working with water as a convenient reference substance, one finds that (to first order of accuracy) equal masses of water, initially at different temperatures, end up at the mid temperature when mixed in thermal isolation from their surroundings. Unequal masses end up at a final temperature such that the ratio of the two temperature *changes* is the inverse ratio of the two masses. Thus, one verifies the possibility of using the masses and temperatures of quantities of water as a convenient way of assigning numbers to amounts of heat transferred. This leads to definition of the "calorie" and to the concept of "specific heat" of substances other than water. (The fact that the specific heat of water itself must be redefined since it varies with temperature, is a second-order discovery, illustrating the continual process of sharpening and refinement of concepts that takes place as the formation of a concept results in more sophisticated insight and more precise quantitative measurement.)

At this juncture, it is important to return to cases such as that of the bucket of hot water cooling in the room. Few students explicitly recognize the room, or the house, as having a heat capacity, very much larger than that of the bucket of water but nevertheless finite. They must be led to recognize that the air in the room undergoes a temperature change very much smaller than that of the water but not zero.

It is worth emphasizing that we neither see nor measure heat transfer *directly*. (Few students notice this explicitly.) The quantities that we observe are masses and temperature changes, and we *infer* the amount of heat transferred from these observations. (Here is another opportunity to discriminate between observation and inference.)

For the sake of clarity and precision in forming and using the energy concepts in subsequent study, it is advisable never to speak of the "heat *in* a body," even

---

[3]As with the kinematical concepts discussed in Sections 2.5 and 2.9, here is still another opportunity to show students that scientific concepts are invented by acts of human imagination and intelligence and are not objects that are "discovered" as existing entities.

in the early stages of development of the concept. The term should be used *only* in connection with the process of *transfer* into or out of a body or system. This is the usage in thermodynamics, and it properly underpins the line integrals and inexact differentials associated with quantities of heat and work transferred.

As will be shown in Section 5.8, speaking of heat as though it resides in a body and implying it to be a function of state, raises severe impediments to clear formation of the concept of conservation of energy, even at an elementary level.

Furthermore, speaking of "converting work into heat" (when work is dissipated through frictional processes, for example) confuses the issue because heat is not actually transferred to the system in such a process. The work done on the system is converted directly into thermal internal energy *without* transfer of heat. The increase in thermal internal energy is found to be exactly equal to the heat transfer that *would* have been necessary to produce the same temperature change in the system. (These questions are discussed in more detail in Sections 5.6–5.10.)

5.6 THE IMPULSE–MOMENTUM AND WORK–KINETIC ENERGY THEOREMS

Most textbooks start with

$$\mathbf{F}_{net} = m\mathbf{a} \tag{5.6.1}$$

and develop the impulse–momentum and work–kinetic energy relations, either as integrals of both sides of Eq. 5.6.1 with respect to clock reading t and position s respectively (in calculus-based courses) or in simplified, constant-force terms (in algebra-based courses). The initial development is, quite properly, carried out for the simplest possible case, namely particles or point masses, and the equations take the familiar forms:

$$\int_{t_1}^{t_2} \mathbf{F}_{net} dt = m\mathbf{v}_2 - m\mathbf{v}_1 = \Delta(m\mathbf{v}) \tag{5.6.2}$$

or

$$\mathbf{F}_{net}\Delta t = \Delta(m\mathbf{v}) \tag{5.6.3}$$

and

$$\int_{s_1}^{s_2} \mathbf{F}_{net} \cdot d\mathbf{s} = \frac{1}{2}mv_2^2 - \frac{1}{2}mv_1^2 = \Delta\left(\frac{1}{2}mv^2\right) \tag{5.6.4}$$

or

$$\mathbf{F}_{net}\Delta s = \Delta\left(\frac{1}{2}mv^2\right) \tag{5.6.5}$$

Both sets of results are perfectly valid and general for the situation for which they were derived, namely the changes in motion of point masses. Serious difficulties begin to enter, however, when these relations are uncautiously extended to systems of interacting particles or to objects that are treated as continuous but are deformable or have other internal degrees of freedom. [Very common examples of such systems are: motions imparted to our own bodies in walking, running, or jumping; propelling a car; discriminating between elastic and inelastic collisions; and even pushing a (seemingly rigid) box over a floor with friction.]

Equations 5.6.2–5.6.5 are valid dynamical relations providing the velocities and displacements being referred to are those of the center of mass of the system under consideration. If one elects to attack extended systems, the significance of the center of mass and its role in the analysis should be developed at least plausibly if not with complete rigor. The general concept of conservation of linear momentum is properly associated with Eqs. 5.6.2 and 5.6.3.

Although Eqs. 5.6.4 and 5.6.5 are generally valid numerical relations (provided it is clearly maintained that velocities and displacements refer to the center of mass), serious conceptual and verbal problems begin to arise as soon as they are extended beyond application to point masses. The difficulties arise principally because these equations are presented as *conservation* statements for work and energy in analogy to the ways in which Eqs. 5.6.2 and 5.6.3 are conservation statements for linear impulse and momentum, the implication being that Eq. 5.6.4 is an initial version of the First Law of Thermodynamics and needs only refinement and extension to incorporate other forms of energy. Actually, Eqs. 5.6.4 and 5.6.5 are true work–energy statements for point masses but are *not* true energy statements for most applications to extended systems. Trouble arises because work and heat are forms of energy that are *transferred* across the *boundaries* of a system, and the system in question may be the single, deformable object.

That something is amiss begins to emerge as soon as we consider cases involving zero-work forces that impart acceleration to, and thus alter the kinetic energy of, extended bodies in various familiar situations:[4] The force exerted on us by the ground when we jump vertically upward; the force exerted on us by the wall when we stand on roller skates and set ourselves in motion by pushing off from the wall with our hands; the horizontal frictional force exerted by the road on an accelerating car; and the frictional force exerted by the ground on us when we walk or run (in the absence of slipping) are all zero-work forces. Yet, in many texts, it is either said or implied that these forces do work and thus impart kinetic energy to the accelerating bodies, regardless of what might have been said earlier about forces that cannot be considered as doing work.

Another illustration that something is amiss stems from consideration of the very simple situation in which we push a box at uniform velocity along a floor with friction. If we denote the force of sliding friction by f and the displacement of the center of mass of the box by Δx_{CM}, Eq. 5.6.5 yields

$$f\Delta x_{CM} - f\Delta x_{CM} = 0 \tag{5.6.6}$$

implying that zero net work has been done on the box. Yet we know, from our more advanced knowledge of energy transformations, that a net amount of work has in fact been done on the box–floor system and that the equivalent of this amount of work has appeared as thermal internal energy, associated with the

---

[4] Virtually all texts and teachers properly point out, in connection with giving the name "work done by the force F" to quantities such as $F \Delta s$ or $\int \mathbf{F} \cdot \mathbf{ds}$, that zero work is done if there is zero displacement of the force in the direction in which the force acts. This is usually illustrated by appeal to situations such as our pushing on a rigid wall, where the displacement is zero, and carrying a suitcase at uniform velocity, where the displacement is orthogonal to the force. This aspect is, for the most part, competently embedded in instruction. It should be noted, however, that these illustrations are seldom, if ever, developed for situations in which kinetic energy is being imparted to a body.

temperature rise exhibited by the box and floor. In introductory courses we tend to paper over the apparent paradox by saying that the work done by us was "converted into heat," with the implication—to the student—that heat has been transferred to the box. We know, however, that no heat transfer has taken place to increase the box–floor temperatures and that, whatever heat transfer does actually occur, is from the box and floor to the surrounding air.

The origin of these difficulties resides in the fact that Eqs. 5.6.4 and 5.6.5 are not really energy equations for anything but a point mass system without friction and without internal degrees of freedom. This does *not* mean that the equations are invalid or incorrect. They are numerically valid connections among external forces (whether work-doing or not) and the displacement and velocity of the *center of mass* of the system to which $\mathbf{F}_{net} = m\mathbf{a}$ and its integrals are being applied. The contradictions being pointed to arise from a flawed and inconsistent invocation of the energy concepts.

Because Eqs. 5.6.4 and 5.6.5 are not true energy equations (except in a few very special cases), I shall, following Sherwood (1983), drop the "work–kinetic energy" terminology and refer to them as "center of mass" or CM equations. To emphasize this awareness, it might be more appropriate to write these equations in the form:

$$\int_{s_1}^{s_2} \mathbf{F}_{net} \cdot d\mathbf{s}_{CM} = \frac{1}{2}mv_{2,CM}^2 - \frac{1}{2}mv_{1,CM}^2 = \Delta\left(\frac{1}{2}mv_{CM}^2\right) \tag{5.6.7}$$

or

$$\mathbf{F}_{net}\Delta s_{CM} = \Delta\left(\frac{1}{2}mv_{CM}^2\right) \tag{5.6.8}$$

making explicit the role of the center of mass coordinates and velocities and emphasizing the fact that the quantity on the left-hand side is *not* calculated from forces and their displacements around the *periphery* of the system.

5.7 REAL WORK AND PSEUDOWORK

Since the concept of "energy" is not a primitive or intuitive one, students do not come to this aspect of physics with reasonable and deeply rooted preconceptions based on everyday experience—as they do with elementary dynamics. They come as *tabula rasa* and acquire certain misconceptions that are implanted by the majority of text presentations. These misconceptions then become hard to eradicate in later study.

The principal misconception planted in introductory physics is that the "work" quantity (force times center of mass displacement) appearing in the "work–kinetic energy theorem" (Eqs. 5.6.4 and 5.6.5), obtained by integration of Newton's Second Law, is identical with the "work" appearing in the general law of conservation of energy, namely the First Law of Thermodynamics. (The reference to *real* work in the heading of this section is to the latter quantity rather than the former.) That this equating of the two "work" quantities is a misconception has been discussed in some detail in recent years [cf. Erlichson (1977); Penchina (1978); Sherwood (1983); Sherwood and Bernard (1984)], but the necessary awareness has not yet penetrated many text presentations.

Work done on or by a system (other than a single point mass) must be calculated by integrating forces and their accompanying displacements around the entire *boundary* of the system—the *region* to which the conservation law is being applied. Bridgman (1941) describes the operational problem in characteristic fashion:

> *Turn now to an examination of the W of the First Law. This W means the total mechanical work received by the region inside the boundary from the region outside [or delivered to the region outside from the region inside]. As in the case of [heat transfer] Q, this work is done across the boundary, and the evaluation of W demands the posting of sentries at all points of the boundary, and the summing of their contributions. In the simple cases usually considered in elementary discussions, the work received by the inside from the outside is of the simple sort typified by the motion of stretched cords or of simple linear piston rods. Our sentry can adequately report this sort of thing in terms of finite forces acting at points and finite displacements. In general, however, there will be contact of the material outside over finite regions of the boundary, and we become involved in the stresses and strains of elasticity theory. [Bridgman goes on to mention the "infelicities that result when we apply the notion of work to the sliding of two bodies on each other with friction."]*

Penchina (1978), joined by Sherwood (1983), suggests distinguishing between the two work quantities mentioned above by adopting the name "pseudowork" for the quantity connected to displacement of the center of mass and reserving the name "work" for the quantity appearing in the First Law of Thermodynamics. Although this terminology has not yet become standardized, it is convenient because it does not completely sever the connection between the two quantities and because it does not resort to a radically new vocabulary. I shall therefore adopt it in the subsequent discussion.

5.8 THE LAW OF CONSERVATION OF ENERGY

The Law of Conservation of Energy is, of course, *not* derivable from the dynamical laws of motion; it is an independent statement about order in nature—one of the "principles of impotence" as the conservation laws and the Second Law of Thermodynamics are sometimes called. The general conservation law, including heat transfer as well as kinetic, potential, and other energy changes, is a *new* statement that, in most cases, has very little to do with the CM equation.[5] The way in which energy concepts are most frequently introduced in elementary physics courses does not make this fact sufficiently clear to the students.

Let us examine what aspects of the First Law of Thermodynamics might be comfortably incorporated into elementary physics in such a way that the treatment of energy concepts is correct and consistent but does not invoke mathematical and other complexities that are meaningless to students at that level.

The First Law, stated in the familiar forms

$$dE = đQ - đW \qquad (5.8.1)$$

or

$$\Delta E = Q - W \qquad (5.8.2)$$

---
[5]See Section 5.6 for the definition of "CM equation."

where E is called the "internal energy of the system", is considerably more than just a statement of conservation of numbers calculated according to the "recipes" prescribed in the operational definitions of the various energy quantities. We have path-dependent quantities on the right-hand side (inexact differentials) and a new *function of state* (exact differential) on the left. In fact, the mathematical statement of the First Law can be regarded as a *definition* in the sense that it turns out to be a law of nature that (1) there exists an *energy quantity* that is a *function of state* of the system (i. e., its changes are path independent) and that (2) we can calculate changes in this quantity in terms of path-dependent energy transfers that are *not* themselves functions of state. (It must be noted that up to this point we have defined only state variables such as temperature, pressure, volume, mass, density, and so forth, without having any way of including an energy as a state variable.)

I do not propose trying to develop this sophisticated a perspective in elementary physics, but I do join Sherwood (1983) in urging that we adopt an essentially verbal translation that makes what we are actually doing in many commonly treated physical situations clearer to our students.

After we have shown that the dissipation of (real) work produces heating effects that are directly proportional to the amount of work dissipated (as Joule did in his famous experiments) and have suggested that future extensions will include electric, magnetic, and chemical effects in addition to the thermal and mechanical ones, we might say that we are asserting a new law of nature, namely that we shall always be able to find ways of calculating numbers (or "keeping the books") such that the change in what we shall call the "internal energy of the system" is numerically equal to the sum of the quantities of heat and work transferred to it.[6] In symbols, this would be translated as

$$\Delta E = Q + W \qquad (5.8.3)$$

Note that this form alters the sign of W from that in the usual American convention in the First Law.[7] I suggest this convention for the sign of W, at least temporarily, since it would be very messy to give negative signs to work done by the force accelerating an object at a stage when students are having great difficulty with interpretation and use of algebraic signs in general. By the time one gets to thermodynamics with students who continue to that level, the convention can be changed without too much stress.

In preparation for future applications, let us list the internal energy changes with which we shall be concerned and specify the notation to represent each one:

Internal thermal energy change: ΔE_{therm}

Internal chemical energy change: ΔE_{chem}

Internal kinetic energy change: ΔE_{kin}

---

[6] It is in this same spirit that Feynman introduces the energy concepts in the fourth of the Lectures on Physics [Feynman, Leighton, and Sands (1963)].

[7] In the most commonly used American convention, work is taken to be positive when it is done *by* rather than *on* the system. This is very likely motivated by the convenience of associating a positive value of dW with positive values of $p \, dV$ for expanding fluid (or so-called chemical) systems. The convention is the reverse in much European thermodynamic literature.

Internal kinetic energy changes would have subcategories such as translational ($\Delta E_{\text{int,tr}}$) and rotational ($\Delta E_{\text{int,rot}}$)

Internal potential energy change: ΔE_{pot}

Internal potential energy changes would have subcategories such as gravitational ($\Delta E_{\text{pot,grav}}$); springlike in compression or extension ($\Delta E_{\text{pot,sp}}$); electrical ($\Delta E_{\text{pot,el}}$); and so forth

Miscellaneous internal energy changes: ΔE_{misc}, encompassing emission or absorption of sound, radiation, or other messy interactions

The general symbol ΔE in the First Law (Eq. 5.8.3) would then stand for the algebraic sum of the various different internal energy changes specified in the above list:

$$\Delta E = \Delta E_{\text{therm}} + \Delta E_{\text{chem}} + \Delta E_{\text{kin}} + \Delta E_{\text{pot}} + \ldots \qquad (5.8.4)$$

To illustrate the implications of the approach being advocated, let us apply it to a very simple special case, namely the transfer of an amount of heat Q to a system in the absence of any doing of external work and in the absence of any internal energy change other than thermal. Equation 5.8.3, in combination with Eq. 5.8.4, would then give

$$\Delta E_{\text{therm}} = Q \qquad (5.8.5)$$

This readily translates into the verbal statement that, under such circumstances, the amount of heat transferred to the system is equal to the change in thermal internal energy. Note that this statement uses the term "heat" only in connection with its *transfer*, as is systematically done in thermodynamics, and avoids the implication that heat "resides in," or is a "property of" the system. (The subtlety that arises in connection with applying this terminology to everyday situations, such as those invoked in Section 5.5, is discussed in the next section.)

The First Law, in the form of Eq. 5.8.3 or the preceding verbal statement, should be distinctly separated from the CM equation. It can be very neatly and simply applied to all the situations discussed in elementary physics, and, as I shall try to show in the following sections, such application greatly improves the clarity and consistency of the treatment of energy transformations in general, just as it does with the special thermal case illustrated above.

5.9 DIGRESSION CONCERNING ENTHALPY

Critical readers, especially chemists, will have noticed that the discussion, at the end of Section 5.8, of heat transfer and thermal internal energy change in the absence of doing of work lacks complete rigor. What is implied is a transfer of heat at constant volume of the receiving object or system. Although such transfer is perfectly possible with gases, it is extremely difficult, if not actually impossible, with liquids and solids. Liquids and solids that expand on increase in temperature would have to be confined under enormous pressure to maintain constant volume, while water, which contracts in volume as its temperature increases anywhere between 0°C and 4°C, would have to be subjected to hydrostatic "tension."

5.9 DIGRESSION CONCERNING ENTHALPY

Virtually all the transfers of heat that we confront in everyday experience take place at constant pressure rather than constant volume, and some amount of work is inevitably exchanged with the surrounding atmosphere. Thus the amount of heat transferred is not strictly equal to the thermal internal energy change as defined by the First Law. It was for this reason that the concept of "enthalpy" was invented: we define still another energy quantity which is a function of state and the changes of which are rigorously equal to the amount of heat transferred under constant pressure instead of constant volume.

This is achieved by applying a so-called Legendre Transformation to the original expression for the First Law:

$$dE = dQ - p\,dV \tag{5.9.1}$$

$$d(pV) = p\,dV + V\,dp \tag{5.9.2}$$

If we add these two equations, we obtain:

$$d(E + pV) = dQ + V\,dp \tag{5.9.3}$$

The new function of state $E + pV$ is usually denoted by H and is called the "enthalpy of the system." Equation 5.9.3 then becomes:

$$dH = dQ + V\,dp \tag{5.9.4}$$

and, rigorously speaking, the amount of heat transferred at constant pressure (as is the case in most everyday phenomena) is equal to the enthalpy change ΔH rather than the internal energy change ΔE. (The so-called heat of chemical reaction, dealt with in chemistry, is an enthalpy change between specified initial and final conditions.) The constant pressure heat capacity c_p is then defined as

$$c_p = \frac{1}{M}\left(\frac{\partial H}{\partial T}\right)_p \tag{5.9.5}$$

where M denotes the mass of the system and T the temperature.

Under normal circumstances, with any system other than a gaseous one, the difference between ΔH and ΔE is extremely small and is legitimately neglected. With gases, one must be careful. (It is interesting to note that Julius Robert Mayer, who shares credit with Joule for discernment of the quantitative "equivalence" between work and heat, estimated the "mechanical equivalent" not from thermal effects resulting from dissipation of work, as did Joule, but from comparison of the constant volume (c_v) and constant pressure (c_p) heat capacities of gases. He correctly interpreted the difference as reflecting the amount of work done on expansion as heat is transferred at constant pressure.)

In presenting this discussion of the enthalpy concept, I do so to clarify the subtleties involved and not to advocate its development in introductory physics courses. In the following sections, I shall use ΔE as an adequate approximation for ΔH. Formal introduction of enthalpy can perfectly well be left to the beginning of the more formal treatment of thermodynamics. (What we have here is still another illustration of how concepts are refined, redefined, and invented as knowledge and understanding deepen through successive approximations. It is worth noting that more than a century elapsed between Joseph Black's investigations of heat and the invention of enthalpy.)

A very few exceptional students occasionally get worried about the difference between constant volume and constant pressure processes. They can be directed to the more sophisticated view without confusing the issue for the others.

5.10 WORK AND HEAT IN THE PRESENCE OF SLIDING FRICTION

To illustrate the approach being suggested, consider the prototypical situation represented in Fig. 5.10.1: A block of mass m is accelerated from rest along a floor or table by an applied force F against a force of sliding friction f. The displacement of the center of mass of the block is denoted by Δx_{CM}, and the final velocity by $v_{CM,f}$.

We have two equations to apply to the changes taking place. One is the CM equation

$$F_{net} \Delta x_{CM} = \Delta \left(\frac{1}{2} m v_{CM}^2 \right) \tag{5.10.1}$$

and the other is the conservation of energy (COE) equation suggested in Section 5.8:

$$\Delta E = Q + W \tag{5.10.2}$$

For the situation in Fig. 5.10.1, the only relevant internal energy changes are thermal and kinetic translational, and Eq. 5.10.2 becomes

$$\Delta E_{therm} + \Delta E_{kin,tr} = Q + W \tag{5.10.3}$$

We can apply the CM equation 5.10.1 to the block itself, and we obtain

$$(F - f)\Delta x_{CM} = \frac{1}{2} m v_{CM,f}^2 \tag{5.10.4}$$

We must now carefully select the *system* to which we are to apply the COE equation 5.10.3. The block alone is *not* an appropriate system for this purpose because, although we can calculate the work done by the force F in displacing the center of mass, we cannot calculate the work done by the frictional force f at the interface. What happens at the interface is a very complicated mess: we have

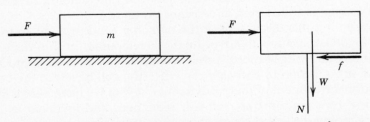

Figure 5.10.1 A block of mass m is accelerated from rest along a floor or table by an applied force F against the opposing force of sliding friction f.

abrasion, bending of "asperities," welding and unwelding of regions of "contact," as well as shear stresses and strains in both the block and the floor. The work done on the block at the interface is *not* simply $f\Delta x_{CM}$, and we are unable to deal fully with the quantity W on the right-hand side of Eq. 5.10.3.[8]

In this circumstance, we can take advantage of a possibility frequently available in application of the conservation laws: that of sweeping an area of ignorance under the rug by judicious choice of system. The system to choose in this instance is the *combination* of block and floor. (We could include the air also, but there is an insight to be gained by excluding it.)

If we apply the COE equation 5.10.3 to the system of block and floor, we no longer have to worry about the intractable situation at the interface since the frictional forces are now internal and no work is exchanged with the surroundings. Work is done on the system by the force F, and *no* heat is received from the surroundings. If there is any heat exchange with the surroundings (air) at all, it would be a *loss* following increase in temperature of the block and floor. Let us represent this by the positive quantity Q_{loss}. Substituting into Eq. 5.10.3, we obtain

$$\Delta E_{therm} + \frac{1}{2}mv^2_{CM,f} = F\Delta x_{CM} - Q_{loss} \qquad (5.10.5)$$

yielding

$$\Delta E_{therm} = F\Delta x_{CM} - \frac{1}{2}mv^2_{CM,f} - Q_{loss} \qquad (5.10.6)$$

Interpretation of Eq. 5.10.6 yields the following insights: (1) The increase in thermal internal energy of the floor–block system is directly equal to that part of the work done by force F that does not go into the form of kinetic energy of the block, providing no heat is transferred to the surrounding air. (2) This work, which is said to be "dissipated", is directly transformed into thermal internal energy; the increase in thermal internal energy does *not* result from a transfer of heat to the system. From this perspective, $F\Delta x_{CM}$ is real work done on the block–floor system, while $f\Delta x_{CM}$ is pseudowork done on the block. If F and f happen to be equal in magnitude and the block moves at uniform velocity, the pseudowork happens to be numerically equal to the real work, but that does not make the two quantities identical conceptually, and it does not make the work done on the system equal to zero.

This approach is desirable because it helps one avoid misleading and confusing locutions about "converting work into heat" when no heat transfer is taking place. (The latter phraseology, inherited from the nineteenth century and never altered when other concepts were refined, helps implant, on the one hand, the misconception that heat resides *in* bodies, and, on the other hand, that heat is transferred to the system when work is dissipated.) One can now point out that the temperature change developed in the system by direct dissipation of work is equal to the temperature change that *would* have resulted from the transfer of an equivalent amount of heat even though no heat was actually transferred.

---

[8]Sherwood and Bernard (1984) attack the problem of work done at the interface by adopting a plausible model and analyzing the consequences. They develop some interesting insights, and I recommend a reading of the paper. I would hesitate to take time for such an analysis, however, in introductory courses.

Given Eq. 5.10.6, one can explicitly idealize the situation by taking heat loss to the air to be zero, and one can apportion the total thermal energy increase between the block and the floor as may appear reasonable or as may yield upper or lower bounds on estimated temperature changes.

The reader might find it helpful to carry out the parallel analysis for the case, still dealing with the system in Fig. 5.10.1, in which the force F is removed and the block coasts to a stop from an initial velocity $v_{CM,i}$. The CM equation becomes

$$-f\Delta x_{CM} = 0 - \frac{1}{2}mv^2_{CM,i} \qquad (5.10.7)$$

and the COE equation becomes

$$\Delta E_{therm} = \frac{1}{2}mv^2_{CM,i} - Q_{loss} \qquad (5.10.8)$$

No real work is done on or by the system, and all the internal kinetic energy is converted into internal thermal energy if there is no heat loss to the air.

Summary comment: The CM equation for a particular body gives an entirely correct numerical relation among dynamical quantities. Some of the terms in the equation are amounts of real work done on or by the body, but others are not and should be described as pseudowork. The proper interpretation of *energy transformations* comes from the COE and not from the CM equation. The COE equation must be applied to a properly defined *system*. In some instances the pseudowork [a quantity that *looks like* an amount of work done (e. g., $f\Delta x_{CM}$), but is not a real work done by (or against) that force over the indicated displacement] is shown by the COE equation to be *numerically equal* to an amount of real work that was done by some other force (e.g., F) and was, say, dissipated. (Note that this kind of equality is analogous to another one that was discussed in Chapter 3: Although the downward force exerted on our hand by the body we are supporting is, in some circumstances, *numerically equal* to the weight of the body, the force on our hand is not the same force as the weight. The gravitational force exerted by the earth on the body and the contact force exerted by the body on our hand are two entirely different forces conceptually even when they are numerically equal.)

5.11 DEFORMABLE SYSTEM WITH ZERO-WORK FORCE

Consider the situation in which a person of mass m jumps vertically upward (Fig. 5.11.1). The average normal force exerted by the ground on the person is denoted by $\overline{N}$. The center of mass of the person starts from rest and acquires a change in elevation Δh_{CM} and a velocity $v_{CM,f}$ at the instant the feet leave the ground. For this situation the CM equation gives

$$(\overline{N} - mg)\Delta h_{CM} = \frac{1}{2}mv^2_{CM,f} \qquad (5.11.1)$$

The quantity $\overline{N}\Delta h_{CM}$ is a pseudowork: it looks very much like an expression for work done on the person by the normal force $\overline{N}$, but it cannot be a real work done on the body since $\overline{N}$ is a zero-work force. The COE equation may show us that this expression is *numerically equal* to an amount of work done by some other force, but that interpretation has to come from the COE and not from the

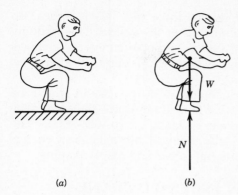

Figure 5.11.1 Forces acting on person jumping vertically upward

CM equation. The quantity $mg\Delta h_{CM}$ can be interpreted, on the other hand, as an actual amount of work done by the jumper *against* the force of gravity if the jumper is taken as the system. It becomes an internal potential energy change in the jumper–earth system.

Let us first apply the COE equation to the jumper–earth system. No external work is done on the system, so $W = 0$. The only heat transfer might be a loss Q_{loss} from the jumper to the surrounding air. The relevant internal energy changes are: chemical (the origin of the biological effects in the muscles), thermal (we warm up in such exercise), kinetic translational ($(1/2)mv_{CM,f}^2$), kinetic rotational (flailing of arms), and gravitational potential ($+mg\Delta h_{CM}$). Hence,

$$\Delta E_{chem} + \Delta E_{therm} + \frac{1}{2}mv_{CM,f}^2 + \Delta E_{kin,rot} + mg\Delta h_{CM} = -Q_{loss} \quad (5.11.2)$$

If we elect to ignore the flailing of the arms and the thermal effects, this reduces to

$$\Delta E_{chem} = -\frac{1}{2}mv_{CM,f}^2 - mg\Delta h_{CM} \quad (5.11.3)$$

Thus the source of the upward kinetic energy of the jumper and of the increase in the potential energy of the jumper–earth system resides in the decrease of chemical energy within the body of the jumper. This chemical energy is transformed through internal work-doing forces and displacements that cannot be described in quantitative detail. The CM equation 5.11.1 shows that the right-hand side of Eq. 5.11.3 happens to be numerically equal to $-\overline{N}\Delta h_{CM}$ despite the fact that $\overline{N}$ is a zero-work force.

A useful insight is gained if we apply the COE equation to the system of the jumper alone rather than to the jumper–earth combination. In this case, external work $W = -mg\Delta h_{CM}$ is done by the gravitational force, and ΔE_{pot} must be taken as zero; otherwise this quantity would be entered twice. Ignoring thermal effects and internal rotation, the COE equation again gives Eq. 5.11.3

$$\Delta E_{chem} = -\frac{1}{2}mv_{CM,f}^2 - mg\Delta h_{CM}$$

but the implications are different from what they were for the jumper–earth combination: the $mg\Delta h_{CM}$ term appeared, from the beginning, on the right-hand side of the equation as a work done by the gravitational force and did *not* appear on the left-hand side as an increase in potential energy of the *system*. This strongly reinforces the contention that we should only speak of potential energy changes of the interacting system (jumper–earth) and not speak of potential energy as residing in the elevated object (jumper) alone.

The reader is urged to set up the parallel analysis for the case in which the person stands on roller skates and pushes him- or herself off from a rigid wall. The normal force $\overline{N}$ exerted by the wall on the skater is again a zero-work force, and the source of the kinetic energy of the skater is the chemical internal energy change mediated by work done by unquantifiable internal forces and displacements.

5.12 ROLLING DOWN AN INCLINED PLANE

The treatment of an object rolling down an inclined plane, with its usually sudden introduction of rotational kinetic energy and obscure condition of rolling without slipping, offers very great difficulty to many students. The usual consequence is memorization without understanding. The difficulties cannot be reduced to zero for the introductory level, but the physics can be significantly clarified by explicit separation of the CM and COE equations and careful interpretation of the content of the latter.

Consider the familiar situation in which a spherical or cylindrical object with mass m, radius R, and moment of inertia I rolls down an inclined plane (Fig. 5.12.1), starting from rest and acquiring center of mass translational velocity v_{CM} and angular velocity ω after center of mass linear displacement Δs_{CM} along the plane. The frictional force between the rolling object and the plane is denoted by f. Applying the CM equation yields

$$(mg\sin\theta - f)\Delta s_{CM} = \frac{1}{2}mv_{CM}^2 \qquad (5.12.1)$$

Since W and Q are both zero, applying the COE equation to the object–plane–earth system gives

$$\Delta E_{therm} + \Delta E_{kin,tr} + \Delta E_{kin,rot} + \Delta E_{pot} = 0$$

which, on appropriate substitution, becomes

$$\Delta E_{therm} + \frac{1}{2}mv_{CM}^2 + \frac{1}{2}I\omega^2 - mg\sin\theta\,\Delta s_{CM} = 0 \qquad (5.12.2)$$

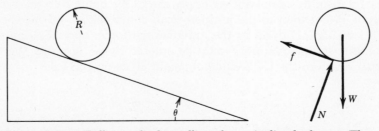

Figure 5.12.1 Ball or cylinder rolling down inclined plane. (The plane is assumed to be undeformed in the region of contact with the rolling object.)

Combining Eqs. 5.12.1 and 5.12.2 and solving for the thermal energy change gives

$$\Delta E_{\text{therm}} = f \Delta s_{\text{CM}} - \frac{1}{2} I \omega^2 \qquad (5.12.3)$$

Thus the extent of thermal dissipation is determined by the comparative values of the two terms on the right-hand side of Eq. 5.12.3.

The additional information needed comes from the dynamical equation

$$\tau_{\text{net}} = I \alpha \qquad (5.12.4)$$

where τ_{net} denotes the net torque acting around the axis through the center of mass of the rotating object, and α denotes the angular acceleration.

Integration of both sides of Eq. 5.12.4 with respect to angular displacement ϕ gives the rotational analog of the translational CM equation:

$$\tau_{\text{net}} \Delta \phi = \Delta \left(\frac{1}{2} I \omega^2 \right) \qquad (5.12.5)$$

and specializing to the situation in Fig. 5.12.1 gives

$$f R \Delta \phi = \frac{1}{2} I \omega^2 \qquad (5.12.6)$$

where $\Delta \phi$ denotes the total angle through which the object has turned since the rolling started.

Now let us examine what the energy equation 5.12.3 says about different conditions of rolling:

1. If the rolling takes place without slipping, the circumferential displacement (or unrolling) $R \Delta \phi$ must be equal to the linear displacement Δs_{CM}.[9] Thus, for rolling without slipping, we have the condition

$$\Delta s_{\text{CM}} = R \Delta \phi \qquad (5.12.7)$$

and Eq. 5.12.6 becomes

$$f \Delta s_{\text{CM}} = \frac{1}{2} I \omega^2 \qquad (5.12.8)$$

Putting Eq. 5.12.8 into Eq. 5.12.3 gives

$$\Delta E_{\text{therm}} = 0 \qquad (5.12.9)$$

showing that, for rolling without slipping, there is no thermal dissipation and that the decrease in gravitational potential energy of the system must be entirely

---

[9]Many students have a dreadful time seeing that rolling without slipping implies that $v_{\text{CM}} = R\omega$. Much of the trouble resides in their shaky understanding of the velocities. Students are helped if led to look at the problem concretely in terms of the total displacements, i. e., by comparing the linear displacement Δs_{CM} of the center of mass with unrolling of length of arc $R \Delta \phi$. Many students are unable to see the relation in the abstract and must be led to roll a cylindrical object themselves.

reflected in the total kinetic energy of the rolling object. This is easily verified algebraically by going back to the earlier relations.

2. If we consider the case in which

$$\Delta s_{CM} > R\Delta\phi \qquad (5.12.10)$$

the object must have been sliding along the plane without rotating as fully as it does in case 1. Equation 5.12.6 still holds, but it now indicates that

$$f\Delta s_{CM} > \frac{1}{2}I\omega^2 \qquad (5.12.11)$$

Under these circumstances, Eq. 5.12.3 indicates that

$$\Delta E_{therm} > 0 \qquad (5.12.12)$$

and some of the decrease in potential energy has been dissipated thermally. This is the nature of ordinary rolling with some slipping.

3. Now consider the case in which

$$\Delta s_{CM} < R\Delta\phi \qquad (5.12.13)$$

(This means that the object is *spinning* faster than it would in nonslip rolling.) The pattern of the preceding analysis indicates that we will now obtain

$$\Delta E_{therm} < 0 \qquad (5.12.14)$$

This is, of course, a result forbidden by the Second Law of Thermodynamics. Such spinning could occur only through input of work not included in the specification of the present system.

5.13 INELASTIC COLLISION

Consider the case of a ball striking, and perhaps bouncing from, a rigid wall, neglecting effects in the vertical direction and assuming zero spin (Fig. 5.13.1). Let us apply the CM and COE equations to the interval between first contact between ball and wall and the instant at which the center of mass velocity of the ball is zero. With center of mass subscripts implied on displacements and velocities, the CM equation for the ball gives

$$-\overline{N}\Delta x = 0 - \frac{1}{2}mv^2 \qquad (5.13.1)$$

where $\overline{N}$ denotes the average force exerted by the wall on the ball, Δx the displacement of the center of mass of the ball between contact and the instant of zero center of mass velocity, and v the center of mass velocity at instant of initial contact.

The $\overline{N}\Delta x$ term is a pseudowork, and we have no way of calculating the actual work done by the wall on the ball in the deformation accompanying the collision.

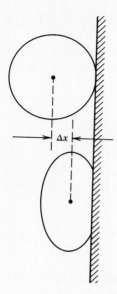

Figure 5.13.1 Inelastic collision of ball with wall.

If we consider the ball–wall system, however, both W and Q are zero, and the general COE equation becomes

$$\Delta E_{\text{therm}} + \Delta E_{\text{kin,tr}} + \Delta E_{\text{pot,sp}} = 0 \tag{5.13.2}$$

where $\Delta E_{\text{pot,sp}}$ denotes any compressional potential energy that might be stored in the ball–wall deformations.

Rearranging Eq. 5.13.2, we obtain

$$\Delta E_{\text{therm}} = \frac{1}{2}mv^2 - \Delta E_{\text{pot,sp}} \tag{5.13.3}$$

Although Eq. 5.13.1 tells us virtually nothing, Eq. 5.13.3 is highly informative. It says that there is no increase in thermal internal energy, and therefore no thermal dissipation, if the increase in compressional potential energy in the system is equal to the initial kinetic energy of the ball. This defines the perfectly elastic collision (with coefficient of restitution of unity) since the center of mass speed (and kinetic energy) would be restored as the motion is reversed. (One could follow the reversal with the same equations we have written above but with appropriate changes in algebraic signs.)

At the other extreme, if the potential energy increase is zero (a ball of putty striking the wall), the increase in compressional potential energy is zero, and the increase in thermal internal energy of the system is equal to the initial kinetic energy of the ball (providing there has been no transfer of heat to or from the ball–wall system.) This defines the perfectly inelastic collision with coefficient of restitution of zero. In such instances, if one wishes to pursue the issue further (e.g., will the lead bullet with a given mass and initial velocity melt on striking the wall?), one must make explicit assumptions about the distribution of the thermal internal energy among the components of the system. The articulation of such assumptions becomes clearer and easier when one has explicitly defined the system and written an equation such as Eq. 5.13.3. The CM equation, on the other hand, is of no help at all.

Equation 5.13.3 also makes it possible to say something about partially elastic collisions, the cases intermediate to the two extremes examined above, if appropriate information is available or is assumed.

5.14 SOME ILLUMINATING EXERCISES

Because of the ubiquity of the experience, a very worthwhile exercise for the students is examination of real-work, pseudowork, and energy transformations associated with the propulsion of a car. If we accelerate a car of mass m from rest to a final velocity v, the CM equation gives, for the horizontal direction

$$(f_{dr} - f_{res})\Delta x = \frac{1}{2}mv^2 \qquad (5.14.1)$$

where f_{dr} denotes the frictional force exerted by the road on the driving wheels, and f_{res} lumps together all the other forms of dynamical resistance to the motion. (Center of mass subscripts are implied on displacement and velocity.) Although f_{dr} is the accelerating force, it is a zero-work force, and the terms on the left-hand side of Eq. 5.14.1 are pseudowork, rather than real-work, terms.

The COE equation for the car as the system, with W equal to zero, gives

$$\Delta E_{therm} + \Delta E_{chem} + \Delta E_{kin,tr} + \Delta E_{kin,internal} + \Delta E_{misc} = -Q_{loss} \qquad (5.14.2)$$

where the right-hand side denotes heat loss to the surrounding air. One can make various idealizations and apportion various amounts to various terms. The main point to bring out, however, is that chemical internal energy, initially resident in the fuel, is transformed through work done by *internal* forces, into kinetic energy of the car, thermal energy increases, and so on and that no external work is done on the system despite the existence of an external accelerating force.

Sherwood (1983) suggests a number of other simple problems that are very illuminating conceptually and help point up the profound distinction between the CM and COE equations. The following problems are borrowed from his paper.
1. Sherwood ascribes the following to Michael Weissman: Consider the situation in Fig. 5.14.1 in which two frictionless pucks, connected by a string, are accelerated by the force F. (Note that this is a deformable system, that the force is displaced farther than the center of mass point, and that the pucks are assumed to undergo an inelastic collision on contact.)

The CM equation gives (for starting from rest)

$$F\Delta x_{CM} = \frac{1}{2}(2m)v^2 \qquad (5.14.3)$$

Taking the system to be that of the two pucks, the external work done on the system is $F\Delta x$, and the COE equation gives (in the absence of heat transfer)

$$\Delta E_{therm} + \Delta E_{kin,tr} = F\Delta x \qquad (5.14.4)$$

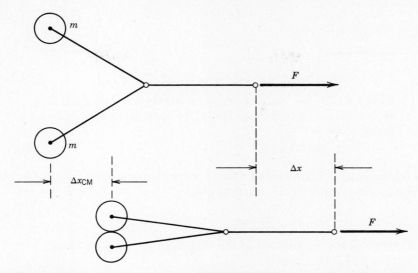

Figure 5.14.1 Top view of two frictionless pucks on an air table. The pucks are connected by strings as shown, and the external force F accelerates the system.

which, rearranged, yields

$$\Delta E_{\text{therm}} = F\Delta x - \frac{1}{2}(2m)v^2 \qquad (5.14.5)$$

Note that the displacement Δx of the force F is greater than the displacement Δx_{CM} of the center of mass point and that the real work done on the system is greater than the pseudowork appearing in the CM equation. The difference between the two is dissipated in the inelastic effects and is equal to the increase in thermal internal energy. (A numerical problem could be posed by asking for the final velocity v and the amount of dissipated energy, given F, Δx, m, and the length of the string connecting the pucks.)

An alternative version of this problem would be that of starting with a bunched-up chain lying on a table. A horizontal force applied to one end of the chain stretches the chain out to full length and accelerates it along the table. Here again the displacement of the applied force is greater than the displacement of the center of mass of the chain, and thermal internal energy is increased through inelastic effects even in the absence of friction between the chain and the table.

2. Consider two identical blocks of mass m initially at rest on a frictionless table and pulled away from each other by equal forces F (Fig. 5.14.2). If we take the system to be that of the *two* blocks, the CM equation gives

$$(F - F)\Delta x_{\text{CM}} = \frac{1}{2}(2m)v_{\text{CM}}^2 = 0 \qquad (5.14.6)$$

Because of the zero on the left-hand side, this might seem to be a trivial result, but that is not so for the students. It emphasizes the fact that the center of mass of the system is not displaced and acquires no velocity despite the fact that work is done on the system and its individual parts acquire kinetic energy. In treating this situation, we normally apply the COE equation intuitively rather than formally, but a formal treatment in very simple cases does much to clarify the formalism

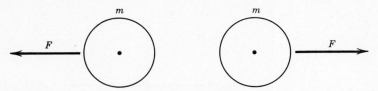

Figure 5.14.2 Top view of system consisting of two frictionless blocks on an air table. The blocks are accelerated by equal and opposite forces F.

for subsequent use in complicated cases where intuition fails us. In this instance formal application of the COE equation gives

$$\Delta E_{kin,tr} = W = 2F\Delta x \tag{5.14.7}$$

where Δx denotes the magnitude of the displacement of each block.

Equation 5.14.7 becomes

$$2F\Delta x = 2\left(\frac{1}{2}mv^2\right) \tag{5.14.8}$$

where v denotes the velocity acquired by each block. This is, of course, the result that we would write down intuitively, but it is actually justified only by the COE concept.

5.15 SPIRALING BACK

An element referred to repeatedly throughout these discussions of physics teaching is the desirability of spiraling back so as to allow students to review or reencounter important ideas and lines of reasoning in increasingly rich or sophisticated context. Following is an illustration of an opportunity for such spiraling back through use of momentum change in elastic collision in two dimensions.

The formula for centripetal acceleration of a particle moving at tangential velocity v in a circle of radius R

$$a_c = \frac{v^2}{R} \tag{5.15.1}$$

is usually derived initially in a kinematical treatment such as that in which one evaluates the acceleration associated with continual "falling" from the tangent line to the circle, or that in which one sketches the vector change in tangential velocity between two locations of the particle along the circular arc and then evaluates the rate of change of the velocity as the time interval tends to zero.

A simple dynamical derivation is given by Newton in Proposition IV of Book I of the *Principia*. Newton's derivation has been accorded very little attention, possibly because it is given entirely in words citing proportional relations and is unaccompanied either by a geometrical figure or by algebraic equations. (In the Scholium following several corollaries to the proposition, Newton acknowledges that " . . . by such propositions Mr. Huygens, in his excellent book, 'De Horologio Oscillatorio,' has compared the force of gravity with the centrifugal forces of revolving bodies. " Perhaps he did not play up the derivation because of Huygens's priority.)

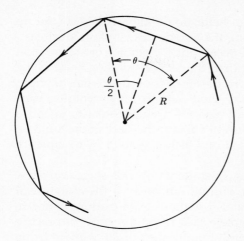

Figure 5.15.1 Particle of mass m moves at velocity v along sides of a polygon inscribed within a cylinder of radius R. The particle collides elastically with the walls of the cylinder at each vertex of the polygon. No gravitational effects are present.

Newton's proof, translated algebraically and geometrically, runs as follows (see Fig. 5.15.1): Consider a particle of mass m moving at velocity v inside a rigid cylinder of radius R. Suppose at first that the particle moves along the sides of a polygon inscribed in the cylinder, undergoing successive elastic reflections from the wall of the cylinder at each vertex of the polygon. The momentum change Δp at each collision is given by

$$\Delta p = 2mv\sin\frac{\theta}{2} \tag{5.15.2}$$

The time Δt to traverse one side of the polygon is given by

$$\Delta t = \frac{2R}{v}\sin\frac{\theta}{2} \tag{5.15.3}$$

Since Δt is the time interval between successive impacts, the average rate of change of momentum is obtained by dividing Eq. 5.15.2 by Eq. 5.15.3, and, since the average force $\overline{F}$ acting on the particle is equal to the average rate of change of momentum (Second Law)

$$\overline{F} = \frac{mv^2}{R} \tag{5.15.4}$$

(It is interesting to note that no small angle approximation is involved.) Newton thinks in terms of the collisions with the wall coming closer and closer together until, in the limit, the force exerted by the wall on the particle becomes continuous rather than consisting of a series of discrete impulses.

This analysis has very substantial pedagogical value—not as a *substitute* for the earlier kinematical derivation but as a powerful *supplement*. It gives the opportunity to come back to the centripetal acceleration and centripetal force ideas from an entirely new point of view after some time has elapsed and after the momentum concept and its vector nature have been developed. The context is rich and conceptually significant: one makes use of the vector change of momentum in bouncing from the wall to obtain an important result and reinforce an earlier physical insight; the bouncing does not take place at normal incidence; the bouncing from the wall is not left as a seemingly sterile exercise in an end-of-chapter homework problem unconnected to other phenomena; a better foundation is laid for future use of such momentum change in formulating the kinetic theory of an ideal gas.

The *order* in which the different derivations of centripetal force or acceleration are developed is probably far less important than the fact of spiraling back. If the momentum concepts were developed prior to the discussion of circular motion, one could give the above derivation first and then reinforce it by coming back to the kinematical treatments later. It should also not be necessary to spend a great deal of valuable class time on the second treatment; the derivation could be assigned as a homework problem the solution of which is guided by a sequence of Socratic questions.

The Socratic questioning is needed by most students. At this stage, very few are ready to proceed with such an analysis without guidance. If the assignment is too cryptic, they will simply sit and stare at it without even putting pencil to paper. This disability is largely due to complete lack of opportunity to practice such thinking. Opportunities that are sufficiently simple analytically without being trivial conceptually are not very easy to find. That is what makes the opportunity outlined in this section especially valuable.

APPENDIX 5A

Sample Homework and Test Questions

1. The net force acting on a glider on an air track (essentially frictionless system) varies with time as shown in the following diagram:

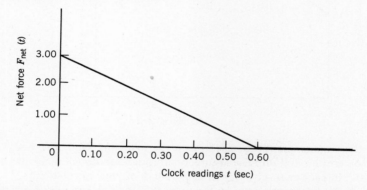

The glider has a mass of 0.850 kg. When the force is applied to the glider at clock reading $t = 0.00$, the glider has an initial instantaneous velocity of 0.150 m/s in the positive direction.
   (a) Describe, in words, what happens to the glider over the time interval between $t = 0.00$ and $t = 0.60$ s. (Use the vocabulary of impulse, momentum, work, and kinetic energy rather than that of Newton's Second Law.)
   Perform the following calculations by making use of the concepts and relations referred to in your verbal description. In each calculation, describe your reasoning briefly.
   (b) Calculate the momentum of the glider at $t = 0.00$ s (i. e., the initial momentum.)
   (c) Calculate the net impulse delivered to the glider by the applied net force between $t = 0.00$ and $t = 0.60$ s.
   (d) Calculate the *change* in momentum of the glider over this time interval.
   (e) Calculate the momentum of the glider at clock reading $t = 0.60$ s (i.e., the final momentum.)
   (f) Calculate the instantaneous velocity of the glider at clock reading $t = 0.60$ s.
   (g) Calculate the initial and final values of the kinetic energy of the glider.
   (h) Calculate the *change* in kinetic energy of the glider.
   (i) Calculate the work that must have been done by the net force acting on the glider.
   (j) Calculate the potential energy change of the glider–earth system over the interval under consideration.

142 MOMENTUM AND ENERGY

2. A block with mass 12.0 kg is being pushed along a horizontal floor by a force **P** as shown in the following diagram. The kinetic frictional force opposing the motion of the block is constant at a value of 15.0 N. At clock reading $t = 0.00$ s and position $x = 0.00$ m the block has an instantaneous velocity of 2.00 m/s.

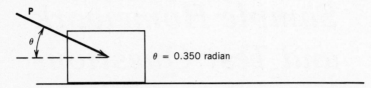

The force **P** is applied to the block at the instant it is in position $x = 0.00$ m. The direction of **P** remains fixed, but its magnitude varies with position as shown in following graph:

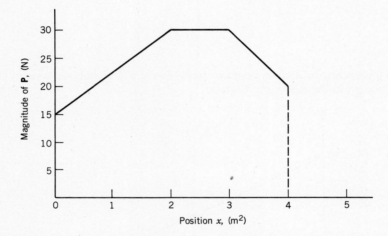

Calculate the various quantities asked for in the following sequence. Use work, kinetic energy, momentum, and impulse arguments throughout; do *not* make use of Newton's Second Law. In each case, give a brief explanation of your calculation.
(a) Calculate the total work done by the force **P** in the displacement from position $x = 0.00$ to position $x = 4.00$ m.
(b) Calculate the work done against the frictional force in the same displacement.
(c) Calculate the work done by the *net* force acting on the block during this displacement.
(d) Calculate the *change* in kinetic energy of the block.
(e) Calculate the *final* kinetic energy of the block, that is, the total kinetic energy it possesses on reaching position $x = 4.00$ m.
(f) Calculate how far the block will slide beyond position $x = 4.00$ m if the force **P** abruptly drops to zero at this position and remains zero from there on.
(g) Calculate and describe any internal *potential* energy changes of the block–earth system that take place over the entire history of the motion dealt with above.
(h) Calculate and describe any internal *thermal* energy changes of the block–floor system over the entire history of the motion dealt with above.
(i) Calculate the velocity of the block at position $x = 4.00$ m.
(j) Calculate the momentum of the block at position $x = 4.00$ m.
(k) Calculate the momentum *change* imparted to the block over the interval between positions $x = 0.00$ and $x = 4.00$ m.
(l) Calculate the *net* impulse that must have been delivered to the block over this interval.

NOTE: The following suggests a lecture or class demonstration aimed at helping students establish concrete connections between observed motions and the symbolic statements embodied in the impulse–momentum and work–kinetic energy theorems. [See Lawson and McDermott (1987) for a report of student difficulties with such connections and for a description of the apparatus used in their interviews.]

3. Gliders on an air track, pucks of different masses on an air table, or dry-ice pucks on a glass plate can be accelerated relatively slowly by means of the air stream emerging from the hose of the inverted vacuum cleaner system usually used to pump air through air tracks and air tables. (Slow acceleration is desirable in order to make the sequence of events more clearly visible and apprehensible.) Attached to the hose opening are small strips of paper that, when blown out by the air stream, serve as spacers for maintaining a constant separation between the hose and the puck. This makes it possible to keep applying a fixed force to different objects.
   (a) Direct the air stream at a puck, started from rest, for a given interval of time, and repeat the demonstration once or twice to allow the students to acquire a clear visual impression of the time interval involved and of the velocity acquired by the puck. Then apply the air stream for shorter and longer intervals of time to the same puck and ask the students to compare the impulses delivered and the momentum changes imparted. (Emphasize that to "compare" means to indicate whether one quantity is greater, equal, or smaller than another—not to give numerical values.)
   (b) Direct the air stream for the same interval of time at two different pucks having clearly different masses and acquiring visibly different velocities. Ask the students to compare the impulses delivered and the momentum changes imparted.
   (c) Perform similar experiments in which the air stream is applied to the pucks over controlled *displacements* rather than over controlled intervals of time. Ask students to compare quantities of work done on the pucks and the changes in kinetic energy imparted.
   (d) Lawson and McDermott (1987) applied the *same* force to two pucks of very different mass over the space interval between two lines drawn on the table. They then asked the students (drawn from algebra-based and from calculus-based physics courses) to compare the momenta imparted to the two pucks and the kinetic energies imparted to the two pucks. The results obtained indicate the need students have for practice with the ideas afforded in this exercise.

   Although it is useful to discuss at least some of the observations and interpretations as the demonstrations are being performed, it is also useful to leave some questions to be answered for homework. Many students need time to dwell on what they have seen and to think about the connections between the visible effects and the unfamiliar *composite* concepts (impulse, momentum, work, kinetic energy) to which they are being asked to accord physical meaning.

CHAPTER 6

Static Electricity

6.1 INTRODUCTION

For those of us who have, over years of experience, become intimately familiar with concepts and phenomena of electricity[1] and magnetism, it is all too easy to lose sight of how highly abstract this part of physics really is and how frustratingly difficult it turns out to be for many students. Its concreteness resides only at the level of observation of noncontact interactions that involve energy transfers through acceleration of objects, through deflections against opposing forces, or through thermal effects. We then construct abstract concepts and models that rationalize the observed effects. Because of the additional layers of concepts (such as "electric charge," "like charges," "unlike charges," "electric current," "potential difference," "Lorentz force," "field strength") that are introduced, this conceptual structure is even further removed from the concrete manifestations than is the conceptual structure of mechanics.

It should not be surprising that students have very substantial difficulty with the most fundamental aspects of this subject: we are asking them to absorb an entirely new range of very sophisticated abstractions before they have fully mastered the related structure of mechanics. It is necessary to recognize this difficulty—to allow time for assimilation of new operational definitions, for confrontation with "How do we know . . . ? Why do we believe . . . ?" questions, and for establishing firm connections between the concepts and the phenomena. It is also important to keep invoking and applying the basic mechanical concepts (velocity, acceleration, force, mass, momentum, energy) at every opportunity.

Because most students have been hearing the language of electric and magnetic effects since childhood, many teachers and textbooks (especially at college level) tend to assume that the rudimentary ideas have been absorbed and understood. Unfortunately this not the case. Many students have never directly observed electrostatic interactions: many have never even seen the commonplace experiment in which a rubbed object attracts bits of paper, and many have never played with simple magnets. Many students, especially women, are afraid to touch

---

[1] I have placed the discussion of static electricity ahead of the discussion of current electricity, not because I consider this order essential, but because it is the order most commonly used in existing courses. One might perfectly well start with current electricity, as many teachers and some texts elect to do, and consider electrostatic phenomena afterwards. The same logical questions must be examined, however, in either case.

small electric batteries because they have never had the chance to play with such objects, and, at the same time, they have been conditioned to fear electricity.

Thus, many of the students coming to us at high school and college level have been hearing the vocabulary without ever having seen the phenomena that motivate it, and very few have examined the operational meaning of the terms with which they have become familiar. It is especially important to reexamine the basic vocabulary in any introductory physics course, whether at high school or college level. The failure to do this, the starting with the assumption that students must already "know" both the phenomena and the terminology, is responsible for a substantial portion of the subsequent difficulties students have with this subject matter.

6.2 DISTINGUISHING ELECTRIC, MAGNETIC, AND GRAVITATIONAL INTERACTIONS

Since discussion of electric, magnetic, and gravitational effects is usually introduced separately and at different times, whether it be in elementary school, high school, or college science, very few students have been afforded the opportunity to compare and contrast these phenomena operationally. The result is that many students in introductory physics courses, at any level, seriously confuse the terms and the effects. Interviews and test questions reveal that they do not distinguish clearly between electrostatic and magnetostatic effects: the terms are frequently used randomly and synonymously; many students will predict that north magnetic poles will repel positive electric charges; many expect other irrelevant and impossible effects; some students have the impression that gravitation is some sort of electric or magnetic phenomenon. Very few students have ever had direct, hands-on experience with the phenomena under circumstances in which they are considering the different manifestations at the same time and can compare the similarities and differences explicitly.

This aspect of learning and understanding is mentioned at this point not in order to advocate such discussion in class *prior* to the experiences outlined in the following sections, but in order to alert the teacher to the importance of raising the question and returning to it periodically as further operational distinctions can be added to the list. The point is to keep the students explicitly conscious of the differences as they observe electric and magnetic phenomena and compare them with gravitational ones. (It should be kept in mind, incidentally, that the only familiar aspects of gravitation are the weight and the free fall of objects. The minute gravitational interactions of ordinary bodies and gravitational interaction on the cosmic scale both involve abstract extension of the concept of "gravity" to situations beyond the realm of direct experience.) Keeping the various phenomena in mind simultaneously enriches the context and enhances understanding of the vocabulary and remembering of the effects.

6.3 FRICTIONAL ELECTRICITY, ELECTRICAL INTERACTION, AND ELECTRICAL CHARGE

Never having personally observed the phenomena and interactions that lead to formation of the concept of "electrical charge," but having heard the words used in and out of school since early childhood, many students use the term without knowing what it means and what it does *not* mean. To many students, charge is some invisible kind of substance that may be smeared on things or that may drip out of household sockets.[2]

In an introductory physics course, students should be led to retrace at least some of the observations that led to the formation of the concept of "electrical interaction" resulting from frictional effects, then to formation of the concept of "charge," recognition of the mobility of charge, and finally to realization that there are two varieties of charge (or two "charge states") and to the operational definitions of the terms "like" and "unlike." Teachers who are not familiar with the historical sequence as it developed in the eighteenth century will find the story very instructive and illuminating even though they might not want to follow all of its intricacy in the class work. [Especially useful references to this end are Whittaker (1951) and Roller and Roller (1941). An abbreviated version is given by Arons (1965)].

Those students who have never observed the attraction of bits of paper or fibers to a charged plastic rod or comb should be led to do so. In making the observations, they should be led to think about gravity at the same time. They should articulate the fact that the interaction is different from the gravitational since gravity has nothing to do with the rubbing of objects and persists unchanged by the rubbing effects. They should also perceive that the electrical interaction is enormously stronger than the gravitational one since the bits of paper are given a high upward acceleration despite the downward pull of the entire earth. Such differences are sufficient to justify a new name, and the term "electrical interaction" can be introduced.

The term "charge" is introduced as the name for the property acquired by the interacting objects—a property that seems to "leak away," can be restored by rubbing, is transferable from one object to another by contact, is highly mobile on metallic objects, and so on. It must be explicitly emphasized that "charge" is an abstract construct and not a substance—that it is the name of a property that we infer from observed interactions and that we never come to know what charge "is" or how it "works" any more than we know what gravity "is" or how *it* "works." (Many students, having heard about electrons in connection with electricity and electrical charge, and having no idea whatsoever as to what the term "electron" means or where it comes from, labor under the misconception that, by knowing about electrons, they must know the nature of electrical charge. They must eventually be disabused of this misconception. One approach that makes a dramatic impression on most students is to lead them to the perception that the only reason we know electrons to be negatively charged is, in the final analysis, that we observe electron beams to be repelled by rubber rods that have been rubbed with cat fur.)

The terms "conductor" and "nonconductor" (or "insulator") should be given explicit operational definition in connection with those demonstrations and experiments that deal with the mobility and transferability of charge. This is more or less competently done in most presentations, and it is not necessary to dwell on it here.

6.4 ELECTROSTATICS EXPERIMENTS AT HOME

Since very few students have observed electrostatic effects at first hand, there is much to be gained by encouraging them to perform relevant experiments them-

---

[2]Readers aware of the humorous writings of James Thurber may note that these views are very similar to those ascribed by Thurber to his mother in *My Life and Hard Times*. The ascription is not idiosyncratic.

selves in addition to passively watching any lecture demonstrations that may be performed. (This is not meant to diminish the importance of the lecture demonstrations; these are very worthwhile and should certainly be performed, but they do not obviate the need for firsthand involvement.)

Fortunately, it is possible to perform effective and useful electrostatic investigations at home, and such homework assignments (as opposed to routine pencil-and-paper work) prove engaging to many students, especially if they are given adequate guidance.

The key to such electrostatic experiments is ordinary white ("Magic™") Scotch® tape. If a strip of tape two or three inches long is stuck down on virtually any smooth, dry surface and is then quickly pulled off, it turns out to be charged—strongly with some materials and relatively weakly with others. (One end of the strip should, of course, be folded over so as to provide a "handle" that does not stick to the surface and allows the strip to be handled without sticking to the fingers.)

If one such strip of tape is stuck to the back of another and the doubled tape is stuck to a smooth surface and pulled off, the two pieces of tape turn out to be oppositely charged when they are pulled apart.

One end of such a charged strip of tape can be stuck to some readily available support (e.g., the edge of a table, the bottom of a lamp shade, the slat of a chair back, etc.), and the suspended strip then becomes an electroscope leaf that can be observed to react to the presence of other charged objects. If the two oppositely charged strips of tape are suspended not far from each other, one has two reference strips capable of interacting with, and checking, the charge on other objects.

In a lecture demonstration, students can be shown how to set up such electroscopes and how to go about conducting an investigation of the interaction of charged objects. The first steps of investigation might involve charging other strips of tape by similar procedures and observing their interactions with the electroscope strips, but it should be made clear that the investigation should not stop with strips of tape. Many other household items (plates, glasses, containers of all kinds, tooth brush handles, combs) can be charged by rubbing with various materials (silk, wool, fur, cloths made of synthetic fibers, plastic bags or wrapping films), and the rubbed objects can be tested for interaction with the electroscope strips. One possible test object is one's own body after scuffing over a rug in dry weather, the test being made by bringing one's finger near each of the oppositely charged electroscope tapes.

A report of the French Academy in 1733 points out that "when an electrified body is brought close to the face [or arm], it causes a sensation like that of encountering a cobweb." This is a sensation students should be led to experience and interpret.

Some of the systematic investigations that might be conducted and questions that students should be led to address are outlined in the following sections.

6.5 LIKE AND UNLIKE CHARGES

Very few students, even those who may have acquired some substantially correct knowledge of electricity in prior schooling, can give an adequate operational definition of what is meant by "like" charges. If asked for the meaning of the term "like," virtually all of them say "like charges repel" and regard that as a definition. Very few textbooks pay any attention to this nontrivial linguistic problem. Many textbooks confound the issue by assuming that everyone "knows" that there are two kinds of charge and that like charges repel while unlike charges attract. They fail to comprehend the necessity of defining the technical terms "like" and "unlike" and of asking why we believe there to be more than one variety (or state) of electrical charge and why we also believe that there are not more than two. Only a very few textbooks handle these questions clearly and correctly.

Furthermore, a few students seem to acquire the (not very explicitly articulated) idea that there is really only *one* variety of charge. When questioned closely, they indicate a belief that repulsion between two bodies reveals that both are charged while other objects, attracted by either of these bodies, are actually without charge. This is, of course, a subtle issue since charged bodies do always attract uncharged ones, and we eventually rationalize this phenomenon by inventing the concept of "polarization." Even though most students do not reveal this misconception openly (they have memorized the dictum that there are two varieties so efficiently that the difficulty does not seem to arise), very few students are able to outline experimental evidence supporting the view that there are at least *two* "charge states" rather than one.

These elementary, but conceptually very significant, questions can be resolved by raising them explicitly while guiding the students through home experiments such as those suggested in Section 6.4. Guidance is best provided through a sequence of Socratic questions tailored to the level of preparation and sophistication of the group. If the students are simply turned loose to "investigate" the electrostatic interactions on their own, very little happens except among an extremely small number of exceptional individuals. The investigation has to be guided with sequences of questions that lead the students to make relevant observations and draw inferences. The questions must be carefully structured, however, so as to lead the students from one insight to another without giving the whole story away by simply asserting the end results.

The first step involves leading students to recognize that two strips of tape pulled off the same surface always repel each other and then to the extension that identical objects, charged in an identical way by being rubbed with the same material, always repel each other. This provides a first cut at the operational definition of "like." (An extension or redefinition of the concept into the completely general assertion that "like charges repel" comes later when we begin to recognize that there seem to be only two varieties of charge.) The recognition of two distinct charge states (or varieties of charge) comes, of course, from the observations that (1) rubber rods that have been rubbed with cat fur always repel each other; (2) that glass rods rubbed with silk always repel each other; and (3) that the glass and rubber rods then always attract each other. Finally one notes (4) that if either set of like rods were uncharged, they would not interact.

One way of approaching these insights in the home experiments is to set up the two electroscope strips of tape that are obtained when first stuck together back to front and then pulled apart after having been pulled off some smooth surface. (As indicated in Section 6.4, the two strips are found to attract each other strongly.) Students can be led through the logical sequence outlined in the preceding paragraph by use of a second set of strips prepared in the same manner as the first two.

Students, in their investigation of subsequent effects, can then bring up other charged objects to each strip and check whether the interaction is attractive or repulsive. Very few students will spontaneously discern and articulate the key, common element in the resulting observations. They must be led into articulating the insight that, if a charged object attracts one of the two strips, it always repels the other one, and vice versa. They must then be led to recognize the importance of what is *not* observed but can be imagined: namely that another object that has been rubbed with a different material is never found to repel *both* of the electroscope strips that attract each other. Neither do we observe a charged object that attracts both of the strips, but here one must be very careful to be sure that the test object is indeed charged, since any large uncharged object does attract both strips. (See further discussion below.)

Because no one has ever found a charged object that either attracts or repels *both* of the two strips of tape that attract each other, we come to believe that there are only two "charge states." It is not possible to "prove" that there are only two any more than we can "prove" the Law of Conservation of Energy. We accept the assertion of a regularity only because a violation has never been observed.

Now that we are limited to two "states" of electrical charge, we can extend the meaning of "like" to embrace all repulsive interactions and invoke the terms "unlike" or "opposite" to cover all attractions (with the exception of the subtlety outlined below). The names "positive" and "negative" can then be logically introduced in any one of the usual ways.[3]

As they make their observations, students should simultaneously be led to pay attention to the qualitative strengths of the interactions they observe. They should become explicitly aware of the fact that they intuitively assess the strength of the interaction by the *force* exerted on one of the interacting objects (e.g., the deflection of the strip of tape). They should recognize that (in the light of Newton's Third Law) there must be an equal and opposite force acting on the other object, but that the effect is usually too small to be observed directly as an acceleration or deflection. (The deflection of both strips of tape is readily observed, however, if a hand-held charged strip is brought near one suspended at the edge of the table.) They should be led to recognize that the force increases markedly with decreasing spacing between the interacting objects. They should recognize that, when each of two charged objects is separately brought near a charged strip of tape, the force exerted on the tape at the same distance of separation is frequently quite different, stronger for one object and weaker for the other. From this they should begin to draw the implication that one might eventually measure the charge state of an object by the force exerted on another "standard" charged object.

In the sequence of electrostatic experiments with the oppositely charged Scotch tape electroscopes, there is a very important subtlety that students must be led to recognize. Any relatively large object that has not been rubbed with

---

[3]From a carefully logical point of view, it is worth noting that there is a subtle distinction between recognizing the existence of two distinct "charge states" and asserting the existence of two distinct "varieties" of electrical charge. As far as macroscopic phenomena are concerned, we can account for the observed phenomena in either of two ways: (1) We can visualize the displacement of a single, conserved, imponderable fluid (as Franklin did) leading to one charge state in which the fluid is present in excess of normal amount and another state in which there is a deficiency, or (2) we can visualize two distinct varieties or kinds of charge, as we have become used to doing.

It is wise to remember that, on the macroscopic scale, we cannot really distinguish between the two models and that either one is thus equally valid. The choice of the "two distinct varieties" model is eventually forced on us by *microscopic* rather than macroscopic phenomena. How much of this detail one should belabor with students at introductory level is something the individual teacher must decide, given the existing constraints of time and coverage. One should not, however, suppress those students who try to pursue the "one fluid" model initially. They are not wrong, and they are in good intellectual company, given the history of the concept. They should be led to see how the modern "two varieties" picture came to be accepted rather than the "excess" and "deficiency" picture.

something else (and hence is presumably uncharged) will attract both charged strips of tape. One's own hand if brought close, for example, will attract both of the electroscope strips. Students must become aware of the fact that, although the repulsive interaction is always clear cut, an attractive interaction must be carefully checked and cross-checked to determine whether it is really an interaction between unlike charges or whether it is the ever-present attraction between a charged object and an uncharged one.

In the home experiments, there is a good way of showing that, if an object acquires one variety of charge on being rubbed, the rubbing material acquires the opposite charge. Normally (as with the hand-held fur and silk of the conventional lecture demonstrations), charge leaks off the rubbing material so rapidly that it is almost impossible to show that it becomes charged. If one puts one's hand, however, into a small plastic sandwich bag and charges some object by rubbing, the plastic bag retains its charge far more effectively than do the silk and fur. When tested against the electroscope strips, the bag will show a charge opposite to that of the rubbed object. One must, however, take the hand out of the bag before approaching the electroscope; otherwise the attraction of the large uncharged object (the hand) will overwhelm the interaction one seeks to observe.

There is ample room in these apparently simple home experiments for careful and skillful experimental work and for keen observation. As one performs the investigations oneself and devises ways of guiding students of different levels of sophistication and preparation through a fruitful sequence, one begins to see very vividly how subtle the unraveling must have been for the eighteenth century investigators and why it took them so long to recognize explicitly that there were two distinguishable charge states and not just one.

6.6 POSITIVE AND NEGATIVE CHARGES; NORTH AND SOUTH MAGNETIC POLES

Many students labor under the misconception that the names "positive" and "negative" for the two varieties of electrical charge are in some way necessary or inevitable, or that they literally refer to something in excess and to something missing. If the teacher desires, it is worth explaining how the names came into use historically through Benjamin Franklin's plausible (but, from the modern point of view, mistaken) model [see references such as Cohen (1941) and Roller and Roller (1957) for the details.] It should be made clear to the students in any case, however, that the names are perfectly arbitrary and could just as well have been chosen to be "red" and "blue" or "George" and "James" or "charming" and "revolting" (to parody some of the tongue-in-cheek choices that have been made in recent years in high-energy physics.) The rubber rod, rubbed with cats' fur, can be established as the primary classical reference for negative charge in lecture demonstration and can be used to calibrate the students' Scotch® tape electroscopes for home use.

Once the positive–negative terminology has been established in this way, it is well worth making a brief digression to the behavior of ordinary magnets. Some students have played with magnets in their earlier years, but it is surprising how many have not. Furthermore, even those who have played with magnets have rarely done so under circumstances in which they made systematic and guided observations rather than engaging in random manipulations that were never ordered or organized conceptually. Many students, for example, have

6.6 POSITIVE AND NEGATIVE CHARGES; NORTH AND SOUTH MAGNETIC POLES

played with magnets only to the extent of lifting tacks, nails, or paper clips with a single magnet but have never explored the interaction between two magnets and are astonished when they first observe repulsion. It is also a fact that some students may have seen such effects demonstrated under various circumstances, but the experience has been entirely vicarious, and they have never set up the interactions themselves or felt the forces with their own muscles. It is important for such students to feel for themselves the attraction between two magnets on being pulled apart in appropriate orientation or the repulsion when pushed toward each other.

Here again, students can be guided into systematic home-based experiments after a few illustrations in lecture and under the guidance of notes that supply a suitable sequence of Socratic questions. Students can be encouraged to purchase their own magnets for home experiments or suitable kits can be issued from the classroom.

The term "magnetic pole" can be developed operationally from observations of the differences between interactions associated with the ends of bar magnets and their middles. It should be made clear, preferably by direct demonstration, that magnetic poles cannot be separated by breaking the magnet and that all simple magnets, however large or small, are always bipolar. The question can then be posed as to whether there might be a valid role for the terms "like" and "unlike" as there was with electrical charge. The best reference frame for developing the operational definition is the historical reference frame, namely the earth itself.

Many students are aware of the interaction between magnets and a compass needle, but initially it is best to stay away from the compass because the compass needle is itself only an ordinary magnet. Many students do not realize this, and others all too easily lose sight of the fact that the primary reference in magnetic nomenclature is the earth rather than the compass.[4]

The students should be led first to establish the north–south direction wherever they are located. They can then take two rod or bar magnets (first having checked that they are actually both magnets and that one of them is not simply an unmagnetized object) and suspend them on strings, with a suitable yoke to maintain a horizontal orientation. The students should observe for themselves that each of the magnets, after coming to rest, is approximately lined up with the north–south direction. They should then mark the "north seeking" poles in some way and check the interactions between all of the combinations of magnet ends. From such observations they can be led to give an operational meaning to "like" and "unlike" poles and to articulate the observation that like poles always repel each other while unlike poles attract. They can also see the origin of the names "north" and "south" for the respective north- and south-seeking ends.

Following such a sequence, most students will readily conclude that the compass is simply a pivoted magnetized needle and will recognize that the earth itself must be a huge magnet. Most teachers are aware of the confusion students exhibit in connection with nomenclature concerning the poles of the earth, but this confusion stems principally from the fact that students have never had the opportunity to go through a sequence of observation, reasoning, and invention of

---

[4]Teachers should be alert to the fact that many students do not know how the directions "north" and "south" are established relative to the surface of the earth (see Section 1.14), and that many labor under the misconception that the compass *defines* these directions. It must be made clear that the directions are defined by *astronomical* phenomena and that the compass turns out to be a convenient secondary device only because it was discovered that the needle tends to align itself approximately in the astronomically defined north–south direction.

names such as that outlined above. They have been constrained to passive reading or listening and have only tried to memorize the vocabulary without sharing the experiences on which it is based. When they make the observations themselves, they can see through the tricks of the vocabulary and can recognize that the so-called north magnetic pole of the earth must be the south magnetic pole of the earth-magnet.

Playing with magnets provides an opportunity to extend the list of operational distinctions among electrostatic, magnetic, and gravitational forces. Students should be led to try the effects of magnets on their Scotch® tape electroscopes and to ascertain that there is no interaction other than the previously noted attraction observed in the presence of any sizable object. This paves the way for eliminating misconceptions such as repulsion between a north magnet pole and a positive electric charge, and so on.

Students should be led to make detailed lists of both the similarities and the differences among the effects, for example, there are two kinds of magnetic poles and two kinds of electrical charge, but there is no known bipolarity in gravity; the two varieties of electrical charge can exist separately, but magnet poles cannot; magnetism, as it is being explored with common permanent magnets, is confined to certain metals and has nothing to do with frictional effects as does electrical charge; magnetism is not "drained away" from an iron magnet when it is handled (as charge is immediately drained away on touching any isolated charged conductor); gravitational interaction is ever present, in all materials, without having anything to do with either frictional effects or magnetization; and so on. Even fairly knowledgeable students have never thought of making such a list and, when the question about specific operational distinctions is raised, have trouble giving illustrations of behavior and articulating the differences.

6.7 POLARIZATION

At about this juncture, it is helpful to direct students' attention to the fact that accumulating experience with electrical phenomena points to a very deep role for electrical charge in the structure of matter, even though experiments up to this point do not reveal just what the role is: (1) if we accept the notion that the architecture of matter is discrete (atoms and molecules) rather than continuous, the apparent presence of both varieties of charge in all materials hints that attraction between positive and negative might be what holds the discrete entities together; (2) sparks jumping through air (and, in the process, discharging charged bodies) hint at some sort of breakdown of structures ordinarily neutral and containing both kinds of charge—the breakdown resulting in conduction not present initially; (3) conduction, charging by frictional contact, and charging by induction all testify to different degrees of mobility of electrical charge within various materials.

Given awareness of these implications of the observations being accumulated, one can fruitfully go back to the observation that there is always an attraction between charged and uncharged objects. On the basis of what we now know about the ubiquity and mobility of electrical charge, can we provide a plausible description of how this interaction might arise?

The plausible description is, of course, provided by visualizing displacement of charge within the neutral body when it is in proximity to the charged one: opposite charge within the neutral object is then closer to the charged body than is the like charge, and the net result is attraction between the two objects. We bring to bear here almost everything we have learned so far about electrostatic phenomena, and we give the name "polarization" to the displacement of charge that is induced in the neutral object.

This well-known set of ideas is summarized in such detail at this point, not because it is in some way obscure or difficult, but because of the unfortunate manner in which it is developed in many existing texts. A very common approach is to assert that attraction between charged and uncharged bodies arises "because" of the phenomenon of polarization, and this is usually done without first reviewing and summarizing the existing knowledge as is done in the first paragraph of this section. Few students are prepared, at this stage, to unscramble the logical sequence on their own. Furthermore, they take "because" statements very literally: when the scientist says that such and such happens "*because* of so and so," many students take such statements as an absolute accounting of "*why*" the phenomenon occurs. The scientist has come forth with an a priori *reason* that the student feels he or she could not possibly have conceived, and the revelation is obediently memorized without comprehension of whence it came.

I have, on various occasions, characterized such a presentation as "backwards science"; it forgoes the opportunity to give students a more realistic view of how science works, and it obscures the intelligibility and motivation of the model.

The appropriate starting point is not the model but the observed *fact*, and the observed fact is that of attractive interaction between charged and uncharged bodies. We seek to account for this interaction in terms of what we now know, and the concept of "polarization" is deliberately *invented* to provide a plausible account. The concept is invented a posteriori and not a priori. This is the "forward" rather than the "backward" sense of the development.

This context can be enriched by leading students to consider the analogous effect in the case of permanent magnets: the lifting by a magnet of a string of tacks or paper clips and the general occurrence of attraction between a magnet and any unmagnetized piece of iron. The logical sequence is essentially the same as that in the case of electrical polarization, but students should also be led to note the differences as well as the similarities in the observable phenomena.

Without some direct help, many students find the *process* by which electrical polarization occurs very difficult to visualize and comprehend. They should be assisted in visualizing possible processes in both conductors and nonconductors. Since charge is visualized to be nonmobile in nonconductors, the appropriate model is that of inducing electric dipoles. (Many texts introduce this idea but usually so cryptically that students are mystified and tend to memorize without comprehension.) Since charge is visualized to be mobile in conducting bodies, polarization is to be visualized in terms of actual displacement or separation of charge.

In the latter case, it is important to lead students to the realization that the model accounts for attraction by a charged body regardless of whether one imagines positive charge to be mobile, or negative charge to be mobile, or both varieties to be mobile simultaneously. Many students are puzzled by this and are very reluctant to accept the indeterminacy. They expect to have the "right answer" immediately, and they also find it hard to accept the idea that one cannot obtain a unique answer from the experiments and observations. (In the same way, students, when they begin to do some naked eye astronomy and observe the behavior of the sun, moon, and stars, expect to "see" that the earth and planets revolve around the sun and that the earth rotates on its axis. They find it disconcerting to confront the fact that the heliocentric and geocentric models account for the observations equally well and that it is impossible to distinguish between them at this level of observation.)

Furthermore, the electrical situation is heavily confounded by the fact that many of the students have heard about "electrons" (without understanding anything about the origin and meaning of the term), and they have been misled into explaining virtually all electrical effects in terms of motion of "electrons" whether or not the concept is relevant. Students should have the opportunity to see that macroscopic observations and experiments never do resolve the question of what variety of charge is displaced in conductors and that, for the time being, any one of the three possible models is equally valid. Also, students who labor under the fixed delusion that *all* conduction is by electrons should be apprised of the fact that, in ionized gases and in electrolytes, there exist both negative ions (that are *not* electrons) and positive ions, and that conduction in such systems takes place through migration of both kinds of ions.

6.8 CHARGING BY INDUCTION

Many students in introductory physics courses find charging by induction to be a difficult and mystifying process and have great trouble visualizing it correctly and making correct predictions. Most of this difficulty stems from inadequate grasp of the concept of polarization discussed in the preceding section. If the idea of polarization is clearly motivated and developed, much of the difficulty with charging by induction goes away (providing it is made clear that the *entire* system, object plus other body in contact, becomes polarized in the presence of the charged object causing the induction.)

Exercises with charging by induction should, however, require that students visualize the effects in terms of the mobility of either kind of charge, or even in terms of mobility of both. Such practice is readily cultivated if it is made clear that test questions might require use of any one of the models chosen at random.

In analyzing situations involving charging by induction, students should be required to draw pictures of charge distributions at various successive stages in the sequence of operations and to describe, in their own words, what is happening. It is under the requirement of describing each step in words that understanding is most effectively generated.

6.9 COULOMB'S LAW AND THE QUANTIFICATION OF ELECTRICAL CHARGE

Given the fact that electrical charge is not a ponderable substance and, up to this point in the sequence of development, is simply the name for a certain state in which objects attract or repel each other, it is far from obvious to students that the "quantity" of electrical charge is measurable. Eighteenth-century investigators (prior to Coulomb) had no way of measuring the quantity of charge, although they qualitatively recognized more and less highly charged states by the intensity of the observed effects, such as sparks and shock, and they compared charge states roughly by the number of turns that had been given the electrostatic generator prior to having the generator make contact with the Leyden jar or other body accepting charge.

It was Coulomb who first showed convincingly that charge might be quantified by measurement of the force of interaction. In the famous torsion balance experiments in which he established the inverse square law, he also examined the interaction between charged conducting spheres at fixed separation as the charge state was altered in a systematic way: having set up a situation in which the charged spheres on the torsion balance were, say, repelling each other, he brought an identical uncharged conducting sphere in contact with one of the two interacting spheres and observed that the force between the charged spheres was now half of what it was initially. On bringing the uncharged sphere in contact with the second of the charged spheres, he observed that the force of interaction was now one quarter of what it had been initially.

Arguing from symmetry (namely, that identical spheres should share charge equally if charge is indeed a systematically measurable quantity), one supposes that the quantity of charge on each of the interacting spheres was successively cut in half in the above procedure. Since the force of interaction was also cut in half at each step, it is plausible to infer a direct proportionality between observable force and the product of the two interacting quantities of charge. Thus charge becomes truly quantified not by qualitative observation of intensity of sparks and shocks or by number of turns of an electrostatic machine but by means of a previously established numerical scale, namely that of force. This leads directly to the usual statement of Coulomb's Law *induced* from these observations: that the force of interaction between point charges is directly proportional to the product of the two quantities of charge (and inversely proportional to the square of the separation). In the old c.g.s. system of units, each of the interacting spheres was defined as carrying one unit of charge when the force between them was one dyne at a separation of one centimeter between centers.

Thus, in electrostatics, prior to the connection to current electricity and its chemical effects, support for the notion that quantity of electrical charge was measurable came from the halving of the force in Coulomb's experiments. The PSSC film "Coulomb's Law" (narrated by Eric Rogers), presents an excellent demonstration of these effects on large scale, showing that the force between charged spheres varies inversely as the square of their separation and also showing that the force at fixed separation is cut in half each time one of the spheres is brought in contact with an identical uncharged sphere.

These effects can also be demonstrated quantitatively and very effectively by setting up a Coulomb's Law experiment on the platform of an overhead projector. (The warmth supplied by the projector helps maintain dryness that minimizes leakage of charge during the course of the demonstration.) A transparent graph paper on the platform supplies a scale. The system consists of one pithball on a bifilar suspension so that the ball can swing away from equilibrium position, the displacement from equilibrium position being a measure of the force acting on it. A second pithball is fastened to the top of a short insulating rod on a small stand of transparent plastic that can be moved along the projector platform. The stand can be moved around, bringing its pithball to within different distances of the ball on the bifilar suspension. The two balls are set up so as to be quite close to the platform of the projector, and their images, as well as that of the graph paper, appear on the screen. The inverse square effect can be nicely demonstrated up to the point at which small separation of the spheres leads to appreciable distortion of the charge distributions by induction. The halving of the force can be shown by touching one of the charged spheres with an identical uncharged one.

In developing this story for students, it is important to emphasize the fact that Coulomb's Law is not "derived" mathematically or "proved" physically. Postulating the law on the basis of the very limited observations described above is a matter of *inductive* and not deductive reasoning. Ultimate acceptance of the law and belief in its validity reside in the fact that it has been tested over a vast variety of situations and applications and has always been found to "work" and never found to fail. Furthermore, the quantification of charge via Coulomb's Law turns out to be consistent with quantification through definition of electric current and through observation of chemical effects (i. e., deposition of material in electrolysis). In this context it is also illuminating, for example, to show students why it is that careful investigators have taken great pains and invested substantial effort in testing the accuracy of the inverse square relation. Williams, Faller, and Hill (working at Wesleyan University in 1971) showed that, if the exponent deviates from exactly 2, it does so by less than 2×10^{-16}.

Finally, this is an opportune point at which to remind students that, just as Coulomb's Law is induced rather than derived or proved, so also are the Newtonian Laws of Motion, the Law of Gravitation, and the conservation laws. When one is traversing these developments for the first time, it is very easy to lose sight of subtle logical aspects of this variety; yet the awareness is crucial to understanding the nature and limitations of scientific knowledge, as well as the essential difference between science and mathematics.

It is also worth noting that Coulomb, in the infancy of the science, talks not about electrical "charge" but about the electrical "masses" of his charged spheres. It is obvious that "electrical mass" has nothing whatsoever to do with "inertial mass." By analogy, one can dramatize the idea and help students discriminate operationally between inertial and gravitational mass by temporarily referring to the latter as "gravitational charge." (See Section 3.9)

6.10 ELECTROSTATIC INTERACTION AND NEWTON'S THIRD LAW

The encounter with electrostatic and magnetostatic forces provides a valuable opportunity for "cycling back" of the variety advocated repeatedly in these notes, namely a reencounter with Newton's Third Law after some time has elapsed since its use in mechanics. Those teachers who are inclined to believe that the issue has been settled through clear exposition, and through the numerous exercises assigned in the first encounter, would do well to try the following problem on students who have begun the study of electrostatics:

> *The following diagram [Fig. 6.10.1] shows uniformly charged spheres firmly fastened to, and electrically insulated from, frictionless pucks that ride on an air table. One sphere carries a charge of $+2.0 \times 10^{-8}$ coul while the other sphere carries a charge of $+6.0 \times 10^{-8}$ coul. The pucks, with the spheres they are carrying, are free to accelerate along the table. Draw separate free-body force diagrams of the spheres and of the pucks, showing all the forces acting on each of the four objects. When forces are equal, show the arrows as equal in length. When forces are not equal, show the larger forces with longer arrows to some reasonable scale.*

When I gave this problem on a diagnostic pretest to a group of 40 second-year students who were beginning an introduction to modern physics after having completed a full year of introductory calculus-based physics, 65% of the group showed the electrical force acting on one sphere to be three times as great as the electrical force acting on the other, and 85% showed no horizontal force of interaction between the spheres and the pucks (60% also had something wrong with at least some of the vertical forces). I submit that this experience dramatically

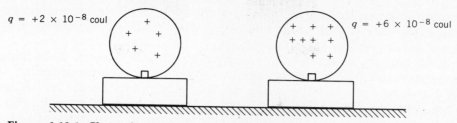

Figure 6.10.1 Charged conducting spheres are pinned to nonconducting frictionless pucks and are allowed to accelerate freely on an air table.

illustrates the importance of giving students repeated exposure to qualitative questions of this variety. It is only after several encounters, in altered context, spread out over time, that the concepts register with the large number of students who do not operate at the front edge.

In addition to recognizing that the forces are still equal and opposite even when the charges are different, the students should also recognize that, when the charged rod they are holding attracts bits of paper or repels a charged pithball, forces are being exerted on the rod even though they cannot be sensed directly. This fact is commonly overlooked, and the situation is not connected with the one in which forces between, and accelerations of, small interacting objects are clearly revealed.

Still another question involving Newton's Third Law can be fruitfully raised at this juncture: Suppose we have two charged spheres attracting or repelling each other; in accordance with the Third Law, the forces are equal and opposite. Suppose we now suddenly displace one of the spheres. Does the force exerted on the other sphere change instantaneously (i.e., does electrostatic interaction operate as "action at a distance" in the manner Newton assumed that gravity did)? If the change is *not* instantaneous, and an interval of time elapses before the change, it must be the case that the Third Law does not hold during that interval. What are the implications of this line of thought? (See Section 3.13 for comments on the initiation of this discussion.) It should be noted that "closure" need not be provided immediately. Many students have been conditioned into expecting an immediate pat answer for any question raised, however subtle and sophisticated it might be, and, in the case of the present question, the answer was long in coming. Faraday spent the last years of his life vainly trying to observe such time delays, and the question came to be resolved only with Maxwell's invention of field theory and the recognition that the Law of Conservation of Momentum has primacy over Newton's Third Law.

6.11 SHARING CHARGE BETWEEN TWO SPHERES

The sharing of charge between identical conducting spheres was made use of in the discussion of Coulomb's Law and the quantification of charge (Section 6.9). Distribution of charge between spheres of different radii is usually discussed in connection with the concept of capacitance. In all these instances, contact is made between the outside surfaces of the spheres, and it is shown that the ratio of the charges acquired by the spheres is equal to the ratio of the radii.

After it has been shown that there is no charge residing on the *inside* of a hollow conductor, and students should have this idea firmly registered, it is illuminating to give them a question such as the following: Suppose we charge a large hollow sphere and then insert a small uncharged sphere into the space within the large one through an opening in the latter. Now we connect the small sphere to the interior of the large one with a wire. What will be the final charge distribution between the two spheres?

It has been my experience that many students blindly go back to the relation for the charge distribution between the two spheres making external contact and apply the same formula to the small sphere making contact with the inside of the large one. They fail to invoke what has been learned about the interior of the hollow conductor, and they do this even after vivid demonstrations of the "ice pail" experiment, especially if a bit of time has elapsed since the discussion regarding zero electrical field in the interior of a hollow conductor. They show better retention of the latter concept after having made the mistake indicated above. This is another illustration of the value of "spiraling back" in a later context.

6.12 CONSERVATION OF CHARGE

If one seeks to cultivate and enhance "critical thinking" among the students, an important avenue for doing so resides in leading them to address the "How do we know . . . ? Why do we believe . . . ? What is the evidence for . . . ?" questions that arise at the most fundamental levels of concept formation and of perception of order in natural phenomena. The questions raised in the preceding sections underpin genuine understanding of, and critical thinking about, electrical phenomena. Adoption of the view that electrical charge is conserved is an integral part of the structure. What is the evidence for this principle? The insight that electrical charge is conserved is far from obvious in the performance of macroscopic electrostatic experiments, and it should not be surprising that this realization came rather late in the historical sequence.

In 1747, Benjamin Franklin conducted a series of simple, ingenious experiments with Leyden jars. Two identical jars were charged in the same way by simultaneously touching the "hooks" (contacts to the internal foil) to the glass tube of an electrostatic generator while the external foil was grounded. Franklin observed that if the outer coatings were connected to each other (by holding the jars with his fingers, his body being the conductor), nothing happened when the hooks were brought in contact. However, when he held one jar by the hook and the other by the outside coating, and then brought the coating of the first in contact with the hook of the second, a spark occurred, and the jars were completely discharged.

This experiment, together with variations of it, led Franklin to surmise that the "electrical fluid"[5] was merely shifted from one body to the other in frictional contact and that the total quantity was always conserved. It is true that the experiments are crude and qualitative, but Franklin's intuition was sound; the conjecture influenced other investigators and proved to be very fruitful in subsequent research.

The Leyden jar demonstrations are easily done in class or lecture, and questions regarding their implications make excellent homework assignments—assignments that give students a chance to consider the interpretation of real physical phenomena, with all the attendant caveats and uncertainties, in addition to the conventional numerical problems with precanned parameters.

---

[5]Franklin held a one-fluid model of the charging process and (fallaciously) convinced himself that glass acquired the excess of the fluid while the material with which it was rubbed acquired the deficiency. From this heuristic model came the terms "positive" and "negative."

6.13 ELECTRICAL FIELD STRENGTH

The concept of "electrical field strength **E** at a point in space" as defined by

$$\mathbf{E} = \frac{\mathbf{F}}{q_{\text{test}}} \tag{6.13.1}$$

is well known to offer difficulty to many students. Most of the difficulty appears to reside in the definitional aspects, but, since there are several of these, the effect is compounded.

First, the use by most texts of the ordinary equals sign (=) instead of the identity (or definition) symbol (≡) deflects attention from the fact that the *idea* being invented resides on the right-hand side of Eq. 6.13.1 while the left-hand side is simply a *name* for this idea. (See discussion of different kinds of "equalities" in Section 3.23.) This needs to be emphasized several times before it sinks in, and the identity symbol helps in maintaining the emphasis.

Next, we have the mystique of the ratio. This is still a matter of discomfort to many students, and they fail to ask themselves what the ratio *means*, that is, what is the verbal interpretation? They hope to be able to manipulate the formula without facing the verbal issue. It is necessary to lead them to articulate the fact that we are talking about the force per unit positive charge and not about force alone. Obvious though this may seem, it is frequently lost sight of. Since it helps to have done this kind of thinking more than once and to have started in some more familiar situation, quite a few texts now mention the parallel concept of "gravitational field strength," that is, gravitational force per unit mass, and lead the student to recognize that this is just another way of talking about the familiar concept denoted by g. This parallel illustration is well worth invoking.

Next, there is the question of *why* we want to invent such a definition in the first place. The point, of course, is that we wish to create a number that is an intrinsic measure of the force field to be described and that must therefore be made independent of numerical properties of our detecting device. Although this objective is explicitly stated in many presentations, it is frequently put forth so briefly and casually that most students miss it. Students should be led into participating in the invention of the concept and should have to articulate the purpose and the interpretation in their own words. It is through such articulation that they begin to perceive that a concept is being *created* for a specific purpose and cease to regard Eq. 6.13.1 as a "formula" received through revelation.

Next, students should explicitly summarize and describe the *operations* underlying Eq. 6.13.1, that is, they should be able to describe step by step what they themselves would *do* to obtain the number denoted by **E**. It is in such a description that students are constrained to indicate that a force is to be measured at a *point* in space and that an electrical force cannot be observed without placing a charged particle at that point. This description becomes the best place to emphasize that the test charge must, in principle, be so small as not to disturb or rearrange the charges creating the field under consideration, and that **E** therefore denotes force per unit charge without actually being measured as a force on one whole coulomb.

One might hope that, having been led through the preceding sequence, all students would understand the meaning of "electrical field strength" and would pursue the solving of conventional numerical problems without a hitch. Most teachers are aware, however, that this is not the case. The trouble is that, when starting in on a problem (especially one involving the calculation of a field strength involving superposition of effects from two or more point charges), only a very few students take themselves back through the meaning of **E** on their own initiative. The majority sit staring at the problem and wishing to discern how to substitute in the formula when what they should be doing is asking themselves; "What does **F** stand for? What does q_{test} stand for? What would I do to obtain the respective numbers?" Only then do they proceed, for example, to find the vector sum of the forces exerted on the test charge by other point charges in the vicinity. It is necessary to help students form the habit of stopping and asking themselves such questions whenever they find themselves in difficulty with definitional statements appearing in symbolic form.

6.14 SUPERPOSITION

"Each point charge in a distribution of charges contributes to the total value of **E** at a given point as though it (the point charge) alone were present." This seems like a simple and reasonable statement, and it is—except for the fact that many students are confused by a few normally unspecified aspects.

If one is concerned with a single idealized charge distribution in empty space (e. g., a set of point charges, a line charge, a sheet of charge) nothing more need be said, and the only difficulty that arises is associated with the idea of setting up an integration for the resultant effect of a continuous distribution. Here the difficulty usually resides in the fact that many students do not yet, at this stage of the game, clearly associate integration with the process of addition needed in these circumstances. They have been *told* about limits of sums, but they have rarely, if ever, had to use such language themselves in describing what they are doing in specific integrations. They have executed algorithms of integration without talking about or interpreting them, especially in physical applications. Many do not yet recognize the evaluation of an area as an addition process, and, even if they do, they do not carry the addition concept over to the process of finding the total field strength. They must be led to describe what is happening in their own words. Making a partially successful derivation by parroting a procedure in the text is no indication of understanding.

Significant conceptual difficulties with the superposition concept arise when the effects of more than one simple array of charge are being evaluated. First, students must be helped to absorb the subtle idea that the insertion of a set of charges in a region where mobile charges (especially those on conductors) are already present will, in general, lead to a rearrangement of the charge distributions and that the superposition principle applies only to the final rearranged state. (I am not concerned here with dielectric phenomena or other sophisticated aspects of polarization but only with simple situations of the kind that arise in an introductory course.) For example, if one is dealing with the field produced by two or more charged conducting spheres and is treating them as point charges concentrated at their centers, students should be articulately aware that the latter assumption holds only to the degree that the interactions among the spheres are not sufficiently strong to distort the initially spherical charge distributions residing on their surfaces. They should also be aware that the true resultant field is that due to whatever final static distribution is actually attained.

A second difficulty with superposition is that many students are puzzled by what happens when the situations under consideration imply the presence of physical objects rather than just disembodied charge distributions. In considering

6.14 SUPERPOSITION

the fields produced by the charged spheres mentioned in the preceding paragraph, for example, they find it very hard to believe that they can treat the system as though the metal spheres themselves were absent. They believe that "one sphere would not let the field from the other sphere 'pass' through it." Similarly, after it is shown that the field on either side of an infinite plane sheet of charge is uniform to infinity, they are unwilling to accept the argument that superposition of the fields of the two plates of a capacitor gives an approximately zero field outside the plates. The unwillingness arises, not from the idealizations and approximations involved, but rather from the robustly held idea that the field of one plate could not possibly "penetrate" through the other plate.

Many students hold these misconceptions on an a priori basis without ever having heard of electrostatic shielding. The misconception is reinforced, however, in those students who *have* heard of shielding. They tend to associate the shielding effect with inability of an external field to penetrate the shield rather than with the cancellations due to the redistribution of charge on the shield itself.

Genuine understanding of the superposition principle can be developed only through recognition and careful discussion of these conceptual difficulties. Students must be helped to realize that the principle is arrived at by inductive reasoning, that it is accepted because it "works," that it is a fact of nature that, once the charge distribution has attained equilibrium, intervening spheres and capacitor plates play no role, and that electric field contributions cannot be thought of as "blocked by" or "passed through" material objects.

CHAPTER 7

Current Electricity

7.1 INTRODUCTION

Research has been showing that the most basic concepts underpinning simple direct current (d.c.) circuits offer very serious difficulties to many students and that certain misconceptions are widely prevalent [Arons (1982); Cohen, Eylon, and Ganiel (1983); Fredette and Clement (1981)]. As in the case of static electricity, the learning problems are aggravated by the remoteness of the underlying phenomena from direct sense perception. The observable effects are not easily linked to abstractions such as "electrical charge," "current," and "energy." Since students are aware that batteries "run down" and that one "uses" household electricity, they believe that "something is used up" in electric circuits, and, to many of them, the most reasonable thing to be "used up" is "electricity" itself. Furthermore, the concept of "potential difference" is difficult enough in its application to electrostatic fields; its relevance to, and applicability in, electric circuits is even more obscure for most students.

The structure of the concepts underlying Ohm's Law, and the operational definitions entailed, are just about as intricate as those underlying Newton's Second Law,[1] but there has been far less epistemological discussion of the electrical case. Textbooks show a very wide range of differences in order and mode of presentation. Research on student learning difficulties is still in its infancy. At the present time, no one mode or sequence of presentation emerges as pedagogically superior to any other. Teachers are on firm ground if they choose any logically sound mode which is most congenial and with which they feel most secure.

Regardless of the logical structure of presentation, however, conceptual difficulties arising in the treatment of current electricity tend to be glossed over much too rapidly and superficially in introductory courses at all levels. As a result, very few students (even in engineering-physics courses) develop understanding of the phenomenology of simple circuits. They are usually tested on numerical manipulations involving Ohm's Law, while the ability to solve the conventional problems has only very weak correlation with understanding of the physical phenomena taking place [Arons (1982)]. This situation can be remedied only by leading the students to think more carefully and more frequently about the qualitative

---

[1] I do not mean to imply that there is a close analogy between Ohm's Law and the Second Law of Motion, although some authors attempt to exploit such an analogy. In logical structure and range of validity, Ohm's Law is much more like Hooke's Law for ideal springs than it is like Newton's Second Law.

aspects of the phenomena and by deepening their concern about the ever-present "How do we know . . . ? Why do we believe . . . ?" questions.

If one accepts this objective, it is unwise to force a completely rigorous formulation on the students from the very start. Just as it is wise to introduce concepts in kinematics and dynamics by starting with fairly primitive, intuitive ideas, and then to refine and redefine them as one penetrates more deeply into the conceptual structure, so it seems wise to follow a similar pattern in forming the picture of simple resistive circuits. Preliminary concepts of "circuit," "current," "conductor," "nonconductor," "resistance" can be formed initially, in qualitative form, through observations of phenomena in simple resistive circuits. The ideas can then be redefined, quantified, and joined with the concept of potential difference in a more rigorous discussion, which becomes far more intelligible to beginning students than a fully rigorous discussion asserted *ab initio*.

7.2 WHICH SHOULD COME FIRST, STATIC OR CURRENT ELECTRICITY?

There is debate among teachers about this question. Some prefer to introduce static electricity first (as do the majority of textbooks); others prefer to introduce current electricity first and go on to static electricity as a special case afterwards. In this book, static electricity has been discussed first because that is the more common approach. It has always seemed to me, however, that the teacher's own preference is the legitimate determining factor. One can develop the relevant concepts equally soundly and logically either way. What is frequently glossed over without adequate care and attention, however, is the *connection* between the two sets of phenomena, that is, how do we know that the voltaic battery or the electromagnetic generator maintains continuous transport within conductors of the same "electrical charge" manifest in static frictional phenomena? (This question is discussed in more detail in the following section.)

The fact that the concepts of electricity developed out of electrostatic phenomena historically need not be the pedagogical determinant unless the teacher prefers to follow the historical sequence. Selected portions of the historical sequence [as treated, for example, by Roller and Roller (1957) in the *Harvard Case Histories*] can be instructive and illuminating, but more detailed study of the actual historical ups and downs and controversies [as analyzed, for example by Heilbron (1979)] would be hopelessly confusing at an introductory level and would not enhance learning and understanding in most students (however much such study might expand the teacher's own insights).

Teachers interested in seeing sound developments that start with current rather than static electricity would do well to examine the treatment given by Rogers (1960) and the excellent elementary school unit "Batteries and Bulbs" in *Elementary Science Study (ESS)* (1968ff). A purely qualitative treatment for college level students (owing much to *ESS*) is to be found in Arons (1977), and a continuation, which effectively quantifies the "Batteries and Bulbs" activities, is given by Evans (1978).

7.3 HOW DO WE KNOW THAT CURRENT ELECTRICITY IS "CHARGE IN MOTION"?

Many textbooks, after forming the concept of "charge" and examining electrostatic phenomena in the context of frictional electricity, make a discontinuous jump to current electricity by simply asserting that electric circuits containing batteries involve "charge in motion." To most students, however, it is far from obvious, at this early stage, that batteries and household outlets on the one hand, and frictional phenomena on the other, have any connection with each other. They are dumbfounded if the "How do we know?" question is raised explicitly. The visible effects are radically different in the different situations, and the students have not had the benefit of the long series of subtle investigations carried out by Galvani, Volta, and their contemporaries, investigations in which the invention of the voltaic pile came very gradually and out of initially direct connection with electrostatic effects.

Furthermore, the students have, since childhood, heard the words "charge" and "electricity" applied indiscriminately to batteries, household outlets, and electrostatic effects. It has never occurred to them to raise the question as to how one knows that there is any connection among these seemingly disparate systems; the application of the same name from very early on has suppressed the questions that should have been raised. Whether one starts with static or with current electricity in a given course, it is important to stop when the transition is made from one subject to the other and examine the evidence that establishes the commonality asserted in the names.

That this is not a trivial intellectual matter, and that there must have been serious questions about it in the scientific community as late as the 1830s is testified to by the attention given it by Faraday himself. In the *Electrical Researches* [Faraday (1965a)], one finds a paper, dated 1833 and titled "Identity of Electricities Derived from Different Sources," in which Faraday describes the experiments summarized in Table 7.3.1. The "different electricities" referred to are voltaic, common (frictional), magneto (from electromagnetic induction), thermo (from a thermocouple), and animal electricity from the torpedo and the gymnotus (electric ray and electric eel, respectively). He shows that each of these different sources produces an array of identical effects: physiological effect (shock), magnetic deflection (of a compass needle as in the Oersted experiment), magnetization of a needle, spark, heating effect (in a conducting wire), chemical action (electrolysis), attraction and repulsion, and discharge by hot air.

In the paper accompanying the table, Faraday describes exactly how each of the observations was made. Some of these required the full exertion of his

Table 7.3.1
Faraday's Table of Experimental Results Showing the Identity of Electricities Derived from Different Sources. The x's denote positive results he obtained himself; the +'s denote positive results filled in somewhat later by other investigators.

	Physiological Effects	Magnetic Deflection	Magnets made	Spark	Heating Power	True chemical Action	Attraction and Repulsion	Discharge by Hot air
1. Voltaic Electricity	x	x	x	x	x	x	x	x
2. Common Electricity	x	x	x	x	x	x	x	x
3. Magneto-Electricity	x	x	x	x	x	x	x	
4. Thermo-Electricity	x	x	+	+	+	+		
5. Animal Electricity	x	x	x	+	+	x		

7.3 HOW DO WE KNOW THAT CURRENT ELECTRICITY IS "CHARGE IN MOTION"?   165

legendary experimental skill. Students are in good company if they initially fail to perceive the identity of the various effects and if they are unable to cite evidence for the interconnections.

"The general conclusion which must, I think, be drawn from this collection of facts," writes Faraday, "is that *electricity, whatever may be its source, is identical in its nature*. The phenomena in the five kinds or species quoted differ, not in their character but only in degree; and in that respect vary in proportion to the variable circumstances of *quantity* and *intensity* which can be made to change in almost any one of the kinds of electricity, as much as it does between one kind and another."

Although one certainly need not try to duplicate all of the episodes in Faraday's table, it is highly desirable to lead students to think about the phenomena involved and to perform a few of the experiments (or at least see them convincingly demonstrated.) The most important links to establish at an early stage are probably those among electrostatics, voltaic batteries, and the household outlet; the connection to other sources will subsequently be plausible without belaboring the issue. Showing Faraday's table and telling the story that goes with it are very helpful in this connection.

Assuming that electrostatic effects have been explored, that the concepts of "charge," "conductor," "nonconductor," "leakage of charge," "electrostatic induction," and "polarization" have been developed, and that the electroscope is now a familiar qualitative device, one can appeal to the following set of experiments with the apparatus sketched in Fig. 7.3.1.

The capacitor plates can be charged in four ways: (1) by touching one plate with a plastic or glass rod that has been rubbed with fur or silk and touching or grounding the other plate (i.e., charging it by induction); (2) by connecting the plates to the two terminals of a Wimshurst machine or other electrostatic generator; (3) by connecting the plates to the terminals of a voltaic battery of sufficiently high voltage (three or four B batteries of 90 V each usually suffice); (4) by connecting the plates to the terminals of a high-voltage source (rectifier) running off the 120-V line or directly to an electromagnetic generator.

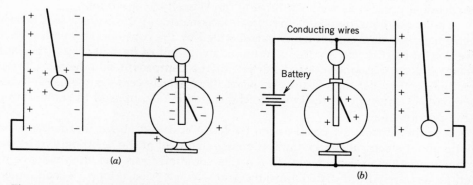

Figure 7.3.1 (a) Capacitor and electroscope charged by induction through contact with charged rod or by connection to Wimshurst machine. Suspended light ball with conducting coating swings back and forth, transporting charge between the plates. System runs down if charge is not resupplied to the plates. (b) Capacitor and electroscope connected to voltaic battery or to a high voltage source running from 120-volt (V) outlet or to an electromagnetic generator. Electroscope reveals presence of charge. System behaves in exactly the same way as it does in setup (a).

An electroscope is connected across the plates as shown, and deflection of the leaf monitors the charge on the plates. A light ball with a conducting coating is suspended as a pendulum between the plates. When the plates are charged, the ball oscillates back and forth, transporting charge between them. (Students should have seen a demonstration of the oscillating ball in the electrostatic case and should have been led to sketch and analyze the charge transport prior to the investigation now being undertaken.)

It can now be pointed out that we have constructed a little "motor" (in principle, one could draw mechanical work from the displacement of the ball) in which charge is transported between the plates. Starting with charging mode 1, we note that, as the ball keeps swinging, the deflection of the electroscope decreases: the plates are being discharged; and the swinging "runs down." We can produce the same final state by connecting a wire (or wet string) between the plates, the discharge taking place so rapidly that we cannot discern a finite time interval, as we can with the swinging ball.

With charging mode 2 we can produce the same swinging of the ball, but the swinging persists, and the electroscope shows deflection, as long as we keep turning the Wimshurst machine. Discharge takes place just as in mode 1 when we cease turning the machine. In other words, the system will maintain continuous transport of charge providing we keep on supplying electrostatic charge to the plates by the frictional process.

With charging modes 3 and 4 the electroscope shows deflection and the ball keeps swinging indefinitely. Since the battery and the rectifier produce the same visible effects (deflection of the electroscope and swinging of the ball) as do the charged rods and the Wimshurst machine, we infer commonality of the underlying electrical effects. (The commonality is, of course, further substantiated by Faraday's other tests, such as spark, shock, chemical, and magnetic effects.) If a wire (high resistance, of course) or a wet string is used to connect the plates, the ball keeps swinging, and the electroscope continues to show deflection. Since, under these circumstances, transport of charge must be taking place continuously through the wire regardless of whether or not the ball happens to be present, this experiment implies that both of these sources have the capacity of continuously supplying charge to the plates, just as charge is continuously supplied through the frictional process as we keep turning the Wimshurst machine.

The continuous nature of the effects originating in the battery (and also in electromagnetic sources) is further corroborated by the continuous chemical effect observed in electrolysis, by the continuous evolution of heat in wires (the Joule effect), and by the continuous deflection of the compass needle in the Oersted experiment. It is also significant that all of these effects cease immediately when the circuit is interrupted, testifying to the dynamic (as opposed to static) nature of the phenomena.

It is the *combination* of all of the evidence adduced above that justifies our talking about electrical charge (and its transport) in *all* of the seemingly unrelated circumstances being considered. Given this common basis and the evidence for continuous transport under circumstances in which a "closed loop" is maintained, we are justified in introducing terms such as "circuit" and "electric current."

Most textbooks introduce the various technical terms so casually and abruptly that it seems worth pointing out how Faraday himself felt about the evolving terminology. "Whether there are two fluids or one," he writes, "or any fluid of electricity, or such a thing as may rightly be called a current, I do not know; still there are well-established electric conditions and effects which the words 'static,' or 'dynamic,' and 'current' are generally employed to express; and with this reservation they express them as well as any other."

It is worth perceiving and respecting the students' "reservations" about these matters and giving them an insight into the observations that force us to the concept of continuous transport of electrical charge. Understanding of physics resides in understanding the connection between the phenomena and the concepts as well as in the solving of subsequent problems.

7.4 BATTERIES AND BULBS (I): FORMATION OF BASIC CIRCUIT CONCEPTS

To illustrate the importance of hands-on experience in connection with the formation of the abstract concepts and models we are now discussing, I cite experience with college-age students in connection with electricity and simple circuits. The subjects in this case were preservice (undergraduate) elementary teachers, undergraduate nonscience majors in a general education physics course, and in-service elementary teachers in a summer institute. Somewhere in their school experience or in other circumstances, all had heard about "electric circuits," most had seen diagrams of electrical configurations in books and on chalk boards, all had been exposed to verbal assertions of facts and concepts of current electricity, even though none had ever taken a physics course.

When these students were given a dry cell, a length of wire, and a flashlight bulb and were asked to get the bulb to light, most started either by (1) holding one end of the wire to one terminal of the cell and holding the bottom of the bulb to the other end of the wire, or by (2) connecting the wire across the terminals (i.e., shorting the cell) and holding the bulb to one terminal. They showed no sense of the functional two-endedness of either the cell or the bulb. Few noticed that the wire became hot when connected across the terminals of the cell, and those that did notice inferred nothing from the observation. It took 20 to 30 minutes for some member of the group to discover, by trial and error, a configuration that lighted the bulb. Then, of course, the message was passed around. Seven-year-old children, incidentally, when given the same task go through exactly the same sequence at very much the same pace.

Absent the synthesis of actual experience into the concept of "electric circuit," the adults, despite the words they knew, the diagrams or pictures they might have seen, the assertions and descriptions they had read or heard, showed no more understanding of the ideas involved than the seven-year-old approaching the phenomena *de novo*. Purely verbal, passive inculcation had left essentially no trace of knowledge or understanding. In the groups referred to above, the only individuals who got the bulb lighted quickly were the few who happened to have had previous hands-on experience with electric circuits; one, for example, had been an electronics technician in the service.

Pursuing the sequence thus initiated is one excellent way of leading students to build up the abstract model and concepts associated with simple circuits. They should be led to sketch *all* the configurations they try out: the ones that do not light the bulb as well as the ones that do. They should be led to separate the two classes and to describe in words what the successful arrangements have in common and how they differ from the other group. Once the intrinsic two-endedness of each object and the necessity of forming the continuous loop have been explicitly recognized, one has arrived at the essential operational description, and the technical term "circuit" can be introduced (note adherence to the precept of "idea *first* and name *afterwards*").

Interposing a variety of different objects made of different materials in the loop (coins, pencils, keys, glass, plastic, rubber bands, string, paper, etc.) leads to the classification of materials that allow the bulb to light or prevent it from lighting even when the necessary loop configuration is provided. The operational definition has been formed, and the technical terms "conductor" and "nonconductor" or "insulator" are appropriately introduced. Air can now be identified as a nonconductor.

The situation, now becoming familiar under repeated exploration, strongly suggests the presence of a dynamic rather than a static effect: All effects (heating of the wire, lighting of the bulb) cease when the loop is interrupted or is not completed with a sequence of conducting objects. The two-endedness of each object and the necessity of a complete loop argue for some continuous, dynamic process, as also does the continuous "running down" of the battery. The fact that the bulb lights with equal brightness regardless of where it is placed in the loop (near either terminal, or in between with use of two wires) as well as the fact that the wire, when connected across the battery terminals, is uniformly heated over its entire length, point to continuity and uniformity of the dynamic effect all the way around the system. The uniformity of heating of the wire is supported by the uniformity of heating of stove and toaster elements in the appliances at home (most students turn out to be aware of this uniformity if specifically asked about it but do not think of appealing to the evidence themselves). It must be emphasized *explicitly* that this uniformity means that whatever effect is taking place invisibly within the conducting loop is *not* stronger near one terminal than near the other, that is, that no directional effect is detectable.

Evidence for a continuous, dynamic effect, equally strong all the way around the loop, motivates the forming of a mental model for the invisible phenomenon: a continuous flow of some sort around the system. (It must be emphasized that the evidence noted and adduced does not "prove" such a flow "exists" nor does it tell us what is flowing. What we are doing is forming a plausible picture based on the qualitative observations, and this picture must be subjected to test and verification under subsequent application in other circumstances.) To the continuous flow we now (provisionally) visualize, we give the name "electric current" or simply "current."

If the concepts of "electrical charge" and "energy" are not yet available from prior study it is, of course, difficult to discriminate what is and what is not being "used up" in an electric circuit. One has no choice but to resort to making some appropriate assertions, perhaps to be substantiated later in the course. If the concepts are available, one can appeal to the prior experience: the law of conservation of charge and the fact that there is no evidence of charge buildup at any points in the circuit support a model in which charge is transported within the system without it being "lost" or "emitted" or "transformed." The emergence of light and heat, however, coupled with "depletion" and chemical change in the battery, or with the supply of mechanical work to a generator, suggest the transformation of energy in association with the continuous transport of (conserved) charge.

One can now exploit the concepts of "circuit," "conductor," "nonconductor," and "current" by exploring the constitution of the bulb itself as well as that of lamp sockets and switches. Using the lighting or not lighting of the bulb as an indicator,

7.4 BATTERIES AND BULBS (I): FORMATION OF BASIC CIRCUIT CONCEPTS

students can be led to explore the construction of each device by identifying the role of conducting and insulating elements.

Most novices have little or no insight into the structure of the bulb itself. In such instances, it is helpful to break the glass envelope and have them examine the structure carefully. Initially, few students are explicitly aware of the two-endedness of the configuration. They do not distinguish between the metallic tip at the bottom of the bulb and the metallic screw base, nor do they recognize the presence of insulating material and the point and function of the latter. All of these aspects should be traced out by using the lighting or not lighting of a test bulb in order to establish what is connected electrically to what (and what is *not* so connected.) Finally the bulb with the broken envelope can be connected to the battery. The quick flash with which the filament burns out causes astonishment and forms, in itself, an instructive episode.

Similar investigations need to be carried out with a switch (even if it is only a simple knife switch) and with a socket. To novices such as those who have never yet formed the necessary concepts and do not light the bulb in the first sequence, these investigations and questions are far from trivial. Such students must be led to describe the point and purpose of the structures in their own words (after having identified the locations of the conducting and nonconducting parts) and they must be led to address the questions of what would, or would not, happen if the objects were made entirely of conducting material or entirely of nonconducting material.

Although the latter questions sound absurdly simple, the fact is that they are not. Most students find the entire sequence surprising and illuminating. It is trivial only to those who have had the prior experience. The latter, of course, should not be subjected to all this detail, but the most common mistake in instruction in the elements of electricity is to assume that these elements are common knowledge and to lose the large number of students for whom this is not the case. A large fraction of students in introductory engineering-physics courses (as many as 30 or more percent in some instances) are unfamiliar with these basic ideas, and the fraction rises to over 90% in courses for nonscience majors. Experience from second grade level on with "Batteries and Bulbs" indicates that the necessary understanding can be readily developed in elementary school through the concrete experience that has been outlined. The understanding is *not* developed, however, because only an insignificantly small number of elementary teachers have the necessary understanding themselves. Not until we equip our elementary teachers with the security that goes with such understanding will we remove the need for such basic concept building at high school and college levels.

A next logical step in concept development and in refining the model can be the exploration of factors influencing the brightness of the bulb, that is, the *intensity* of the effect. Connecting a second bulb end-to-end with the first or inserting increasing length of nichrome wire in the circuit both lead to decrease in brightness under the influence of a given battery. This suggests that the "intensity" or "strength" of the electric current might be decreased by inserting more material end-to-end (in "series") in the loop. Different objects have different degrees of effect. This suggests that different materials and amounts of material are, in some sense, "obstacles" or offer different degrees of "obstruction," to the flow. From this observation, we form the concept of "electrical resistance."

Note that the concept emerges naturally and plausibly from the operational sequence described; it is *invented* to fit the observations and the model being induced. In many presentations, the concept is introduced through the assertion that current is decreased *"because* of the resistance of the material introduced" as though the concept of "resistance" were a necessary a priori and as though the name explained an actual causal mechanism. The student is thus given the impression that "resistance" was something the scientist knew about ahead of time through some process of prior revelation and fails to see it as a concept invented to fit observations he himself can make with everyday objects. I call this mode of presentation "backwards science," and I urge that it impedes learning and understanding and fails to give the student a clear idea of where scientific concepts come from.

In connection with the concept of resistance (or its inverse, conductance), it is interesting and informative to note how Cavendish compared the conductivities of different metals. Charging a frictional generator with a fixed number of turns and using wires of the same diameter but of different metals, he took the shock of discharging the generator through the wire and through his own body. By comparing the lengths of wire that gave the same intensity of shock, he arrived at an essentially correct qualitative ordering of the conductivities. (It is instructive to note that Cavendish performed this investigation some 30 or so years *before* the publication of Ohm's Law. Neither current nor potential difference were quantified, but the concepts were being separated nevertheless. And even Ohm, in his original paper, still measures resistance in terms of length of wire (see Section 7.6). Students can readily see the connection between this story and the invention of the resistance concept in the "Batteries and Bulbs" sequence.

7.5 BATTERIES AND BULBS (II): PHENOMENOLOGY OF SIMPLE CIRCUITS

Research on student learning and concept formation [e.g., Arons (1982)] shows that very few students, even among those in engineering-physics courses, develop sound understanding of what happens in simple d.c. circuits through the conventional text presentations based on formulation and application of Ohm's Law and Kirchhoff's Laws. They may be able to solve conventional end-of-chapter problems, but many are unable to predict qualitatively what will happen to the current at various points in the circuit, or to the potential difference between two specified points, when some change is imposed such as adding or removing a resistive element or shorting two points. This failure indicates serious deficiency in conceptual understanding, but students can be helped to rectify this deficiency if they are given the chance to practice the qualitative reasoning.

Simple configurations of batteries and bulbs lend themselves very nicely to the generation of such practice, and Appendix 7A gives some examples of effective homework and test questions. Many variations on the examples are possible. In connection with questions such as those illustrated in Appendix 7A, it is highly desirable for students to perform the actual experiments, and they should be strongly encouraged to do so as part of their homework. The necessary equipment (flashlight bulbs and batteries, sockets, and hookup wire) is readily available in hardware stores and can be readily acquired by the students themselves. In classes where I have had the necessary support, I have maintained a set of kits of such apparatus and have lent the kits to students to take home for one or two nights of homework.

It is desirable to use only one type of bulb in the initial experiments so as to have identical resistive elements in the simple configurations being explored. Different types of bulbs vary greatly in their intrinsic resistance, and, if used in the same circuit, add undesirable complexity to the observed phenomena in the initial stages. (Fortunately, the fact that the identical bulbs have different resistances when burning at different brightnesses does not seriously affect the qualitative observations and interpretations.)

In connection with such homework, it is desirable to urge students to work in pairs if they can arrange to do so. The opportunity to talk, argue, and explain in the course of observations and experiments contributes greatly to the learning that takes place. An important caution in connection with such homework: Open-ended assignments, in which the students are given the kits and told to "perform some investigations and experiments," prove to be virtually useless. The students "mess around" with the equipment, try various arrangements, and so forth, but very few perform any genuine observations or experiments. They fail to notice systematic changes; they do not impose systematic alterations on a configuration and predict or interpret the resulting effects; they fail to invent interesting and fruitful configurations of their own—at least in the initial stages. The majority of students have to be *guided* into a series of investigations by being supplied with some initial suggestions and leading questions (not cookbook instructions that destroy all the inquiry.) Many then proceed to develop fruitful sequences of their own, but some never achieve this level of inquiry. [One example of the kind of guidance and questioning that many students need is to be found in Chapter 9 of *The Various Language*, Arons (1977)].

There are several very basic aspects of simple resistive circuits to which students should be explicitly exposed in the initial stages of such investigations in order to counter various naive ideas and preconceptions:

1. It is quite plausible to virtually all students that the total effective resistance of a circuit increases as materials (or bulbs) are inserted end-to-end (or in series). It is far from plausible, however, to many students that the effective resistance of a combination of resistive elements *decreases* as more elements are added in parallel. One is "adding more resistance," and the very robustly held naive notion is that the "resistance must increase." Having studied Ohm's Law and developed the formulas for series and parallel combinations does *not* remove this preconception except in a few of the quicker students. It is helpful to lead students into performing batteries-and-bulbs experiments that concretely show the lowering of effective resistance in parallel combinations: a combination of several bulbs in parallel runs a battery down more quickly than does one bulb alone; starting with two bulbs in series, the brightness of the one bulb is increased as additional bulbs are added in parallel with the other bulb; if a short thread of steel wool is inserted in series with a bulb and additional bulbs are then added in parallel with the first one, the steel wool burns out at some point in the sequence; and so on. (The last experiment is one the students usually enjoy, and it also, of course, leads into a discussion of the point and purpose of fuses and circuit breakers.)

2. When two bulbs are connected in series and a wire is then connected across one of the bulbs (short circuit), many students (even among those in an engineering-physics course) are astonished by the fact that the shorted bulb goes out and the other burns more brightly. They have all heard the term "short circuit" but very few have any operational or conceptual awareness of its meaning, nor do they visualize the accompanying effects. All they know is that a short circuit is something "bad." They should be led into forming the concept through "idea first and name afterwards."

172   CURRENT ELECTRICITY

3. Readers will have noticed that the initial experiments being suggested have, at least by implication, been confined to the use of a single battery and to the exploration of the effect of distribution of resistive elements on the current in a circuit, that is, to only the first steps in the elaboration of the current model. A next step, of course, involves the perception that still another factor is involved in determining current anywhere in a circuit, namely a property of the battery itself. A qualitative concept of "strength" of a battery can be introduced through experiments with combinations of batteries in series (both "adding up" and "subtracting" their effects.) This is, of course, a stepping stone to the formation of the concept of "potential difference," but it is not desirable to plunge into the latter concept precipitously. The first step is to show that current is determined by an internal property of the battery as well as by the distribution of resistance in the external circuit. Formation of the concept of "potential difference" is a subtle enterprise in its own right.

7.6 THE HISTORICAL DEVELOPMENT OF OHM'S LAW

Although I do not advocate the historical sequence above other approaches, I find it illuminating to be aware of how Ohm's Law developed and evolved because the steps of evolution reveal how the various subtle insights were originally achieved. Such awareness helps one appreciate difficulties experienced by the students when modern shortcuts are imposed.

Ohm's work, published in 1826 [see translated excerpt in Magie (1935)], did *not* lead directly to the relation we now call Ohm's Law. His investigation was, in a sense, a quantitative version of Cavendish's qualitative comparison of the shock sensed by one's body in taking the discharge from an electrostatic generator through different lengths of metallic wire (see Section 7.4).

Ohm initially attempted to make precise measurements of current supplied, by voltaic batteries with different numbers of plates, to metallic wires of different length and different materials. He made use of a very carefully designed tangent galvanometer (the device had been invented in 1820 as an immediate consequence of the electromagnetic discoveries of Oersted and Ampere), but he found it impossible to obtain reproducible measurements because of erratic fluctuations in the voltaic piles. (It was a long time before the effects of temperature and pressure, cleanliness of surfaces, uniformity of electrolytes, polarization of the electrodes, etc., were brought under control to the extent of producing reasonably stable voltaic batteries.) In consequence, at the suggestion of Poggendorf, he turned to the thermoelectric effect, which had been discovered by Seebeck in 1822.

Ohm constructed a thermocouple consisting of a bismuth–copper junction with the ends maintained at different temperatures. The ends of the thermocouple were connected to cups of mercury, and the wire being investigated had one end immersed in each of the mercury cups. The galvanometer had a torsion suspension made of a thin ribbon of gold leaf (which was shown to be highly elastic in its return to zero), and the steel compass needle suspended at the end of the gold ribbon was carefully held at a fixed distance above one of the rigid conductors running from one end of the thermocouple to one of the mercury cups.

Ohm recorded the torque on the tangent galvanometer as a function of the length of wire inserted (the diameter of the wires was fixed) and found the empirical relation

$$\tau = \frac{a}{b + x} \qquad (7.6.1)$$

where τ denotes the observed torque, a is a constant for a given temperature difference between the ends of the thermocouple, b is a constant property of the "base" (consisting of the thermocouple and all the connections, including the mercury cups), and x is the length of wire being inserted.

Ohm interpreted a as a measure of the "exciting force" driving the electric current and eventually showed it to be directly proportional to the temperature difference between the two ends of the thermocouple (note that no connection is made to the electrostatic concept of "potential difference"). He also, as is to be expected, interprets the quantity $b + x$ as the resistance of the entire circuit with b as the contribution due to the frame (in units of length of wire.) He thus shows resistance of wire of a given material to be a linear function of length if diameter is held constant. He also showed that resistance increases as the temperature of the wire increases.

Ohm went on to show that "an inch of brass wire is equivalent to 20.5 inches of copper wire" and, on using a very long brass wire resulting in a barely detectable current, remarks "we see . . . that the equation fits with experiment very accurately nearly up to the extinction of the force by the resistance of the conductors."

It should be noted, however, that Eq. 7.6.1 is not yet "Ohm's Law" as we usually state the latter in our texts. As indicated earlier, it is much more nearly a quantified version of Cavendish's investigation.

The next step toward the modern version of Ohm's Law was taken by Joule in 1841 through the experiment shown schematically in Fig. 7.6.1. Joule wound

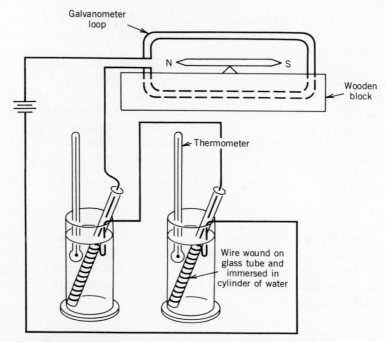

Figure 7.6.1 Schematic diagram of Joule's apparatus for the determination of the evolution of heat in current-carrying wires.

different lengths of wire (the ratio of resistances is thus known) on glass tubes, immersed the windings in cylinders of water, and connected them, in series with each other, to a voltaic battery. He showed that the rate of evolution of heat in each cylinder was in the same ratio as the resistance of the wires, and he observed that the rate of evolution of heat in each cylinder was proportional to the square of the current as measured by the tangent galvanometer, that is, he showed that the rate of evolution of heat in a current-carrying wire is proportional to the product I^2R. It is from these experiments and conclusions that the terms "Joule's Law" and "Joule heat" derive. (It is interesting to note that Joule fully anticipated the relation to current: "I thought the effect produced by the increase of the intensity of the electric current would be as the square of that element . . . arising from increase of the *quantity* of electricity passed in a given time and also from the increase of the *velocity* of the same.")

During the 20 years that followed Ohm's experiments of 1826–27, his results were repeated and confirmed with increasing precision by a number of investigators, but the connection between the law of conduction and other electrical concepts remained obscure until the essential unity of all the various phenomena was grasped through growing appreciation of the Law of Conservation of Energy.

Roget as early as 1832 and Faraday in 1840 had both pointed to the chemical changes taking place within voltaic batteries as evidence that something was being "used up" within the system while electrical charge was being displaced. In Roget's words, "All the powers and sources of motion, with the operation of which we are acquainted, when producing their peculiar effects, are expended in the same proportion as those effects produced; and hence arises the impossibility of obtaining by their agency a perpetual effect or, in other words, a perpetual motion." And from Faraday came the remark that one never found in nature "a pure creation of force; a production of power without a corresponding exhaustion of something to supply it." (It is interesting to note that both of these statements preceded the papers of Mayer and Joule in the 1840s. The idea of conservation of energy was "in the air" as a qualitative "postulate of impotence," albeit not fully formulated or substantiated quantitatively.)

In 1849, just as the accumulated impact of the conservation-of-energy concepts promulgated by Mayer, Joule, and Helmholtz was being felt by the scientific community, the German physicist Kirchhoff pointed out that existing knowledge of electricity fell together beautifully and consistently if the quantity a in Ohm's empirical relation (Eq. 7.6.1) were identified with the magnitude of the potential difference imposed between the ends of a conductor of resistance $R = b + x$. This finally put the relation in the form familiar to us, namely:

$$I = \frac{\Delta V}{R} \qquad (7.6.2)$$

The identification of a with ΔV accounted for the previously established fact that the difference of electric "tensions" between the terminals of a voltaic battery (as measured by means of an electrometer) was proportional to Ohm's quantity a. Furthermore, since the rate P at which work is being done in the displacement of charge must then be given by

$$P = I\Delta V \qquad (7.6.3)$$

it follows, by combination with Eq. 7.6.2, that

$$P = I^2 R \tag{7.6.4}$$

for a resistive element.

The subtle logic of these insights should be carefully noted: There is no a priori reason why all the work done in displacing electrical charge should be converted into thermal internal energy of the system. In fact, there are many instances (e.g., the electric motor) where this is not what happens. In the resistive circuit, however, all the electrical work supplied *is* converted into thermal internal energy. This is what is confirmed by Joule's Law; it cannot be "derived" a priori, and this is why Joule's Law should be recognized as having an essentially independent status and a confirmatory role in establishing and validating the basic circuit concepts.

Because of the way in which most textbooks introduce resistive circuits and concentrate problems and exercises on manipulation of Ohm's Law and its consequences, very few students absorb awareness of the fact that current in a system is *not* always determined by Eq. 7.6.2. It might be determined independently by the power requirement of Eq. 7.6.3, and, in that case, Eq. 7.6.4 is not applicable. (This conceptual aspect is discussed more fully in Section 7.11, where it is suggested that examining the question of why power transmission at high voltages is significantly more efficient than transmission at lower voltages helps students acquire a better understanding of when these relations are applicable and when they are inapplicable.)

7.7 TEACHING ELECTRICAL RESISTANCE AND OHM'S LAW

The teaching of Ohm's Law varies widely among present-day textbooks. Many show at least some care in erecting a motivated and logical conceptual structure intelligible to the level of student addressed, but many assert the concepts of "current" and "resistance" as though they were simple primitives that require no development or explanation. Very few justify the application of the concept of "potential difference,"[2] usually developed in connection with electrostatic effects, to circuits; they simply assert the applicability as though it were obvious to anyone. (Witness how long it took nineteenth-century physicists to perceive the interconnection.)

Among the more careful modes, it seems to me that no single one emerges at the present time as dominant or pedagogically superior, and teachers must, in any case, accommodate themselves as best they can to the approach in the textbook they have selected. In most instances, the concepts and relations are asserted more or less a priori and are eventually justified by the fact that they "work," that is, the results agree with observations. Unfortunately, this ex post facto kind of approach is unsatisfactory to many good students. The logic is unclear to them since it is not explicitly outlined, and, if given the chance, they much prefer to see how something came to be known than to have it asserted ex cathedra and

---

[2]It should be noted at this point that the very widely used term "voltage" generates more confusion than enlightenment. The term is used differently by different authors in different contexts, and no clear, systematic operational meaning has crystallized. The result of this variability is to blur, for many students, the distinction between emf and potential difference. The usage also tends to give naive learners the false impression that "voltage" is a property of a circuit at *one* single location. It is necessary to have the students use the term "potential difference" over and over again to fix the insight that *two* points are always involved. It is for this reason also that I systematically use the symbol ΔV rather than the symbol V throughout this discussion.

then shown to be valid. In the former circumstances, they have the feeling that, in principle, given time and opportunity, they may have formed the ideas on their own; in the latter, ex cathedra, case, they feel they could never have possibly come up with the ideas. And, in the latter feeling, they are on sound ground because the ideas were never, in fact, arrived at in that way.

Iona (1979) gives an excellent survey and commentary on the existing situation in the textbooks. It should be read for its own sake and will not be incorporated here except for the pointing up of a few elements. Iona found that, in the textbooks he examined, the most prevalent approach was to start with the ratio of potential difference to current as the defining expression for resistance, that is

$$R \equiv \frac{\Delta V}{I} \qquad (7.7.1)$$

This is a legitimate approach, especially if the students have been helped to acquire a preliminary, qualitative, nonrigorous notion of "resistance" such as that advocated in Sections 7.4 and 7.5. Equation 7.7.1 then becomes a redefinition and quantification of the concept.

When the consequences of the definition in Eq. 7.7.1 are explored for different conducting materials and systems, it is found that, in simple metallic conduction, R is constant in the sense of being independent of ΔV and of I, although it is, in general, a function of the state (temperature and pressure) of the material. Systems for which R is constant, and for which the relation between ΔV and I is therefore linear, are said to obey Ohm's Law. Systems for which the relation is not linear are said to be "non-Ohmic" or not obeying Ohm's Law, but the definition of R, now as a variable property, still holds.

An alternative approach appearing in some texts is to start with Joule's Law and define resistance as the ratio of power supplied to the square of the current. This, of course, involves prior understanding of the energy concepts and is even more abstract than the approach via Ohm's Law. Connecting it to Ohm's Law also involves the a priori assumption that all of the work done by the source is converted into thermal internal energy, and this tends to prevent the student from clearly separating the elements of the logical structure.

One very fundamental and simple implication of Ohm's Law (as it applies to metallic conduction) is rarely made explicit to the students, yet it illustrates what deep inferences can sometimes be drawn if one recognizes what is *not* the case as well as what is. The fact that the straight line extends all the way down to zero and shows no evidence of a measurable voltage threshold for the onset of current implies that the charge carriers, whatever they may be, are free and unbound in the structure of the metal. This is a kind of physical insight students should be helped to acquire in addition to practice at application of the law itself.

7.8 IS ELECTRIC CURRENT IN METALS A BULK OR SURFACE PHENOMENON?

Very few textbooks pay attention to the question raised in the title to this section, yet it is a question that involves important physics and phenomenology, and consideration of this question can significantly enhance student understanding. We do know that transport of electrical charge in metallic circuits takes place, under ordinary circumstances, through the body of the metal rather than along the surface, but *how* do we come to know this? (That this is not a trivial question is indicated by the fact that, at high frequencies, the so-called skin effect comes into play, and under these circumstances conduction *is* confined to the surface layers.)

Most textbooks simply assume that conduction is a body effect and do not even inform the student that an assumption is being made. The more advanced texts take current density J to be uniform over the cross section of a conductor and assert the relation

$$\mathbf{E} = \rho \mathbf{J} \qquad (7.8.1)$$

where $\mathbf{E}$ denotes the electrical field strength within the body of the conductor and ρ the resistivity of the material. They then go on to show, by invoking Ohm's Law, that the total resistance R will be given by

$$R = \rho \frac{l}{A} \qquad (7.8.2)$$

where l and A denote the length and cross-sectional area of the conductor, respectively.

But what justification is there for assuming bulk flow in the first place? Students should be helped to raise this question and to consider how it might be resolved. The strongest primitive testimony to bulk rather than surface conduction resides, of course, in the *empirical* fact that the total resistance R is found to be directly proportional to the length of a wire and inversely proportional to the cross-sectional *area*, that is, to the *square* of the cross-sectional dimension:

$$R \propto \frac{l}{A} \qquad (7.8.3)$$

The inverse proportionality to square of cross-sectional dimension is what speaks to bulk conduction. Had the resistance been inversely proportional to the *diameter* of the wire, the indication would have been surface conduction. Here the student must confront thinking about what is *not* as well as what *is* the case, and such thinking is deeply conducive to understanding (and to cognitive development as well.) Furthermore, this episode exhibits the power and importance of *ratio* reasoning in circumstances where numbers are not involved but a significant *conceptual* insight is attained.

7.9 BUILDING THE CURRENT–CIRCUIT MODEL

Our picture of electric current, even in "simple" resistive circuits, is very abstract, subtle, and sophisticated. It is cryptically asserted in textbooks with very little discussion of observation of effects—the more elementary the text the more cryptic the assertion—and then teachers wonder why students have wildly erroneous ideas concerning what is happening.

An initial, nonrigorous approach, requiring subsequent elaboration and redefinition, is outlined in Sections 7.4 and 7.5. The closing of a still open question is outlined in Section 7.8. There remain additional questions: Is one variety of charge being displaced or both? Is charge supplied only by the source or is the conductor itself a container of displaceable charge? What are the respective roles of potential difference and resistance in establishing currents in various configurations? Can we visualize what is happening in an electric circuit by appealing to analogy with some more intuitively comprehensible mechanical system (e.g., the fluid model)?

Some of these questions are widely ignored or are treated by assertion without addressing the underlying "How do we know . . . ?" questions. Others are carelessly handled. As shown, for example, by Cohen, Eylon, and Ganiel (1983), many students emerge with the conception that potential difference is a *consequence* of displacement of charge rather than its cause.

Several pilot attempts to evolve more carefully logical and intelligible presentations have been published [e.g., Evans (1978); Steinberg (1983)], but further polishing and work with students are needed. Steinberg, in particular, exploits a batteries-and-bulbs situation combined with newly available inexpensive capacitors of very high capacitance. With these capacitors in the circuits, one can observe transient lighting of the bulbs as the capacitors are charged and discharged. This additional ingredient enormously extends the number of questions that can be asked and answered about location and displacement of charge and greatly enhances the soundness and completeness with which the current model can be constructed. Steinberg outlines a sequence of investigations in which fluid models (including that of a compressible fluid) are tested, found wanting in certain respects, and lead to a final picture of conduction that is fully correct in its own right.

Steinberg's approach makes it possible to attack a common student misconception that circuit elements affect the situation "downstream" while not affecting anything "upstream." Furthermore, the steady-state situation in continuous circuits conceals temporal effects that are important in understanding construction of the electric current model, and continuous circuits do not reveal, as clearly, charge displacements associated with various potential differences.

Teachers would do well to study some of these developments with a view to incorporating into their own teaching those aspects they find congenial and helpful to the students.

7.10 CONVENTIONAL CURRENT VERSUS ELECTRON CURRENT

Especially at precollege, but also at college level, many textbooks are asserting that electric current consists of a flow of electrons. Some texts do not even bother making it clear that they are talking only of metallic conduction. The electron concept is asserted without evidence or basis for acceptance, and students are left with no frame of reference in which to set the term: they are left with no sense of how the properties of this entity (the electron) compare with those of any other on the microscopic scale, how we come to know anything about it in the first place, or what role it plays in the structure of matter in general.[3] All they have is a name (i.e., jargon), and this undermines their capacity to distinguish between jargon and knowledge.

The excuse frequently given for this approach is that the electron picture is "correct" and that students should be given the correct answers as scientists know

---

[3] A few texts (*PSSC Physics* for example) try to provide a sound basis for accepting the electron concept before asserting the electron picture of conduction, but these are rare, and unsubstantiated assertion is the norm.

them. The result is that students are misled in many ways, and they are also robbed of a valuable opportunity for phenomenological thinking and reasoning that enhances understanding and insight not only into this particular concept but into how science works in general.

When electron current is asserted in the manner described above, most students end up with the completely false notion that *all* conduction consists of displacement of electrons and hence of negative charge. They end up with this misapprehension even if it has been qualified to apply only to metals. (This is due to the fact that they are not given the opportunity to encounter any cases of conduction other than in metals, and the restriction is lost sight of and forgotten even if it was ever noticed.) In fact, of course, metallic conduction is a fairly special case, accompanied only by the relatively esoteric instances of conduction under photo or thermal emission. In most electrolysis and in ionized gases, conduction takes place by migration of both positive and negative ions simultaneously. In some cases of electrolysis (e.g., electroplating of copper or silver), conduction takes place by migration of positive ions only. In semiconductors, it is in some circumstances more convenient to visualize the migration of positive "holes" than of electrons. Thus the premature assertion of electron current tends to plant a narrow misconception in student minds rather than liberate them with a "right" answer.

I suggest that it is pedagogically wiser and more fruitful to teach students the positive current convention, and I join Iona (1983) in his forceful argument to this effect. In addition to the fact that only normal metallic conduction and photo and thermal emission are purely electronic, there are a number of other reasons that favor adherence to the positive current convention.

Students, in their initial exposure to electric circuits, tend to be extremely reluctant to accept the fact that it is impossible to determine, through ordinary macroscopic observations (either electrical or electromagnetic), whether it is negative charge, positive charge, or both varieties that are being displaced in metallic conduction. Observed effects can be equally well accounted for and predicted on the basis of any one of the models. It is only the Hall Effect [see, for example, Magie (1935); Resnick and Halliday (1985)] that indicates transport of negative charge, and it is unable to say anything about the carrier. Just as in electrostatics students should confront the indeterminacy as to whether positive charge, negative charge, or both are being displaced, they should first see the same indeterminacy in metallic conduction and recognize that the positive current convention is, in this sense, as good as any other. It is only such experience that registers the insight that positive current is a *convention* and not a fact. This helps them toward far deeper appreciation of the electron picture when it is finally substantiated and makes them far more sensitive to the "How do we know . . . ? Why do we believe . . . ?" questions. They see the course of question asking and development and not just the sterile end result.

The principal reason for retaining the positive current convention is that it underlies: (1) the definitions of electric field strength and potential difference; (2) the treatment of capacitive and inductive circuit elements; (3) all the standard mnemonics of electromagnetism and the Maxwell equations; and (4) the standard notation in diagrams of electronic circuits. This convention has not been, and will not be, changed because of the acceptance of electron current. Students who come to electromagnetism without having had experience of the positive current convention suffer severe and unnecessary confusion without having had any real pedagogical gain.

7.11 NOT EVERY LOAD OBEYS OHM'S LAW

Given the usual presentations of current electricity, very few students, even in engineering-physics courses, emerge with the realization that Ohm's Law does not apply in all circumstances in which a source of potential difference drives an electric current. Just telling the students this, of course, leads nowhere. (I am speaking here not of nonlinearity in resistance but of loads that are not purely resistive, i.e., running a motor, charging a battery.) An effective way of leading students to confront this issue is to raise the question of why it is that power transmission is far more efficient at high voltages than at low. This embeds the idea in a rich and memorable context and and gives it a connection to engineering and societal questions.

Even though power transmission is alternating current (a.c.) rather than direct current (d.c.), there is no serious error in treating this as a d.c. problem at an elementary level. Suppose we wish to supply a fixed amount of power P_L at a potential difference ΔV_L to a given load. The resistance of the power lines from the distant source is denoted by R_T. It should be specified that the nature of the load is not known and that the voltage level is to be chosen on the basis of investigation of the favorable conditions.

The crux of the matter is that the resistance of the load is irrelevant even if it is known. The current drawn through the supply lines is determined by the fixed power requirement:

$$I = \frac{P_L}{\Delta V_L} \quad (7.11.1)$$

regardless of whether or not the load is resistive. This is what students should be led to see, and it is usually a significant obstacle in the minds of many. What they want to do, by habit, is apply Ohm's Law.

Since the current is fixed by the power requirement, the Joule heat loss in the power lines is now fixed: .

$$\text{Resistive energy loss} = I^2 R_T \quad (7.11.2)$$

The efficiency η is given by

$$\eta = \frac{P_L}{P_L + I^2 R_T} \quad (7.11.3)$$

Combining Eqs. 7.11.3 and 7.11.1 gives

$$\eta = \frac{1}{1 + \frac{P_L R_T}{\Delta V_L^2}} \quad (7.11.4)$$

Examination of Eq. 7.11.4 shows that the efficiency rises dramatically as the potential difference is increased. Students should be led to explore and interpret this equation both graphically and numerically. In Rogers (1960), one of the few texts dealing with this illustration at an elementary level, students are led through a good numerical comparison of two cases without the algebraic analysis.

Whether this problem is examined algebraically or arithmetically, it is important to lead students through the reasoning Socratically, having them fill in and explain the steps and interpret the results. Presenting them with the derived result and calling for some algebraic analysis or numerical substitution does not register an understanding of the crucial aspects of the reasoning.

7.12 FREE ELECTRONS IN METALS: THE TOLMAN–STEWART EXPERIMENT

In many textbook discussions of metallic conduction at secondary school and introductory college levels, the existence of electrons and their role as free entities in metals are simply asserted on the "scientists know that . . ." basis. Given a large volume of such asserted end results, many students, especially the slower ones who are still developing abstract reasoning capability, are unable to discriminate what, of "knowledge" they "possess," is based on evidence and understanding and what consists of memorized, unsupported assertions. This is *not* a healthy intellectual condition, and, for those students who will not become scientists and to whom such study is a matter of general education, this condition is destructive of any understanding of the nature, power, and limitations of science. With such background, they do *not* become citizens who comprehend the nature and role of science in intellectual history and in our society. I have long voiced my opposition to such presentations of physics, and I continue to do so.

If one is to discuss the role of electrons in metallic conduction (and this can perfectly well be done, if desired, while still teaching the positive current convention), this can and should be done by first laying some basis for understanding why we believe in electrons, when and how they are detected, what their properties are compared to those of other entities on macro- and microscopic scales. Some texts that deal with electrons do lay such a basis in a plausible and reasonable, even if not fully rigorous, way, and I consider this fair game. It is the bald, unsubstantiated assertion that is destructive.

Given some plausible basis for understanding the evidence underlying acceptance of the electron concept, the next "How do we know . . . ?" question has to do with the justification for believing electrons to be the free carriers of charge in metals. Freedom of the charge carriers is inferred from the observed fact of zero threshold potential difference for the initiation of current in the metallic conductor, that is, from the fact that the straight line in Ohm's Law extends all the way to the origin (see Section 7.7). As often happens, the historical sequence is very illuminating in this respect, and, at the same time, allows the kind of spiraling back in later, richer context that has been advocated throughout this book.

After Thomson's initial (1897) work on the corpuscular nature of the cathode beam (see Sections 10.4 and 10.5), he proceeded to show that particles with the same charge-to-mass ratio appeared outside the metal in both photoelectric and thermal emission. (The cathode beams he studied were ones that resulted from field emission.) These observations led naturally to the widely held supposition that electrons were present as such within metals and were probably the free carriers of charge. Supposition, however reasonable, is nevertheless not evidence. There remained the problem of showing that unbound electrons were indeed present, as such, within metals. This problem was attacked in a very direct and simple way by Tolman and Stewart (1916), (1917).

The procedure adopted by Tolman and Stewart was that of imparting a large acceleration to a metal sample and observing the effect of any potential difference set up between the ends by virtue of the presence within the metal of free charge carriers possessing inertial mass. Given the presence of such carriers, the effect would be analogous to that in a closed, accelerated tube of fluid. In the fluid, a pressure gradient would be set up, the hydrostatic pressure becoming higher at one end than at the other, until all the fluid within the tube had the same acceleration. In the metal, the charge carriers would shift within the lattice, becoming more concentrated at one end than at the other until the electrical field thus set up within the body of the metal would be sufficient to accelerate all the free charge carriers along with the accelerated lattice.

In the actual experiment, Tolman and Stewart used a rapidly rotating spool (diameter about 24 cm) carrying about 450 m of wire, connected to a very sensitive ballistic galvanometer. The rotation of the spool was stopped abruptly, and deflection of the galvanometer revealed the passage of a pulse of current. From this measurement, it was possible to infer the charge-to-mass ratio of the free carriers within the metal, and, within fairly large experimental uncertainty, this ratio was equal to that observed by Thomson and others for electrons outside the metal. At an introductory level, however, complete analysis of this measurement, including interpretation of the behavior of the ballistic galvanometer, would tend to cloud the main issue and is unnecessary. (For more advanced students, in engineering-physics courses for example, the full analysis can be very instructive because of the richness of physical ideas brought together in a significant context.)

Assuming that one can measure, directly, the imparted acceleration and the potential difference set up between the ends of the accelerated conductor, the analysis becomes exceedingly simple. Refer to Fig. 7.12.1.

Applying Newton's Second Law to the individual charge carriers, we have

$$Ee = ma \qquad (7.12.1)$$

and

$$\frac{e}{m} = \frac{aL}{\Delta V} \qquad (7.12.2)$$

Thus the unknown charge-to-mass ratio is expressed in terms of measurable quantities. In the actual experiment, a potential difference does indeed develop

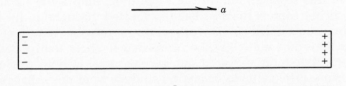

Figure 7.12.1 Schematic diagram of Tolman–Stewart experiment: Metallic conductor of length L is accelerated to the right at known acceleration a. If free charge carriers possessing inertial mass m and charge e are present within the body of the conductor, they will tend to shift, becoming more concentrated at the left than at the right end of the conductor. The result will be a potential difference ΔV between the ends and a field strength $E = \Delta V/L$ within the body of the conductor.

between the ends of the conductor, and the calculated charge-to-mass ratio agrees with that obtained for electrons ejected from the metal. Tolman and Stewart first performed the experiment with copper wire and subsequently showed that the same results were obtained with silver and aluminum. The simplified analysis outlined above is highly instructive for all students because of the opportunity it affords for spiraling back to the use of earlier concepts in setting up the current problem. Furthermore, the context is not that of an artificial end-of-chapter exercise but a real experiment with a significant interpretation.

My colleague Philip Peters informs me that he recently tried this out as a problem for graduate students in an advanced E&M (Electricity and Magnetism) course and found that the students had great difficulty visualizing the situation and dealing with the physics. Such response from the students stems, of course, not from the intrinsic difficulty of the problem, but from the fact that they never had a chance to think about such situations—situations that they should have had opportunity to consider and visualize in undergraduate work.

Students should be led to see the connection between this electrical problem and such mechanical analogs as the pendulum suspended in the accelerating car and the ball on the cart being accelerated by the rear wall of the accelerating cart. A more closely analogous mechanical situation is that of the fluid in an accelerated tube.

In the case of an essentially incompressible fluid of unknown specific volume v in an accelerated tube of length L, the analogous result is

$$v = \frac{aL}{\Delta p} \qquad (7.12.3)$$

where Δp is the measured pressure difference between the ends of the tube and a is the acceleration imparted.

The derivation of this result is an excellent exercise for honors students. Not only does it point up the deep similarity between apparently unrelated physical situations, but it also sets up the basis for understanding, in the light of the pressure gradient developed, how a cork with lower density than the surrounding fluid comes to be displaced in the *same* direction as the acceleration rather than in the opposite direction.

A more sophisticated problem is that of a tube with closed ends containing an ideal gas at initially uniform pressure p_0. When the tube is accelerated, a pressure gradient develops as in the isothermal gas in a gravitational field. One must integrate the differential equation for the pressure, obtaining an exponential as in the case of the isothermal atmosphere, and one must formulate and apply the condition for conservation of mass in the tube. For this situation one could, in principle, find the molecular mass μ of the gas from the measurable quantities since one obtains:

$$\mu = \frac{RT}{aL} \frac{\Delta p}{p_0} \qquad (7.12.4)$$

where ΔP denotes the pressure difference between the ends of the tube. Although this is certainly not a good way of measuring molecular mass, good students find the exercise in mathematical physics very instructive.

APPENDIX 7A

Sample Homework and Test Questions

INTRODUCTION

Following are several sample problems on qualitative, phenomenological aspects of simple d.c. circuits—aspects seriously neglected in most textbooks. Such problems are useful as homework or as open-book test questions. These are simply prototypes; it is obvious that many variations are possible (and desirable).

1. In the circuit shown in Fig. 7A.1, the battery maintains a constant potential difference between its terminals at points 1 and 2 (i.e., the internal resistance of the battery is to be considered negligible).
    Three identical flashlight bulbs A, B, and C are screwed into their sockets and are lighted when the circuit is closed. After each of the changes suggested in the following questions, the system is returned to the initial condition shown in the figure before the next change is made.
    The question "What happens to . . . ?" refers to whether the quantity in question increases, decreases, or remains unchanged. Indicate your reasoning briefly in answering *each* question.
    (a) How do the brightnesses of bulbs A, B, and C compare with each other in the initial condition?
    (b) What happens to the brightness of *each* of the three bulbs when bulb A is unscrewed and removed from its socket? What simultaneously happens to the current at points 3, 4, and 5?
    (c) What happens to the brightness of *each* of the three bulbs when bulb C is unscrewed and removed? What simultaneously happens to the current at points 3, 4, and 5?
    (d) What happens to the brightness of *each* of the three bulbs if a wire is connected from the battery terminal at point 1 to point 4? What simultaneously happens to the

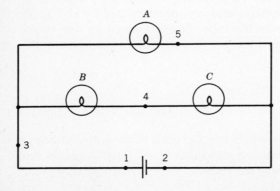

Figure 7A.1 Identical bulbs connected to a battery maintaining constant potential difference across its terminals.

current at point 3? What simultaneously happens to the potential difference across bulb B? What simultaneously happens to the potential difference across bulb C? What happens simultaneously to the potential difference between points 1 and 5?
(e) What happens to the brightness of *each* bulb and to the current at point 2 if a wire is connected from the battery terminal at point 2 to the socket terminal at point 5?
(f) What happens to the brightness of *each* bulb if a fourth bulb (D) is connected in parallel with bulb B alone, i.e., *not* in parallel with both B and C? What happens simultaneously to the current at point 3? What happens simultaneously to the potential difference between points 3 and 4? To the potential difference between points 4 and 2?

NOTE: The following is a variation on question 1. Insert introduction similar to that in question 1 above. Several questions of this variety are necessary before students begin to pick up the lines of reasoning.

2. In the circuit shown in Fig. 7A.2, the battery is sufficiently strong so that all the identical bulbs are visibly lighted. Internal resistance of the battery is negligible, and it maintains a constant potential difference across its terminals.
   (a) Suppose bulb C is removed from its socket. What happens to the brightness of each bulb? How do the final brightnesses compare with each other? What happens to the current at point 1?
   (b) Restore initial conditions in the figure. Suppose a wire is connected between points 2 and 3. What happens to the brightness of each bulb? How do the final brightnesses compare with each other? What happens to the current at point 1? What happens to the potential difference across bulb C? What happens to the potential difference across bulb B?

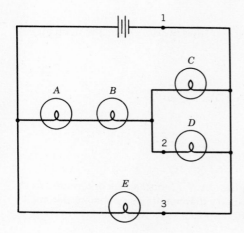

Figure 7A.2 Circuit consisting of identical bulbs connected to a source of constant potential difference.

NOTE: The following problem is a version of the preceding type of question in a multiple choice format. Since it allows rapid correcting, it is useful as a test question after questions of the preceding form have been used. Without the requirement of verbal explanation, however, the right answer is still not necessarily indicative of genuine understanding.

3. Consider the resistive circuit shown in Fig. 7A.3. The battery has negligible internal resistance, and the resistors are all identical.
   Circle the correct choice of word or words in the statements following each suggested change in the circuit.
   (a) Another identical resistor is connected between points E and G.

1. The current at point A

    (increases   decreases   remains unchanged)

2. The current at point F

    (increases   decreases   remains unchanged)

3. The potential difference between points E and F

    (increases   decreases   remains unchanged)

(b) Return to the initial condition shown in the figure. An identical resistor is connected between points C and B.

1. The current at point A

    (increases   decreases   remains unchanged)

2. The current at point D

    (increases   decreases   remains unchanged)

3. The potential difference between points A and C

    (increases   decreases   remains unchanged)

4. The potential difference between points F and G

    (increases   decreases   remains unchanged)

(c) Return to the initial condition shown in the figure. A wire is connected from point E to point C.

1. The current at point F

    (increases   decreases   remains unchanged)

2. The potential difference between points C and D

    (increases   decreases   remains unchanged)

3. The current at point H

    (increases   decreases   remains unchanged)

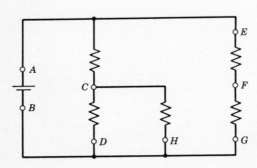

Figure 7A.3 Identical resistors connected to a battery that maintains a constant potential difference across its terminals.

NOTE: As indicated in Section 7.9, many students, including those in engineering-physics courses, hold beliefs about "downstream" effects in electric circuits. The following problem is a way of confronting such beliefs and helping the student rectify them by making a mistake and revising the thinking.

4. Consider the circuit shown in Fig. 7A.1. Suppose the connection is broken at point 4, and a resistor is introduced in series with bulbs B and C. What will happen to the brightness of each of the two bulbs B and C? Explain your reasoning briefly.

NOTE: The following may seem to be a trivial question, but it is not. Most students have very substantial difficulty with it. This forms an excellent preparation for discussion of the Wheatstone bridge. It can readily be extended to the case in which the resistances are not identical.

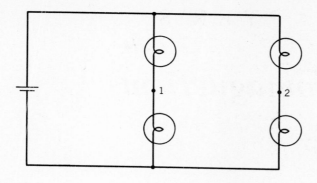

Figure 7A.4 Identical bulbs connected to a battery that maintains a constant potential difference.

5. The circuit shown in Fig. 7A.4 consists of identical bulbs and a battery with negligible internal resistance.

   Suppose a wire is connected between points 1 and 2 in the circuit. What happens in this wire? What happens to the brightness of each bulb? What happens to the current drawn from the battery? What happens to the potential difference across each bulb? Explain your reasoning briefly in each case.

NOTE: The following question affords the opportunity to spiral back to earlier thinking, reasoning, and concept formation. The question may sound trivial to the instructor, but many students, even among those in engineering-physics courses, are astonished to recognize the household meter as a device that measures area under a curve and also to see the connection to apparently unrelated exercises dealt with earlier in the course.

6. The household electric meter is essentially an integrating device—one that measures the area under a curve. It measures the instantaneous power $P(t) = (\Delta V)I(t)$ supplied to the household circuit and calculates the total energy supplied over a period of time.
   Shown in Fig. 7A.5 is a hypothetical graph of $P(t)$ versus clock reading t. The potential difference ΔV is kept constant at 120 V.
   (a) What would have been the appearance of the $I(t)$ versus t graph?
   (b) From Fig. 7A.5, obtain a numerical value for the total energy supplied to the household circuit (i.e., the reading in kilowatt-hours that would have been registered on the meter) over the eight hours shown.
   (c) Compare what you have just done in this problem (and the reasoning involved) with an exercise in which you calculate the displacement of a body in rectilinear motion when you have been given a graph of velocity versus clock reading.

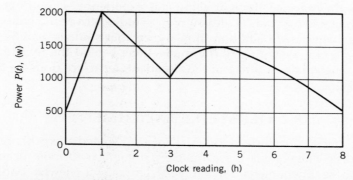

Figure 7A.5 Graph of instantaneous power $P(t)$ versus clock reading.

CHAPTER 8

Electromagnetism

8.1 INTRODUCTION

It was pointed out in Section 6.2 that many students have not developed, in prior experience, sufficient familiarity with electric and magnetic phenomena to distinguish operationally between them. If care is taken to form the distinction at the earliest introductory levels, as recommended in Chapter 6, understanding of electromagnetism is greatly facilitated. If the operational distinctions have not been clearly formed, however, the electromagnetic phenomena become a source of profound confusion. The recourse becomes, as usual, desperate memorization without comprehension. Among students subject to such confusion, one finds expectations such as the repulsion of positive electric charges by north magnetic poles, attraction or repulsion of stationary charges by the magnetic field around a current carrying wire, and a variety of other nonexistent effects.

Our own knowledge of the phenomena comes not from deduction from abstract principles but from observing the actual effects, discerning the systematic order and relations that are maintained, and memorizing the facts observed. The only way to help students acquire this knowledge is to provide ample opportunity for the visual and kinesthetic experiences that facilitate the necessary memorization. Among the chief sources of failure of instruction in this area of physics are the failure to pay attention to the most basic operational underpinnings and in the failure to allow time for sufficient concrete experience to embed the phenomena in the memory.

Major emphasis in this chapter is placed on ways in which to cultivate experience, and connected thought about this experience, so as to generate the grasp of phenomenology that eludes so many students over the entire spectrum of introductory physics. In addition, it is argued that students should be helped to acquire some understanding of the motivations behind the invention of field theory in the nineteenth century and the attendant departure from the Newtonian action at a distance view.

8.2 OERSTED'S EXPERIMENT

Although the physical circumstances making up Oersted's experiment are very simple, the basic phenomena are completely unfamiliar to most beginning students, and these phenomena remain very tenuous if viewed only as rapid demonstrations in a lecture room. Fortunately, the magnetic effect around a current carrying wire can be easily studied in simple experiments conducted at home. One needs only a small compass and a circuit consisting of flashlight batteries, bulbs, and wire. Such "take-home" assignments can be readily coupled with the batteries-and-bulbs homework suggested in Section 7.5, and they greatly enrich study that otherwise consists almost exclusively of pushing numbers into end-of-chapter problems. Opportunities for studying real phenomena in homework with simple apparatus are not very plentiful, and there is substantial educational profit in capitalizing on them when possible.

Very few students will conduct a meaningful investigation without guidance, however. The homework assignment should guide them Socratically into: (1) investigating the compass deflection both above and below the current carrying wire; (2) investigating the effect of reversing the connection to the battery terminals; (3) ascertaining the pattern of the effect all the way around the wire—not just above and below; (4) qualitatively noting the effect of changing the distance between the wire and the compass needle; (5) qualitatively studying the strength of the effect on the compass needle (held at fixed distance from the wire) when additional bulbs are inserted in the circuit either in series or in parallel with an initial single bulb; (6) studying the effect of introducing an additional battery in series with the first; (7) forming, from synthesis of the observations, the right-hand-rule mnemonic for the direction of magnetic field around the current carrying wire.

Out of this simple experience, it is easy to generate understanding of the tangent galvanometer—historically, the first device for comparing the strength of electric currents.

Finally, one can lead the students to study (and thus assimilate a genuine understanding of) the superposition of magnetic effects around current carrying wires. Superposition is rarely discussed explicitly in the textbooks; it is asserted or taken for granted without even a hint that experimental investigation and substantiation are required. Fig. 8.2.1 illustrates how simple this investigation can be made, using the compass at fixed distance from the wire as a crude tangent galvanometer.

Observing the same intensity of effect on the compass needle at positions A and D reinforces the concept of the continuity of electric current all the way around the circuit (i.e., the model initially formed from the batteries-and-bulbs observations outlined in Chapter 7). The exercise on superposition (comparing the effects at positions A, B, and C) is necessary in its own right, but it is rarely outlined to the students. One can speculate about superposition and hope that

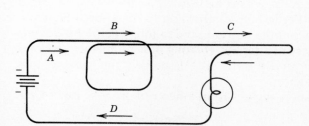

Figure 8.2.1 A battery-and-bulb circuit with a long wire bent and laid out in the form shown. At positions A and D, we have a single wire carrying the same current. At position B, we have the effect of two wires carryng current in the same direction. At position C, the two wires carry current in opposite directions. A tangent galvanometer, exhibiting a given torque at position A and D, exhibits twice the torque at position B and zero torque at position C.

nature is simple and linear in this context, but speculation is not fact. One must appeal to experiment for ultimate sanction. Validation of simple superposition then provides the logical basis for using the right-hand rule to predict the direction of magnetic field around various configurations of the current carrying wire (e.g., coils and solenoids).

The story of Oersted's serendipitous discovery, in the course of a lecture demonstration, of the effect that now bears his name has become such a commonplace in physics instruction that one tends to lose sight of the fact that deliberate, albeit unsuccessful, search for a connection between electricity and magnetism had a long prior history. Franklin had tried to magnetize a needle by electrical discharge. Whittaker reports in the *History of the Theories of Aether and Electricity* that "In 1774 the Electrical Academy of Bavaria proposed the question 'Is there a real and physical analogy between electric and magnetic forces?' as the subject of a prize." In 1805 two French investigators attempted to determine whether a freely suspended voltaic pile orients itself in any fixed direction relative to the earth. Proponents of Naturphilosophie in the early years of the nineteenth century hoped to find interconnection among all of the "forces" of nature. Oersted himself was deliberately investigating such connections before his own discovery. It is conducive to insight and learning if students are invited to speculate on these matters prior to abrupt introduction of electromagnetism. Few students have had the intellectual experience of engaging in informed speculation in which one must carefully discriminate speculation from observed fact.

It should be noted that Oersted's experiment introduces an interaction that is profoundly different from interactions students have encountered up to this time. In all the interactions previously encountered (contact, gravitational, electrical, even frictional), the two forces of interaction are colinear and lie along the line connecting the interacting particles. In the Oersted experiment, the forces on the poles of the compass needle are orthogonal to the radial line from the current carrying wire to the point at which forces are being observed. The interaction is that of torques rather than colinear forces. Very few texts or presentations call this fact to the attention of the students; yet this is a deeply significant extension of previous experience.

Finally, students should be led to articulate the question as to whether moving electrostatic charges would produce the same effects as current in a wire. That this is not a trivial question is indicated by the discussion that attended it during the nineteenth century. Although most physicists anticipated an affirmative answer, and although Faraday firmly asserted in 1838, that "if a ball be electrified positively in the middle of a room and be then moved in any direction, effects will be produced as if a current in the same direction had existed," the effect remained to be confirmed experimentally. This was, of course, not an easy task because of the weakness of the effect—eloquent testimony to the relatively enormous amounts of charge transported in electrical conduction. The direct test was finally performed by H. A. Rowland in 1875 while he was spending a year in Helmholtz's laboratory in Berlin before returning to the United States to assume the professorship of physics at the newly founded Johns Hopkins University. Rowland charged the periphery of a rapidly rotating nonconducting disk and showed that the magnetic field around the rim was essentially the same as that around the current carrying conductor.

An effective homework question evolves from a description of some of Oersted's original experiments:

Among his many experiments and observations, Oersted reports the following: (a) "The kind of metal forming the conductor does not alter the effects, except, perhaps as regards their intensity. We have employed with equal success wires of platinum, gold, silver, copper, iron, bands of lead and tin, and a mass of mercury." (b) The effects on the compass needle remain virtually unaffected when rock, wood, glass are placed between the wire and the needle, and when the needle is encased in a copper box full of water. (c) "Needles of copper, glass, and resin, suspended like the magnetic needle, are not affected by the current carrying wire."

Interpret Oersted's experiments: Why did he perform them? What conclusions are to be drawn from the reported observations?

Students are rarely afforded the opportunity to confront questions dealing with the point and interpretation of qualitative observations such as those described by Oersted. Here is a valuable learning experience to be added to that afforded by the conventional problems.

8.3 FORCES BETWEEN MAGNETS AND CURRENT CARRYING CONDUCTORS

In the usual performance of Oersted's experiment, the use of stiff or massive wires and a small, sensitive compass focusses all attention on the compass and tends to conceal the fact that forces exerted on the compass are accompanied by forces exerted on the conductor. Oersted himself was among the first to point out that: "As a body cannot put another in motion without being moved in its turn, when it possesses the requisite mobility, it is easy to foresee that the current carrying wire must be moved by the magnet." Although this was "easy" for Oersted to foresee, it is the very rare student (only a budding research physicist) who thinks of the possibility without prompting.

Socratic prompting can be very effective in this context, especially if very little is given away. Students derive considerable satisfaction from bridging a gap even if the latter is small, and such satisfaction should not be deprecated. Some students are even able to predict the direction of force on the wire starting from the Oersted effect.

Lecture demonstrations of the force on the current carrying conductor in the field of a magnet are almost universally performed, but, in many instances, their effectiveness is limited or diluted for a number of reasons. Sometimes the link to the Oersted experiment is not made clear. In other instances, the demonstrations are performed far too rapidly, and students do not have time to register the multiplicity of directions involved. The greatest lack, however, tends to be the lack of opportunity for the individual student actually to feel the tug on the wire. Despite the fact that they have seen the wire jump and accelerate when the switch is closed, students show astonishment, bordering on incredulity, when they hold the wire and feel the pull. Making this opportunity available to all students is most desirable if at all possible. The usual mnemonic (the cross-product, or second right-hand, rule) is most effectively registered if students have the opportunity to put it together out of their own observations and sensations close up rather than from the remoteness of the lecture demonstration.

A useful and impressive demonstration that does not seem to be widely performed is that of showing the interaction between a magnet and a current carrying electrolyte, that is, an interaction with electric current not confined to a

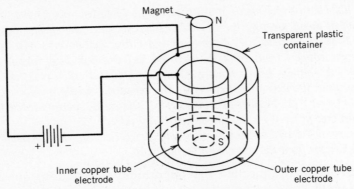

Figure 8.3.1 Interaction of magnet with current carrying electrolyte: Copper sulfate solution is placed in a plastic container between two cylindrical copper electrodes, which are connected to a battery or other power supply. When the end of a rod or bar magnet is inserted at the center of the system (within the inner electrode), the liquid is observed to circulate in the annular space between the electrodes. The circulation can be made visible by sprinkling wood filings or some floating powder on the surface. The direction of circulation is consistent with the right-hand rule devised in connection with the force on current carrying wires.

metallic conductor. The experiment is quite simple, and a basic arrangement is sketched in Fig. 8.3.1.

8.4 AMPÈRE'S EXPERIMENT

Another consequence of Oersted's discovery is the expectation that two current carrying conductors might exert forces on each other since the magnetic effect produced by one would interact with the current of the other, and vice versa. It should be noted, however, that this is merely a plausible argument and does not constitute a reliable inference. As was pointed out by Arago, a magnet exerts a force on each of two unmagnetized pieces of iron, but the pieces of iron, when separated from the magnet, exert no force on each other. The situation with the current carrying conductors could be analogous: namely that the presence of a magnet is essential and that, without the magnet, the current carrying conductors are as neutral as the pieces of iron. Experimental verification is necessary, and it was supplied by Ampère within a week after the news of Oersted's discovery reached France.

Ampère reported to the French Academy the results of the experiment now universally reproduced in lecture demonstrations: the attraction between parallel wires carrying current in the same direction and the repulsion with currents in opposite directions. There is a profound difference, however, between the tone of Ampère's report and that of most text and lecture presentations of the same phenomenon. Ampère was reporting the discovery of a new phenomenon, and, since batteries and electricity were involved, he had to provide convincing evidence that the interaction was not simply electrostatic but justified the designation "electromagnetic," which he introduced. It is this aspect of the phenomenology that is largely ignored in modern presentations: the effect is immediately asserted to be "electromagnetic" without further inquiry and at substantial cost in the opportunity for students to think and learn.

It is far better pedagogy to exhibit the interaction and then to lead students into raising the question as to whether the effect is new and different or simply electrostatic. Once the question is raised, students must confront the actual observations in full detail, suggest additional experiments, and outline in operational terms the differences between what is actually observed and what would be expected of ordinary electrostatic interaction. This engages them in purely phenomenological thinking, without formulas and number grinding, and forces them to review and reconsider earlier experiences. As has been argued repeatedly, such spiraling back in richer and more sophisticated context is at the leading edge of genuine learning and understanding. Homework can be greatly enriched by consideration of such questions since they induce thinking about all aspects of the related (and unrelated) phenomena—not just dealing with the restricted end results.

Ampère regarded the question as far from trivial and argued his case in the following way:

These attractions and repulsions between electric currents differ fundamentally from the effects produced by electricity in repose. First, they cease, as chemical decompositions do, as soon as we break the circuit [i.e., he argues that the effect is dynamic rather than static]. Second, in ordinary electric attractions and repulsions, opposite charges attract, and like charges repel; in the attractions and repulsions of electric currents, we have precisely the contrary. It is when two conducting wires are placed parallel in such a way that their ends of the same sign are next to each other that there is attraction, and there is repulsion when the ends of the same sign are as far apart as possible. Third, in the case of attraction, when it is sufficiently strong to bring the movable conductor in contact with the fixed conductor, they remain attached to each other like two magnets and do not separate after a while, as happens when two conducting bodies, oppositely electrified, come to touch [we tend to perform our demonstrations with insulated wires, but Ampère was using bare conductors that could have transferred charge between each other.]

One might add to this list the observation that *three* parallel wires, carrying current in the same direction, all attract each other. This is an effect that cannot possibly occur electrostatically. (It is interesting that Ampère was aware of this effect but does not adduce it along with the other three listed in the preceding paragraph.)

The phenomenological arguments that the interaction between parallel current carrying wires cannot be electrostatic lend themselves to formation of a very effective homework problem, especially if Socratic guidance is provided without giving the whole story away.

8.5 MNEMONICS AND THE COMPUTER

Virtually all courses and textbook presentations introduce some form of the simple mnemonics that help one keep track of electromagnetic phenomena: the right-hand rules for the Oersted effect and for the Lorentz force on moving charges (and the force on a current carrying conductor in the field of a magnet) and some form of left-hand rule for Faraday's law of electromagnetic induction. Many students, however, fail to master these rules and fail to use them correctly either in concrete or in pencil-and-paper situations.

This failure is clearly not one of logical reasoning or of concept formation; it is simply a matter of practice and memory. Textbooks do not, in general, provide a sufficient number of exercises for application of the rules, and this form of knowledge is not frequently tested. Students do not generate their own exercises and practice—even though they are perfectly capable of doing so if guided into the process and if they know they will be tested on the material.

What is needed here is a certain amount of drill, and the term "drill" is not being used pejoratively. It is only through drill that such knowledge can be fixed in the memory. Drill should take the form of applying the rules in different orientations of the motions and field directions in space. Drill should also lead the student to confront all possible variations of the rules: Given the direction of current in a straight wire (or in a coil), what must be the direction of the magnetic field at various indicated points? Given the direction of magnetic field at some point in the neighborhood of a straight wire or of a coil, what must be the direction of conventional current in the conductor? Given the direction of motion of a charged particle and the direction of the Lorentz force, what must be the direction of the magnetic field? (And the other two combinations of knowns and unknown.)

Students should also have the opportunity to sketch trajectories of moving charged particles in magnetic fields. They frequently say what sound like the correct words in describing such motions, but they proceed to draw trajectories that are physically impossible and fail to recognize what is wrong with their drawings.

It is in such areas of drill that the computer can provide very effective assistance to learning. Interactive dialogues with relatively simple graphics can provide students with as many exercises as are needed to master the rules, and they can help save a great deal of time. Unfortunately, at the time of writing, such computer-based exercises are not readily available. One would hope that this gap might soon be filled. It is one of the potentially cost-effective areas of computer-based instruction.

8.6 FARADAY'S LAW IN A MULTIPLY CONNECTED REGION

Fig. 8.6.1 and 8.6.2 show a simple and very fundamental physical situation that is rarely, if ever, considered in introductory physics. Yet the phenomena arising in this system are deeply significant and greatly enhance a learner's understanding of essential aspects of electromagnetism.

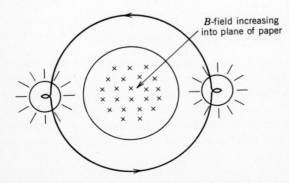

Figure 8.6.1 A long solenoid, axis perpendicular to the plane of the paper, carries a varying current. An emf and a resulting current are induced in the conducting loop surrounding the solenoid. Flashlight bulbs in this loop will light; potential drops will develop across resistors.

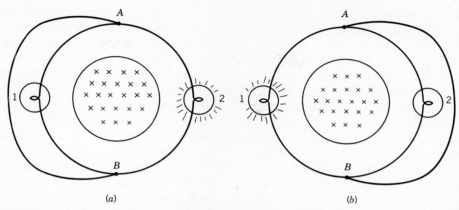

Figure 8.6.2 When points A and B are shorted by an additional wire as in (*a*), bulb 1 goes out while bulb 2 remains lighted and burns more birghly than before. When the same points are shorted as in (*b*), bulb 2 goes out while bulb 1 remains lighted and burns more brightly.

A long solenoid (Fig. 8.6.1), axis perpendicular to the plane of the paper, carries a varying current. Since the magnetic flux through any conducting loop surrounding the solenoid changes with time, an emf is induced in the loop in accordance with Faraday's Law. Flashlight bulbs in this loop will light; potential drops will develop across resistors.

Suppose that a wire is now introduced in such a way as to short circuit points A and B in the loop (Fig. 8.6.2). Students' prior experience with simple batteries-and-bulbs circuits has led to the expectation that short circuited bulbs will always go out. The situation now being considered is, however, very different because the region outside the solenoid is multiply, rather than simply, connected. Only one bulb goes out, and the other burns more brightly, depending on whether the shorting is effected as in Fig. 8.6.2*a* or as in Fig. 8.6.2*b*. Thus, in this situation, naive expectation is contradicted, and one must think through what happens in terms of the change of flux through the various conducting loops in the system.

Romer (1982) discusses this problem in complete and sophisticated detail in terms of Maxwell's equations applied to various regions in the system and in terms of interpretation of the readings on voltmeters connected between points A and B. Readers interested in pursuing, in full detail, the issues raised here would do well to refer to Romer's lucid paper. The presentation given above is merely a qualitative, first stage, introduction to a significant problem. This introduction, even if incompletely rigorous, extends the range of student understanding by broadening and enriching the context.

8.7 FARADAY'S CRITICISM OF ACTION AT A DISTANCE

Among Faraday's experimental achievements were the recognition of several new chemical compounds; work on the liquefaction of gases; discovery of the laws of electrolysis; discovery of the phenomena of electromagnetic induction, diamagnetism, and the rotation of the plane of polarization of a beam of light by a magnetic field. Faraday's great influence on subsequent scientific thought stemmed, however, not only from the importance of his experimental discoveries but also from the influence of his theoretical speculations. His investigations led him to reject Newtonian action at a distance in connection with electric and magnetic interactions and to assign a role to lines of force in the intervening medium. In doing so, he laid the basis for the modern concept of "field," which was elaborated mathematically by Maxwell and which has become one of the most firmly established elements in all of theoretical physics.

Prior to the success and acceptance of the Newtonian theory of gravitation, the concept of action at a distance was anathema to most natural philosophers. In their revulsion against superstition and against the assignment of occult virtues and properties to inanimate substances, they rejected the thesis that a material object could exert an effect in a place where it was not. They tried to visualize clear-cut mechanisms for all interactions, and where an intervening medium was not directly apparent to the senses, they invented effluvia and continuous or corpuscular ethers through which effects could be propagated by contiguous action of layer on layer or particle on particle. The word "attraction," in certain contexts, for example, implied action at a distance and was rejected as yielding to occult beliefs. When Huygens was informed of the imminent publication of Newton's *Principia*, it is said that he expressed the hope that it would not be based on "attractions," as rumors had that it would be. Huygens and Leibniz never accepted Newtonian theory on this account. The opposition, of course, eventually died out because of the repeated successes of the theory, and action at a distance (the applicability of Newton's Third Law instant by instant throughout a gravitational interaction) came to be accepted uncritically. Max Born remarks in this connection that, " . . . after Newton's theory . . . had been established, the idea of a force acting at a distance became a habit of thought. For it is indeed nothing more than a habit when an idea impresses itself so strongly on minds that it is used as an ultimate principle of explanation."

Newton himself had been careful to avoid commitment concerning models of gravitational interaction—"*Hypotheses non fingo.*" He was clearly aware that he had discovered a mathematical formulation that correctly represented the facts of natural phenomena but that might be subject to a variety of interpretations.[1] Newton himself probably did not believe literally in action at a distance. He engaged in much speculation about ethers through which gravitational and other effects might be propagated, and there exist among his writings chapters that were originally intended for the *Principia* but that he withheld, possibly because he felt their speculative nature to be out of harmony with the Euclidean tone of the rest of the work.

Ampère constructed his mathematical theory of electromagnetic interaction between current carrying elements of wire in the action at a distance framework. Although Faraday in the *Electrical Researches* repeatedly expresses admiration for the "beautiful work" of Ampere and other members of the French school, his own fundamental outlook was quite different. Lacking formal training, especially in mathematics, he was never proficient in the abstract language of analysis (calculus). He tended to interpret his observations and formulate his concepts almost entirely in geometrical and physical terms. It is possibly because of this orientation

---

[1]Situations of this kind have arisen from time to time in the history of science. In 1822, for example, Fourier published a complete and elaborate mathematical theory of the conduction of heat in solids (this was the work in which the Fourier series was presented). Fourier thought and wrote in terms of the caloric theory of heat, which was soon to be rejected, but his differential equation correctly represents the conduction of heat regardless of what model we hold as to the nature of heat itself. His theory was therefore in complete agreement with the observed temperature distributions.

that the geometrical patterns formed by iron filings impressed him so deeply. Faraday began to refer to "lines or curves of magnetic force" (a term that had already been used in connection with magnetism at least as early as the beginning of the seventeenth century): "By magnetic curves, I mean lines of magnetic forces, however modified by the juxtaposition of poles, which could be depiced by iron filings; or those to which a very small magnetic needle would form a tangent." Later he extended the idea to include lines of electric force.

Faraday's observations and experiments convinced him that the totality of electric and magnetic phenomena could not be explained in terms of action at a distance between particles, and that the intervening space must somehow be involved. Among the pieces of evidence that seem to have weighed most heavily with him are the following:

1. The lines of force are curved in space and are not simply straight lines connecting interacting magnetic poles or charged particles. Faraday could not conceive the curvature of lines of force in terms of action at a distance, especially in the case of lines around the current carrying wire. He felt that the curvature either indicated a physical existence of the lines themselves, or else stemmed from a state or condition of an intervening ethereal medium.

2. When a slab of dielectric material (Faraday used discs of sulfur and shellac) was inserted between the plates of a charged capacitor, the quantity of charge on the plates was found to be different from what it had been when only air intervened. To Faraday, this experiment indicated that interaction between electrical charges on the plates was not the exclusive factor and that the intervening medium also played an important role. He was also aware of similar effects in magnetism.

3. In connection with induction of electric current in a wire loop when a portion of the wire is moved so as to cut lines of magnetic force (motional emf), he says, "The mere fact of motion cannot have produced this current: there must have been a state or condition around the magnet and sustained by it, within the range of which the wire was placed."

The fact that one might perhaps take issue with some of these views and rationalize the observations in some other way is not the point. These were the factors that influenced his thinking.

The most crucial question, that to which Faraday returned repeatedly, was whether or not finite time intervals were required for the propagation of changes in electric and magnetic effects: When current was abruptly changed in a wire, did a time interval elapse before the magnetic lines of force changed some distance away? When quantity of charge was abruptly changed on an electrically charged body, did a time interval elapse before a change in the force exerted on a test particle some distance away? If a charged particle were abruptly displaced, would a time interval elapse before a change in the force exerted on a stationary test particle some distance away?

An affirmative answer to these questions would have lent the strongest possible support to Faraday's views concerning the dubious status of action at a distance models since it would have implied a propagation of action, perhaps in wavelike fashion, through the intervening space. (Although Maxwell's theory, developed over the decade 1856–65, did answer these questions in the affirmative, direct experimental verification of the theory did not come until Hertz's famous experiment in 1887, 20 years after Faraday's death.)

By the time Faraday was pursuing this line of thought, Young, Fresnel, and others had shown that light behaves like a transverse wave. It was far from clear just what was waving, but many physicists suspected and searched for a connection between light and electricity and magnetism. Faraday himself discovered the first direct physical interaction: the rotation of the plane of polarization of light when propagating in a direction parallel to magnetic lines of force. He entertained the idea that light might itself be the transverse vibrations of lines of force. He speculated on whether the velocity of propagation of magnetic effects might be of the same order as the velocity of light, and these thoughts provided further motivation for rejecting action at a distance and tentatively assigning a fundamental role in electromagnetic phenomena to lines of force or to an ethereal medium.

In his biography of Faraday, Tyndall writes,

During the evening of his life, be brooded on magnetic media and lines of force; and the great object of the last investigation he ever undertook was the decision of the question whether magnetic force requires time for its propagation. How he proposed to attack this subject we shall never know. But he left some beautiful apparatus behind; delicate wheels and pinions, and associated mirrors, which were to have been employed in the investigation.

In retrospect, we know that the technology of the time was not ready for such an experiment.

Readers will note that, in earlier chapters, I have repeatedly advocated directing the attention of students to questions concerning Newton's Third Law, action at a distance, and time intervals associated with interactions. The point, of course, is to prepare students to understand and appreciate this bit of intellectual history—to comprehend Faraday's questions and the motivation for invention of field theory.

8.8 INFANCY OF THE "FIELD" CONCEPT

Faraday, in his writings, carefully separated speculations regarding ethers and lines of force from factual reports of the results of his experiments, and he hedged these speculations almost apologetically:

It is not to be supposed for a moment that speculations of this kind are useless or necessarily hurtful in natural philosophy. They should ever be held as doubtful and liable to error or to change, but they are wonderful aids in the hands of the experimentalist and mathematician; for not only are they useful in rendering the vague idea more clear for the time, giving it something like a definite shape, that it may be submitted to experiment and calculation; but they lead on, by deduction and correction, to the discovery of new phenomena, and so cause an increase and advance of real physical truth, which unlike the hypothesis that led to it, becomes fundamental knowledge not subject to change.[2]

---

[2]This is an especially valuable paragraph to have the students read, think about, and discuss. One can hardly find a more cogent or more lucid description of the power and utility of a heuristic device in scientific thought. On the other hand, the end of the paragraph uses phrases that very few scientists would use today. Faraday was by no means the only scientist of his time to believe that science produces "real physical truth" and "knowledge not subject to change." With only a few dissenting voices influenced by the positivistic movement in philosophy (Mach, Ostwald, Duhem, for example), many nineteenth century scientists would have reflected similar attitudes. Their confidence is quite understandable. So wide in scope, so convincing were the successful applications of Newtonian mechanics and the new theories of thermodynamics and electromagnetism, that they indeed seemed to have led to "knowledge not subject to change." The warnings of the skeptics went largely unheeded. When one reaches the point of discussing the early twentieth-century revolutions in relativity and in atomic and quantum physics, it is very effective to lead students back to this paragraph and to contrast our modern view of the provisional nature of our knowledge with Faraday's confident assertion.

8.8 INFANCY OF THE "FIELD" CONCEPT

Nevertheless, it is evident in much of Faraday's subsequent work that lines of force meant more to him than just a heuristic device. He used this idea so much and with such success that he clearly came to believe in "physical" lines of force. In this conception the lines filled all space and had distinct properties and modes of behavior. They acted like rubber bands that were under tension longitudinally and that repelled each other laterally. William Thomson (Lord Kelvin), in papers published in 1847 and 1854, called attention to mathematical analogies that exist between theories of fluid flow, heat flow, and elasticity on the one hand and electrostatics and magnetism, as described by lines of force, on the other. Faraday, taking mathematical analogy to other physical phenomena as evidence of physical reality, felt his view of lines of force to be strongly supported.

James Clerk Maxwell, then a young Fellow at Trinity College, Cambridge, was deeply impressed both with Faraday's conception of lines of force and with Thomson's revealing mathematical analogies. Gifted with great mathematical talent and with intuitive physical sense on a par with Faraday's, he embarked on an attempt to synthesize into one unified theory all the known phenomena of electricity and magnetism. In his first two papers on this subject, published in 1856 and 1861, he developed an elaborate fluid model of Faraday's lines of force:

> *By referring everything to the purely geometrical idea of motion of an imaginary fluid, I hope to attain generality and precision, and to avoid the dangers arising from a premature theory professing to explain the cause of phenomena. If the results of mere speculation which I have collected are found to be of any use to experimental philosophers in arranging and interpreting their results, they have served their purpose, and a mature theory, in which physical facts will be physically explained, will be formed by those who, by interrogating Nature herself, can obtain the only true solution of the questions which the mathematical theory suggests.*

Note the similarity of this remark to the first part of the statement by Faraday quoted above. In his first paper, Maxwell used an elaborate concrete model involving fluid cells, vortices, and "idler wheels." A diagram of such a system is given in the paper. In these papers, Maxwell also began to use the terms "electric field" and "magnetic field" in the essentially modern sense. In 1865 he published his final version of the theory, explicitly eschewing action at a distance:

> *I have preferred to seek an explanation [of electric and magnetic phenomena] by supposing them to be produced by actions which go on in the surrounding medium as well as in the excited bodies, and endeavoring to explain the action between distant bodies without assuming the existence of forces capable of acting directly at sensible distances.*
>
>   *The theory I propose may therefore be called a theory of the 'Electromagnetic Field' because it has to do with the space in the neighborhood of the electric and magnetic bodies, and it may be called a 'Dynamical' theory because it assumes that in that space there is matter in motion, by which the observed electromagnetic phenomena are produced . . . [The space] may be filled with any kind of matter, or we may endeavor to render it empty of all gross matter, as in the case of Geissler [electrical discharge] tubes and other so-called vacua.*

In this paper (as well as in the subsequent *Treatise on Electricity and Magnetism* published in 1873), the elaborate fluid model of cells and vortices has disappeared. There remain only the mathematical equations and the concept of "field" as a condition or state of an ethereal medium. General acceptance of Maxwell's theory toward the end of the nineteenth century marked the transition from an era dominated by action at a distance philosophy to the present era of field theories in which momentum, energy, and other conserved quantities are propagated through the "field."

Students who go on to more advanced studies in physics and engineering should certainly be helped to acquire some of this phenomenological background so as to better understand and appreciate the real point and purpose behind the introduction of the "field" concept. Such background makes the introduction and the consequences of Maxwell's equations much more understandable and intelligible. However, even students who never go on to more advanced study and will never see or use Maxwell's equations can benefit from this qualitative exposure. Given the build-up and tie-ins urged in earlier chapters, they can acquire at least a qualitative insight into the revolutionary shift in point of view that accompanied the invention of field theory and have some comprehension of what questions and points of view motivate the modern search for evidence of gravity waves.

8.9 LABORATORY MEASUREMENT OF A VALUE OF B

One of the few measurements worth making just for the sake of the measurement itself is that of the strength of a B-field. The reason for this is that very few students develop confidence in the meaning of B directly from text or lecture presentations. If questioned, many actually reveal doubts about the "reality" of such numbers. Such doubts and reservations are markedly reduced by the concrete experience of making a direct measurement of the force acting on a current carrying wire in a magnetic field and calculating the magnitude of B from F/IL. (Flip coil or Hall effect devices do not have the same impact since the connection to B is more abstract and more remote.)

Most apparatus companies offer a device for determining B by "weighing" the force on a known length of current carrying wire. *PSSC Physics* used to exploit a simple homemade device that was quite effective. A simple, flexible setup offers the advantage of allowing students to explore the variation of the force with the angle between the field direction and the wire—something worth doing in addition to making the force measurement when the two directions are orthogonal.

After making such observations themselves, students hold a markedly different attitude toward the meaning of B than they hold in the absence of the concrete experience.

CHAPTER 9

Waves and Light

9.1 INTRODUCTION

The teaching of wave phenomena in introductory physics rightly concentrates on kinematic aspects and leaves most of the dynamics to later, more advanced consideration. One of the pedagogically best and soundest treatments is still that of the *PSSC Physics* text in its six editions. Fortunately, many other texts have drawn heavily on the *PSSC* treatment, especially the fine photographs that have been made widely available. When combined with demonstrations, laboratory experience with ripple tanks, and the collection of excellent film loops showing reflection, refraction, and interference of ripples, such text material is quite effective in generating understanding of wave behavior and the distinction between wave and particle motion. Excessively rapid coverage of this material, however, frequently negates the potential effectiveness, especially when direct laboratory experience with strings, slinkies, and ripple tanks is not made available, and when accompanying Socratic questioning is not provided.

Concrete experience is still an essential factor in cultivating understanding of the phenomena and grasp of the extensive vocabulary that is generated. Furthermore, this experience must be guided by phenomenological questioning of a kind that is missing in many text and lecture presentations. The following sections contain some examples of what might be done to fill a few of the more serious remaining gaps. Examples are also given of insights that enrich the context and lead students to become aware of deep connections between seemingly disparate phenomena.

9.2 DISTINGUISHING BETWEEN PARTICLE AND PROPAGATION VELOCITIES

Although, in the case of the transverse wave on a string, the distinction between particle velocities in the medium and propagation velocity of the disturbance seems quite obvious visually, some students still exhibit residual confusion despite demonstrations they may have seen. They readily admit that the velocities are orthogonal to each other, but they fail to discern that the magnitudes are quite different. Some of this confusion is associated with failure to perceive that the particle velocities vary in both magnitude and direction and do not possess a single unique value (as does the propagation velocity). Some confusion stems from the fact that the maximum particle velocities increase and decrease together with increases and decreases in propagation velocity as the tension in the string is varied.

Lecture demonstrations on these matters tend to go by too rapidly for slower individuals, and such students should be helped to confront these aspects in observations of their own, preferably as part of home experiments with strings or ropes. Such homework, however, needs to be structured so as to guide students into genuine observation rather than vacuous "playing around." They should, for example, be led to see that the particle velocities differ from the propagation velocity in magnitude and that the velocity of the particle is zero at maximum deflection. They should sketch the variation in particle velocity through positive and negative pulses traveling in both possible directions. Initially, it is sufficient that such sketches be qualitatively adequate; to require that they be rigorously correct is asking too much at so early a stage.

In the case of longitudinal waves, many more students fail to discriminate between particle and propagation velocities because of the colinearity of the two. Rapidity of coverage is again a frequent obstacle. Most teachers are aware of how helpful the soft coil spring called the "slinky" can be in this context. Students master these ideas if given a reasonable amount of time to handle and observe the slinky and if they are required to sketch the variation of particle velocity in both compression and rarefaction pulses. The insight does not develop, however, unless suitable *qualitative* questions and problems are provided, and texts, by and large, do not supply the needed guidance. The task devolves on the teacher.

9.3 GRAPHS

As in the case of rectilinear kinematics (see Sections 2.6 and 2.10), the sketching and interpretation of graphs can play a key role in developing the student's understanding of wave kinematics—as well as understanding of many text presentations that are not otherwise assimilated.

Since most texts concentrate almost entirely on sinusoidal wave forms (which, of course, happen to have a simple analytical representation) and tend to ignore arbitrary pulse shapes (which do not), many students fail to develop a clear distinction between graphs in which the abscissa represents clock reading (i.e., passage of time at a fixed location in the medium) and graphs in which the abscissa represents position along the x-axis at a fixed clock reading. The simple way to deal with this is to add homework problems and test questions such as the following: (1) given the "photograph" (y versus x graph) of an *asymmetric* pulse on a string, sketch the corresponding y versus t graph; (2) given the y versus t graph, sketch the y versus x graph. (It is important that the shape be asymmetric because otherwise the distinction between the two representations is lost.) The corresponding particle velocity graphs should also be sketched. (See Fig. 9.3.1 for an illustration of what is being described.)

Problems of this variety are relatively simple for the case of transverse waves on a string because particle displacements are directly represented by ordinates of the graphs (i.e., the shape of the y versus x graph is the same as the shape seen on the string except for a change in scale). Even so, many students have difficulty with the transformation to the y versus t graph as well as with the velocity graphs, especially if they did not have adequate practice with kinematic graphs earlier in the course.

The difficulties are compounded, however, and affect many more students, when the transition is made to longitudinal waves. A major source of difficulty is now the fact that the ordinate in the initial graphs (before one makes the

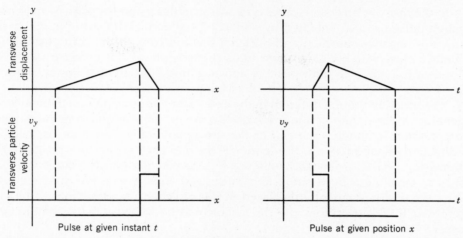

Figure 9.3.1 Corresponding graphs of an asymmetric pulse on a spring: (a) as a function of position along the string at a fixed instant of time; (b) as a function of clock reading at a fixed position along the string.

transition to particle velocity) is some variable such as pressure, or density, or particle displacement from equilibrium position, and the "shape" of the pulse is no longer directly visible as it is on a string.

Again, students' understanding of longitudinal waves can be greatly strengthened if they are supplied with qualitative questions that present asymmetric pulse shapes and call for (1) interpretation of the graph by means of a picture showing the corresponding variations in the medium (i.e., closer or wider spacing of coils of the slinky; higher or lower pressure or density indicated by closer or wider spacing of dots representing molecules of gas); (2) transformation of x into t graphs and vice versa.

The sketching of corresponding particle velocity graphs helps register one of the most significant physical distinctions between longitudinal and transverse waves: although the particle velocity is zero at the point of maximum deflection in the transverse pulse, the particle velocity is at a maximum at the point of maximum compression in a longitudinal pulse.

After such exercises have been performed with asymmetric pulses, they should be performed with the sinusoidal wave trains that become the principal burden of subsequent discussion, but the pulses are important because they emphasize physical aspects that are blurred in the sinusoidal trains. The pulses are also helpful in cultivating better understanding of transverse and longitudinal wave reflections at free and rigid boundaries, a subject dealt with in Section 9.5.

9.4 TRANSVERSE AND LONGITUDINAL PULSE SHAPES

Section 9.3 pointed out the value of invoking asymmetric pulses, in addition to sinusoidal wave trains, in helping students visualize important physical details of wave phenomena through exercises in graphing. Qualitative consideration of the actual generation of simple pulse shapes helps bring out additional physical aspects that are rarely made clear in introductory physics but that play a significant role in developing better understanding of the phenomena.

In the case of transverse waves on a string, either a purely positive or a purely negative pulse can be generated by deflecting the end of the string in either the positive or negative direction and bringing the end back to the zero position. To generate a pulse with both a positive and a negative phase, it is necessary to swing the end of the string to both positive and negative positions before returning to zero.

The situation is quite different in the case of longitudinal pulses. If one wishes to generate a pure compression pulse on the slinky, for example, one must create a compression by moving the end of the spring and *leaving* the end displaced in the direction of compression. If one moves the end back to the zero position, the compression phase is inevitably followed by a rarefaction phase. Similarly, if one wishes to generate a pure rarefaction pulse, one must create the rarefaction by leaving the end permanently displaced in the direction of rarefaction; if the end is returned to zero position, the rarefaction will be followed by a compression phase.[1]

In observing students doing laboratory work with wave phenomena, I have seen significant gains in security and confidence when these aspects of pulse generation become part of their basic understanding.

One aspect of understanding of acoustic compression and rarefaction pulses deserves special attention. When asked to describe in their own words what is happening in the fluid medium as such pulses are generated and then propagate, many students respond with description at a microscopic (atoms and molecules) rather than at a macroscopic level. Although there is nothing intrinsically wrong with such a description (except for the fact that students rarely, if ever, visualize the chaos of thermal motion superposed on the organized wave behavior), recourse to it usually reveals a deeply seated reluctance to deal with the macroscopic properties of pressure and density. The jargon about atoms and molecules has been picked up in earlier schooling, frequently with distorted and misleading overtones, while the macroscopic properties—especially pressure—are not well understood.

Hydrostatic pressure is, in fact, a subtle and difficult concept. Physical understanding of phenomena involving pressure and pressure variations hinges on an understanding of Pascal's Law—the fact that pressure at a point in a fluid is uniform in all directions. Without explicit grasp of this concept, many aspects of what happens in fluid media are imperfectly understood and visualized. With the modern tendency to omit or shortcut the study of fluid phenomena in introductory physics, many students emerge with very weak understanding of the nature of fluid pressure. (Witness the failure of many practicing *physicists* to recognize that, when an oil–water mixture separates on standing, the pressure changes on the bottom of the container if the container has sloping sides.) The question of understanding of fluid pressure is discussed in more detail in Section 11.3.

9.5 REFLECTION OF PULSES

Reflection of waves at boundaries is a very different physical problem from that of particles bouncing off walls. With waves, one is dealing with a boundary value problem in a dynamical system, and it includes all the subtlety associated with

---

[1]These qualitative observations are, of course, directly related to a very basic theorem regarding wave propagation, namely that the net *impulse* carried by a wave must be zero if there is zero final displacement at the point of origin. I do not advocate developing this theorem quantitatively in an introductory course, but the qualitative insights being suggested here provide a firm basis for deeper understanding at later, more advanced levels.

partial reflection and transmission and with absorption, cases of complete reflection being only idealized limits. Simultaneous reflection and transmission is one of the intrinsic properties separating wave and particle behavior in classical physics.

In the introductory course, one cannot proceed to develop the formal mathematical solution of the wave equation at the boundary even for the idealized cases. Visualization of the reflected wave requires something of an ad hoc argument, and the argument tends to be glossed over in many text presentations. As a result, many students are mystified by the approach—that of visualizing a reflected wave propagating out of the "never never land" on the other side of the boundary and having a shape such as to satisfy the boundary condition. They feel this technique to have a touch of black magic and to be something they themselves could never have conceived.

It may not be possible to allay such doubts completely, but it helps to say something explicit and to motivate the approach. The starting point for such motivation is simply the *observed fact* that reflections *do* occur and that the direction of propagation is opposite to that of the incident wave. Since, in the ideal case, the wave in one dimension propagates without change in shape, one can visualize the incident wave as having started or come from *anywhere* along the string or spring or water surface, even from a region into which the medium in question does not actually extend physically. In just the same way, it becomes legitimate to visualize the reflected wave as having been propagating without change in shape from anywhere. This helps justify visualizing the reflected wave as propagating in the region beyond the reflecting boundary (whether there is any medium there or not) and arriving at the boundary in such phase as to maintain the free or rigid boundary condition.

One must confess explicitly that this is a purely *kinematic* approach that helps "save the appearances" (to use an ancient locution), one that accepts the existence of reflection as an observed fact and serves to satisfy the condition at the boundary. It is not a dynamic approach, and it therefore contains no description of a "mechanism" by which the reflection is generated at the boundary. (Most attempts to provide a mechanical explanation of the form of the reflected wave are misleading or specious, and it is better that they be avoided.) Most students are willing to accept the procedure being adopted when one is frank about its justification and limitations.

Although many texts, especially those making use of the excellent *PSSC* illustrations, do give good, clear presentations of what shapes are to be observed when waves are completely reflected from free or rigid boundaries (as well as dealing with partial reflection and transmission), students are rarely given adequate opportunity to sketch the shapes of reflected and transmitted waves under various circumstances. Without such opportunity, many students fail to develop significant understanding of the effects.[2]

Here again, asymmetric pulse shapes play a valuable role because they force the student to consider the sequence of events—a sequence that tends to be obscured in the symmetry of sinusoidal wave trains. Figures 9.5.1, 9.5.2, and 9.5.3 illustrate some of the points at issue. In the case of the transverse wave on a string, incident at a rigid wall (Fig. 9.5.1), the boundary condition of zero particle velocity at the wall requires a reflected pulse inverted in phase relative to the incident pulse. This is relatively clear to students, but what is far less clear

---

[2]It is being taken for granted here that the discussion of reflections is preceded by adequate demonstration and discussion of the superposition of wave trains and pulses. Without prior consideration of superposition, the treatment of reflections is meaningless. Again, one of the best available treatments, with excellent illustrations, is that of the *PSSC Physics* text.

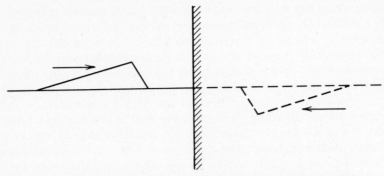

Figure 9.5.1 Transverse pulse on string is incident from the left at a rigid wall. To satisfy the boundary condition of zero particle velocity at the wall, the reflected pulse, imagined as originating in the fictitious region to the right of the wall, must be the phase inverted *mirror image* of the incident pulse.

is the fact that the reflected pulse must be a *mirror image* inverted in phase, and this aspect becomes apparent only if an asymmetric pulse shape is employed as an example. The graphing exercises recommended in Section 9.3 pave the way for the thinking and visualization that are required here.

Another important physical aspect of this situation is the fact that zero particle velocity on the string is maintained *only* at the rigid boundary itself. Elsewhere along the string the particle velocities are *not* zero as the reflected pulse propagates through the incident pulse. This is a point that eludes many students unless

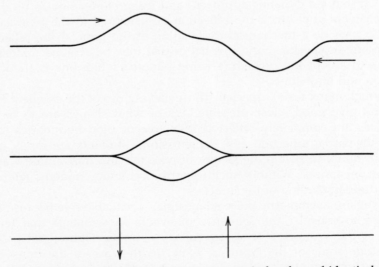

Figure 9.5.2 Superposition of two symmetrical pulses of identical shape but opposite phase, traveling in opposite directions on a string. At the instant of coincidence of the peaks, the instantaneous *deflection* is zero everywhere, but the particle velocity is zero only at the center and has maxima (one with upward and one with downward velocity) on either side of the center.

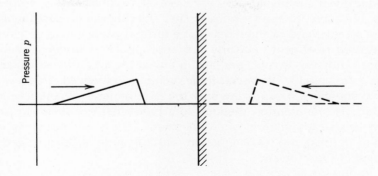

Figure 9.5.3 Compression pulse in a fluid (or on a slinky) is incident from the left at a rigid wall. To satisfy the boundary condition of zero particle velocity at the wall, the reflected pulse, imagined to originate in the fictitious region to the right of the wall, must be the mirror image of the incident pulse.

they are led to sketch the overlapping of the incident and reflected pulses at two or three successive stages and to consider the attendant particle velocities at various locations. The insights acquired here are enriched if students are led to return to the case of superposition of two symmetrical pulses passing through each other in opposite phase and in opposite directions, as illustrated in Fig. 9.5.2. At the instant the pulses "cancel" each other, the net deflection is zero all along the string, but the particle velocities are *not* everywhere zero. The particle velocity is zero in the central region where the maximum deflections cancel, but there are two regions of maximum particle velocity at the sides, one upward and one downward as shown.

With this exposure, students are being prepared to understand the motions that will occur in standing waves resulting from reflection of sinusoidal wave trains.

In Fig. 9.5.3, we consider the reflection of a compression pulse (longitudinal wave) at a rigid wall. Here the boundary condition of zero particle velocity at the wall requires that the reflected pulse be a mirror image of the incident pulse *without* inversion in phase. If students have been exposed to the exercises recommended in Section 9.3, they are far better prepared to understand the reflection now under consideration.

Homework and test questions should, of course, pursue variations on the examples given above: incident negative transverse pulses and rarefaction pulses; incident wave coming from the right rather than from the left; free boundary instead of rigid boundary; sinusoidal wave trains instead of pulses; and also the *reverse* line of reasoning in which the shape of the reflected pulse is given and the shape of the incident pulse is called for. (As has been pointed out in earlier chapters, leading students to traverse a line of reasoning in both possible directions is highly conducive to learning and understanding. The reversal may seem trivial to an expert, but it is far from trivial to the learner.)

9.6 DERIVATION OF PROPAGATION VELOCITIES

The powerful, elegant, and rigorous way of showing that a disturbance will propagate under given circumstances with a velocity established by properties of the system is, of course, to apply the basic laws (mechanical or electromagnetic) governing the system, and to show that some form of wave equation will be obeyed. Such advanced treatment is clearly not appropriate or understandable in most introductory courses—except, perhaps, a few at the second year calculus-physics level. It is quite possible, however, to derive the wave velocity in a few interesting mechanical situations by applying only the requirements of conservation of mass and the impulse–momentum theorem to a pulse that is *assumed* to propagate in a more or less steady state in one dimension. This is not nearly as rigorous an approach as deriving the wave equation, but it is quite reasonable and acceptable as a first cut at the problem.

An important advantage of such derivations is that they give students a chance to see basic laws, encountered earlier, actually employed in a powerful way to obtain significant results in new situations. So far, the only use students have seen for the basic laws (conservation of mass and Newton's Laws of motion) has been in the highly restricted end-of-chapter examples arising in homework. Furthermore, students have not really seen general *derivations* using the basic laws; they have only seen examples of direct application to individual cases, used as separate exercises in homework.

Derivations of wave velocities are presented in the next three sections. I do not mean to recommend the introduction of these derivations in all introductory courses. They are best used at the discretion of the teacher. They might, for example, be made available to front-running students who would benefit from the deeper analytical insight, or they might be offered as opportunities for extra credit or independent study. In some college level courses, however, they are appropriate for an entire class, and, under such circumstances, they provide an opportunity for overview and synthesis that can come only from the cycling back that is entailed.

The derivations outlined in the next three sections are certainly not new. They are presented in order to show how an essentially identical approach can be taken in three disparate cases in order to display the unity of the phenomena. It is this unity, and the continual cycling back to the same set of fundamental ideas, that make the intellectual experience impressive and lasting for the students.

9.7 VELOCITY OF PROPAGATION OF A KINK ON A STRING

Consider a string stretched horizontally under tension T as shown in Fig. 9.7.1. One end of the string is displaced abruptly to a new position, and the corner or kink then propagates along the string to the far end. The action is assumed to be performed without change in the tension, and the angle θ between the original line of the string and the deflected portion is assumed to be small.

The basic (unproved) assumption is that the kink will propagate along the string at some velocity V determined by properties of the string. Given this assumption, we proceed to apply the restriction of conservation of mass and the impulse–momentum theorem to a small chunk of string just encompassed by the wave front in the small time interval Δt.

9.7 VELOCITY OF PROPAGATION OF A KINK ON A STRING

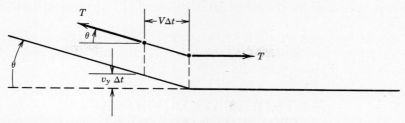

Figure 9.7.1 End of taught string is displaced abruptly so that a kink propagates to the right along the string.

The length of string encompassed will be $\Delta x = V\Delta t$, and, if we denote the mass per unit length of the string by μ, the mass Δm of string encompassed by the wave in time interval Δt is given by

$$\Delta m = \mu V \Delta t \tag{9.7.1}$$

This is, effectively, the continuity (or conservation of mass) equation and should be explicitly identified as such.

If we denote the particle velocity imparted to elements of the string by v_y, the change of momentum imparted to the chunk Δm in the time interval Δt is $\mu v_y V \Delta t$, and the impulse delivered in the y-direction is $T \sin\theta \, \Delta t$. Hence, by the impulse–momentum theorem

$$T \sin\theta \, \Delta t = \mu v_y V \Delta t$$

and

$$T \sin\theta = \mu v_y V \tag{9.7.2}$$

It is clear that one must now say something about the geometrical connection among $\theta, v_y,$ and V. From Fig. 9.7.1 it is apparent that

$$\frac{v_y}{V} = \tan\theta \tag{9.7.3}$$

[Those familiar with the formalism will note that, for a wave shape $y = f(x - Vt)$ propagating in the positive x-direction and obeying the small amplitude wave equation, the particle velocity v_y is given by

$$v_y = \frac{\partial y}{\partial t} = -Vf'(x - Vt) = -V\frac{\partial y}{\partial x} \tag{9.7.3a}$$

which is the counterpart of Eq. 9.7.3]

Combining Eqs. 9.7.2 and 9.7.3 gives

$$V^2 = \frac{T}{\mu} \cos\theta \tag{9.7.4}$$

and, for small θ, with $\cos\theta$ close to unity

$$V = \sqrt{\frac{T}{\mu}} \tag{9.7.5}$$

the familiar expression for the propagation velocity of a small-amplitude wave on a string.

It should be noted that the small amplitude approximation not only takes θ to be small, but also ignores stretching and contracting of the string and any consequent small motions back and forth along the x-axis.

9.8 PROPAGATION VELOCITY OF A PULSE IN A FLUID

Figure 9.8.1 shows a tube of fluid in which a compression pulse is initiated by a rapid displacement of a diaphragm somewhere off to the left. It is assumed that the pulse travels to the right at a velocity V that depends on properties of the fluid. The initially undisturbed fluid has a pressure p_0, a density ρ_0, and a zero particle velocity ($u_0 = 0$). The disturbed fluid behind the leading edge of the pulse has a pressure p, a density ρ, and a nonzero particle velocity u to the right. The cross-sectional area of the tube is denoted by A.

In a small time interval Δt, the leading edge of the pulse engulfs a mass of undisturbed fluid $\rho_0 A V \Delta t$. In the region behind the leading edge, this material will be compressed into the volume $A(V - u)\Delta t$. Conservation of mass requires that

$$\rho A(V - u)\Delta t = \rho_0 A V \Delta t$$

yielding

$$\rho(V - u) = \rho_0 V \tag{9.8.1}$$

Since the net impulse delivered to the chunk of fluid engulfed in the time interval Δt is given by $(p - p_0)A\Delta t$, and since the change of momentum of this chunk is $\rho_0 V u A \Delta t$, the impulse–momentum theorem requires that these two quantities be equal, and the result reduces to

$$p - p_0 = \rho_0 V u \tag{9.8.2}$$

Eliminating u from Eqs. 9.8.1 and 9.8.2 yields

$$V^2 = \frac{p - p_0}{\rho - \rho_0} \frac{\rho}{\rho_0}$$

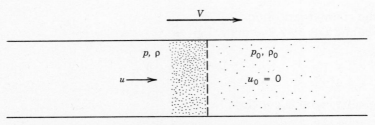

Figure 9.8.1 A compression pulse initiated off to the left propagates to the right in a tube of fluid. The cross-sectional area of the tube is denoted by A.

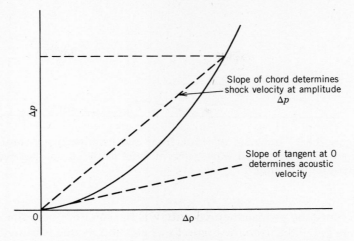

Figure 9.8.2 Schematic graph of Δp versus $\Delta \rho$ for any fluid subject to large adiabatic compressions and corresponding density changes.

or

$$V = \sqrt{\frac{\Delta p}{\Delta \rho} \frac{\rho}{\rho_0}} \qquad (9.8.3)$$

From Eq. 9.8.2, the particle velocity is given by

$$u = \frac{\Delta p}{\rho_0 V} \qquad (9.8.4)$$

It should be noted that none of the equations set down so far contain any small amplitude approximations. They are valid for large amplitude and are limited only by the steady-state assumption concerning the propagation. The change of state, although adiabatic, is inherently irreversible. The compression that takes place is not perfectly elastic; energy dissipation occurs in the process. The medium, after passage through the pulse and return to initial pressure level, has a higher thermal internal energy than it had previously. There is a net entropy increase associated with propagation of a finite amplitude pulse, and the final temperature is higher than the initial temperature on return to initial ambient pressure.

In the limit of small amplitude, Eq. 9.8.3 reduces to the familiar expression for acoustic velocity

$$V = \sqrt{\left(\frac{\partial p}{\partial \rho}\right)_S} \qquad (9.8.5)$$

where the constancy of entropy S indicates reversible adiabatic compression in the limit of vanishingly small amplitude.

Equation 9.8.3 contains interesting implications and merits further examination. A graph of Δp versus $\Delta \rho$ for a fluid subject to large pressure changes would, in general, have the qualitative appearance shown in Fig. 9.8.2. The graph would be concave upward, indicating decrease in compressibility of the fluid with increas-

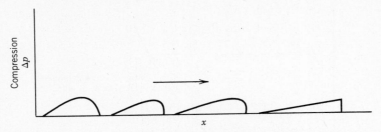

Figure 9.8.3 Formation of shock front and lengthening of pulse as an initially rounded wave pulse of finite amplitude propagates through the medium.

ing pressure. Since the propagation velocity V is the square root of the slope of the chord drawn in Fig. 9.8.2 from the origin to whatever Δp value is being considered, it is clear that higher compressions imply higher propagation velocities.

Thus, in a pulse initially having a rounded front such as that illustrated in Fig. 9.8.3, the higher pressure regions keep overtaking lower pressure regions ahead of them until the leading edge becomes the discontinuity in pressure and density called a "shock front."[3] As the higher pressure region overtakes the lower pressure ahead, it also "runs away" from the lower pressure region behind. Thus, the spatial length (and the duration) of the pulse continually increase as the pulse advances.

It is at the point of formation of the stable shock that the preceding equations become more rigorously applicable since, once the shock front is formed, the propagation is essentially steady except for the dissipation inherent in the finite amplitude transition. In other words, Eq. 9.8.3 is the fully correct equation for the velocity of a shock wave. Similarly, Eq. 9.8.4 is the fully correct equation for the particle velocity behind a shock front.

As an illustration of the importance of leading students to confront what is *not* the case as well as what it is, consider the spurious pulse shape shown in Fig. 9.8.4. Very few students, even those in advanced courses, have the courage to declare this diagram to be physically meaningless and impossible when it is first presented to them.

A note about a more advanced level for anyone who might be interested in pursuing it: With the introduction of an equation for energy conservation through the leading edge, one has a set of three equations that are called the "Rankine–Hugoniot Relations." With the third equation, one can eliminate $\Delta \rho$ in terms of

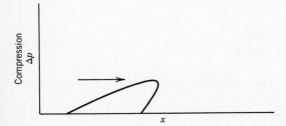

Figure 9.8.4 Meaningless Δp versus x graph of a pressure pulse.

---

[3]The shock is, of course, not a *mathematical* discontinuity. The region of pressure and density change has a width of the order of mean free paths of molecular motion.

the independent variable Δp [analytically for an ideal gas and numerically for any substance whose equation of state data (p, v, T, and c_p) are tabulated] and thus obtain the propagation velocity directly in terms of the pressure amplitude. This, however, involves the use of thermodynamic relations well beyond the level of an introductory physics course.

9.9 PROPAGATION VELOCITY OF SURFACE WAVES IN SHALLOW WATER

The term "shallow" in the present context refers to a layer of water whose depth D is small relative to the wavelength of the surface gravity wave.[4] Under these circumstances, the propagation velocity for small amplitude waves is given by the familiar, simple relation $V = \sqrt{gD}$. This relation will be derived below in a procedure exactly parallel to that applied in the two preceding sections.

The velocity relation indicates that the surface wave travels faster in deeper water, and I have, on various occasions, been asked by both students and colleagues how one can account for this physically. Given the greater inertia one naturally associates with deeper columns of water, it is quite reasonable to expect that the velocity would *decrease* with increasing water depth. The derivation gives insight into this nontrivial physical question.

Restricting the problem to one dimension as in the previous instances, consider a channel having a width Y and containing water to depth D (Fig. 9.9.1). Suppose that a wave pulse having a height h is generated by displacing a vertical wall in the channel somewhere off to the left of the figure. The basic assumptions are that the pulse propagates steadily with velocity V and that the vertical component of particle velocity behind the wave front can be ignored (i.e., that the particle velocity u can be treated as essentially horizontal). The water is, of course, treated as incompressible (ρ = constant).

Conservation of mass (continuity) equation: As the front advances for a small time interval Δt, the volume of initially undisturbed water passing into the disturbed region is $YDV\Delta t$ (sector $ABCE$ in Fig. 9.9.1). After passing through the front, this water is contained in the space $Y(D + h)(V - u)\Delta t$ (sector $GHJC$ in Fig. 9.9.1) providing that the bottom of the channel is rigid and impermeable. Equating these two volumes gives

$$DV = (D + h)(V - u) \tag{9.9.1}$$

and Eq. 9.9.1 reduces to

$$hV = u(D + h) \tag{9.9.2}$$

Impulse–momentum equation: The initial hydrostatic pressure distribution through the depth of the fluid plays no role in acceleration of fluid particles since the attendant forces are balanced throughout. The particle velocity u is imparted by the *unbalanced* force applied over the area YD of initially undisturbed fluid by virtue of the *excess* pressure $\rho g h$, that penetrates the entire depth of the water column. (This is where the restriction to long waves on shallow water enters this

---

[4]When the depth is large relative to the wavelength, the regime is entirely different and significantly more complicated. Under these circumstances, the pressure variations near the surface do not penetrate through the entire water column to the bottom of the layer. The mathematical analysis is much more sophisticated, and the velocity relation is a dispersive one. This regime will not be considered here.

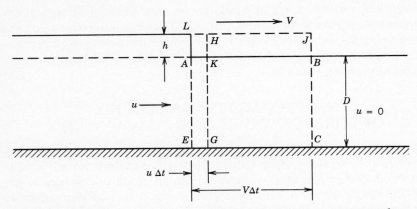

Figure 9.9.1 Positive surface wave pulse of height h propagates along channel containing water of undisturbed depth D and density p. Channel width is denoted by Y and particle velocity behind the wave front by u.

derivation.) Thus, an impulse $\rho g h Y D \Delta t$ is imparted to the mass of water that passes through the front. The momentum change of this water is $\rho Y D V u \Delta t$. Equating the impulse and the corresponding momentum change yields

$$gh = Vu \tag{9.9.3}$$

and, eliminating u from Eqs. 9.9.2 and 9.9.3, gives

$$V^2 = g(D + h) \tag{9.9.4}$$

which, in the limit of small amplitude, gives the familiar

$$V = \sqrt{gD} \tag{9.9.5}$$

One can now begin to see how it is that the wave travels faster in deeper water. First let us rearrange Eq. 9.9.2 as follows:

$$uD = h(V - u) \tag{9.9.6}$$

The two volumes in Eq. 9.9.6 are interpreted in Fig. 9.9.1: The left-hand side refers to the volume of water in the sector $AKGE$ (without the Y and Δt terms) while the right-hand side refers to the volume in the sector $HJBK$ (also without the Y and Δt terms).

Equation 9.9.6 says that, in the time interval Δt, a volume of water $uDY\Delta t$, initially ahead of the wave front, must be transferred into the region $h(V - u)Y\Delta t$ as shown in the diagram. (This is not to say that the first volume is literally lifted into the location of the second. The actual trajectories of the water particles are not being mapped out in this highly restricted analysis.) It is evident from Fig. 9.9.1, however, that the continuity equation requires that the wave front advance far enough in any given interval of time Δt to accommodate the volume $uDY\Delta t$. This means that the front must advance farther if the water is deeper.

This analysis yields the unusual insight that, in the case of the surface wave in shallow water, the progress of the wave is, in an important sense, driven by the continuity requirement rather than by a force. A force, resulting from the pressure difference provided by the wave height, is, of course, necessary in order to generate any disturbance at all, but the velocity is then dictated by the continuity requirement since, in a given time interval, the front must advance farther in deeper water in order to conserve the shifted mass.

Returning to Eq. 9.9.4, we note that, if we admit appreciable wave height h relative to depth D, higher amplitude regions of a pulse propagate at higher velocity than lower amplitude regions and that the pulse therefore tends to build up into a shock front, as does the finite amplitude compression pulse in a fluid. In the case of the water wave, however, the wave "breaks," the most familiar illustration being the running up of waves on a beach. Because of the analogy between finite amplitude surface waves on water and finite amplitude pressure waves in a fluid, some experimentalists have used water waves to model and study the nonlinear effects associated with the intersection and reflection of shock waves.

Final comments: The three preceding derivations of wave propagation velocity have been deliberately designed to dramatize the power of an identical approach applied to seemingly very different physical situations. The results reveal the profound unity underlying the disparate phenomena. This is invariably deeply impressive to students whose previous experience has not exposed them to a comparable synthesis. The repetitive application of the basic laws also plays a very important role in helping the students master the reasoning rather than just memorizing a procedure in an isolated instance.

Another virtue of these derivations is that they entail a very strong physical content, and requirement for phenomenological reasoning, with a minimum of mathematical complexity. Better students, especially those heading for physics or engineering, are very much in need of exposure to such reasoning early on. Many of them come to advanced courses (or even to graduate school) without having had the experience, and this becomes very evident in their performance. Faculty complain bitterly about the inadequacy of the students without identifying the source.

9.10 TRANSIENT WAVE EFFECTS

Once waves and wave pulses have been studied and thought about for a while, student perspectives can be significantly broadened by leading them to reconsider, on a purely qualitative level, a variety of everyday experiences as well as some of the physical situations dealt with in problems earlier in the course. Many of these situations involve transient wave phenomena, but very few students discern this spontaneously since most of the phenomena must be visualized without being directly seen. Students do begin to comprehend the connections, however, if led into thinking about these effects and visualizing them. This, in turn, propels them into a more sophisticated approach to phenomena that transcend direct sense experience.

Some examples:

If we lay a long rod down on the table and displace a box at the far end of the rod by pushing on the near end, what is it that happens at the very beginning of action? Is the force exerted on the rod at the near end equal to force exerted by the rod on the box? Does the displacement of the box begin at the same instant that we push on our end of the rod? What is it that happens in the rod and at the box before a steady state is attained?

An Atwood machine (or any other system of objects and strings dealt with in earlier dynamics problems) is initially held stationary and is then let go and allowed to accelerate. What happens between the instant of letting go and the achievement of the state of steady acceleration? In light of these examples, what do you suppose happens in elevator cables?

Consider the bouncing of a ball from a floor or a wall. What happens in each of the interacting objects during the collision? What happens when two balls collide? What is the origin of the sound we hear when collisions occur?

What happens when a balloon bursts? What is the origin of thunder?

What happens in a crank shaft when a torque is suddenly applied at one end?

You push horizontally at the top of a large box and slide it along the floor against the opposing frictional force. What happens between the application of the push and the time at which the box begins to slide?

It is not usually necessary to supply a long list. Once students are cued to such questions, many of them can begin to invent questions of their own, and many of these turn out to be interesting and ingenious. The object is to get them to discern the presence of longitudinal, shear, and torsion waves in virtually every dynamic interaction despite the fact that these waves elude direct sense perception.

9.11 SKETCHING WAVE FRONTS AND RAYS IN TWO DIMENSIONS

Although most textbooks exhibit diagrams showing transmission, reflection, and refraction of waves in two dimensions, and although excellent film loops are available showing these effects in ripple tanks, the students encounter few, if any, homework problems that lead them to sketch ray and wave-front diagrams of such phenomena themselves. The consequence is that many students are unable to sketch correct diagrams of their own in either representation, especially if the orientation of the interface is changed from that of an illustration given in the text.

This is not a formidable conceptual problem, and most students grasp the ideas fairly quickly. What is missed in instruction is the fact that practice is essential regardless of the clarity of text diagrams and discussion. Numbers are not necessary; purely qualitative questions form the best vehicle.

Questions should include aspects such as the following: Given an interface at which propagation velocity changes and given an incident ray, sketch the reflected and transmitted rays, and then sketch the same situation with corresponding wave fronts instead of rays. Changes in wave length should be indicated in the wave-front diagram. Students should then be led to ring their own changes on this theme: orient the interface differently on the page; reverse the velocity change (i.e., if the initial question had the incident wave in the faster medium, put the incident wave in the slower medium); start with the wave-front diagram and then sketch the corresponding ray diagram; start with the transmitted (or the reflected) rays or wave fronts as given and sketch the rest of the diagram. (Note that students should be led to traverse the line of reasoning in all possible directions, backwards as well as forwards.)

Such situations are usually first encountered with ripples. The sketching should be repeated in the encounter with light. As has been pointed out repeatedly, it is the opportunity to use an idea, after elapse of time and in an altered context, that leads to mastery and retention. In many students, retention is feeble on only one exposure.

The sketching of such diagrams should be required on tests as well as in homework. The phenomenology is important, and since these ideas are readily mastered with a little practice, those students who have done their homework can do well on the test questions, thereby acquiring reinforcement that is frequently lacking in much of our testing.

9.12 PERIODIC AND SINUSOIDAL WAVE TRAINS

Although most texts give reasonably adequate developments of the basic relation

$$V = \nu\lambda \qquad (9.12.1)$$

connecting propagation velocity, frequency, and wavelength for periodic wave trains, many students proceed to use it blindly, as a memorized formula, with no understanding of its origin or justification. Being able to substitute in the formula and obtain correct numerical answers is no indication of understanding.

The learning problem here is a very basic one: there is a profound reluctance among many learners to go back, on their own initiative, to the operational definitions underlying a line of reasoning, articulate the definitions, and use them to obtain a result or draw an inference. They will gloss over text presentations and avoid the sequence of definition and reasoning unless explicitly required to pursue it.

Since, in this instance, the definitions are simple and the reasoning purely arithmetical, this is a valuable opportunity for practice that helps lower resistance to such reasoning. Students should be led to respond to questions such as: (1) State in your own words the definitions of the quantities represented by the symbols V, ν, and λ. (2) Explain in your own words, with accompanying diagrams, how these definitions lead to relations such as $V = \nu\lambda$, or $\nu = V/\lambda$, or $\lambda = V/\nu$.

The representation of sinusoidal wave trains presents not only an opportunity to exploit the various forms of Eq. 9.12.1 but also an opportunity to cycle back to radian measure and the reasons for invoking it (see Sections 1.13 and 4.6). The great majority of students, even among those in engineering-physics courses, have great difficulty grasping the need for the $2\pi/\lambda$ factor in

$$y = A \sin\left[\frac{2\pi}{\lambda}(x \pm Vt)\right] \qquad (9.12.2)$$

They should be led to explain the necessity of this term in their own words.

Finally, an effective kind of homework (and test) question missing from most textbook collections is exemplified in the following (note that only *qualitative* sketching, not numerical plotting, is being called for):

Given the "photograph" (i.e., form at a given instant of time) of a sinusoidal wave train shown in Fig. 9.12.1, sketch the photograph that would be obtained if

(a) the amplitude and the frequency were doubled while velocity remains unchanged.
(b) the frequency and velocity were both doubled while the amplitude remains unchanged.
(c) the wavelength and amplitude are reduced by a factor of three while the velocity is doubled.
(d) and so on and so forth.

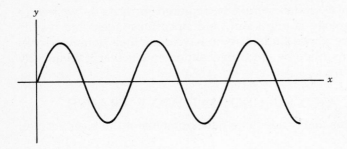

Figure 9.12.1 y versus x graph of a sinusoidal wave train at a given instant.

Students might then be invited to make up, for themselves, a parallel series of questions in which one starts with the y versus t graph at a fixed point in space instead of with Fig. 9.12.1. Students are rarely required to make up questions and problems of their own, and, because of lack of practice, the majority are extremely reluctant to do so. Simple cases of this variety therefore constitute a valuable opportunity, affording practice that helps lower resistance.

Still another aspect with which many students have difficulty is the perception that, in transmission and reflection of sinusoidal waves at an interface, it is the *frequency* and not the wavelength of the wave that is preserved. They must be led to visualize the arrival of crests and troughs at the boundary and to visualize what effect this has on the disturbance transmitted through the boundary. Such exercises are easily coupled to the ones described above and to those suggested in Section 9.11.

9.13 TWO-SOURCE INTERFERENCE PATTERNS

Understanding of two-source interference patterns does not emerge from passively viewed demonstrations or from numerical exercises with the formulas connecting wavelength, spacing between the sources, and the angle between loci of constructive or destructive interference and the principal axis. This is not to say that the demonstrations and the numerical problems are superfluous; both are essential to learning and understanding. The point is that other views are also necessary. It is the overall *mixture* that induces understanding.

The additional ingredients that are helpful, but that are rarely included in exercises provided by the textbooks, involve verbal and qualitative insights. It is effective, for example, to have students look at an actual ripple tank pattern (or a clear picture of one) and confront the question: What is happening at this particular point in the pattern? They must decide, by inspection of the pattern, whether the arbitrarily chosen point is at a node, an antinode, or somewhere in between, and also discern the order of the location. Although this is an exceedingly simple question, it is not trivial, and many students initially have difficulty because they have never been led to examine the pattern from this point of view. The reversal of this question is equally important: Point to a location that lies between a second-order constructive and a second-order destructive interference. Such questions make useful, effective, and easily graded test questions (now that it is easy to reproduce pictures of this kind).

A next level resides in having students make explicit connection between the ripple pattern (usually the first encountered) and analogous acoustic and optical patterns. Texts tend to assume that the connections are obvious, but this is not the case for many students, and it is important to lead them into articulating the similarities and differences. For example, as a student looks at the ripple pattern, one can ask what he or she would have to do to discern a corresponding acoustic pattern: Where would you go? What would you expect to hear at locations that correspond to such and such points in the ripple pattern? How would you map out loci of constructive and destructive interference?

The parallel questions should be raised with optical patterns. Many students fail to connect the ripple pattern (always viewed as a whole from above) with the bright and dark regions of a two-slit optical pattern projected on a screen. They should be led to visualize where they would have to stand and what they would see if the ripple pattern were an optical one in which nothing could be seen from above.

The optical pattern should also be explicitly connected to the ripple pattern. Now that we so glibly show two-slit optical patterns by using lasers, it is easy to exhibit the entire two-dimensional constructive and destructive interference pattern, making it visible from the side by sprinkling chalk dust into the region between the slits and the screen. (Some teachers apparently do this chalk dust demonstration with a grating and imply that the pattern corresponds to the two-source pattern in the ripple tank. It happens that the grating and two-slit patterns are virtually undistinguishable qualitatively in the laser demonstrations. Unfortunately, this is very misleading. The grating effect differs from two-slit interference in a profound and significant way, and a grating pattern in the ripple tank does not look anything like the two-source pattern (see discussion in the following section).

9.14 TWO-SOURCE VERSUS GRATING INTERFERENCE PATTERNS

Mathematical analysis, showing the enormous sharpening of the regions of constructive interference, provides the compelling justification for using gratings instead of double slits in spectrometers. Many students, however, fail to grasp the full force of the mathematical analysis if it is invoked, and, in many courses, the mathematical analysis is inappropriate because the students are not ready for it. It is therefore important to try to develop the difference between these two situations phenomenologically. Unfortunately, modern laser demonstrations tend to conceal the physical difference, and it is advisable to resort to older approaches.

In older texts, students were usually given exercises that involved drawing the Huygens wavelets emerging from each opening when a plane wave is incident on a grating and then identifying the plane wave fronts that are reconstituted in the emerging beam. Such exercises were valuable except that the texts rarely made sufficiently explicit the contrast between the pattern formed by the overlapping circles in the two-source case and the plane wave fronts going off in very sharply defined directions after transmission (or reflection) with a grating. Few modern texts lead students to perform and interpret such drawings, and the physics of the grating effect gets lost through oversight. This is unfortunate because the concept involved is subtle and important.

For this reason, it is well worth leading the students to draw the patterns with compass and straight edge and to describe in their own words the essential differences between the two-source and the grating patterns. It is also well worth setting up a grating pattern in the ripple tank. Although this is a bit tricky, it can be done. One needs to use a low frequency (wavelength of about 2 cm) and a barrier grating with six or eight openings of similar width.

It is necessary to be prepared for what to look for in the ripple tank, since the grating pattern is completely different from the two-source pattern. When the grating pattern is stabilized, one sees only a network pattern, a crisscrossing of straight waves—nothing like the "fingers of a fan" associated with two sources. If one lets the eye sweep with the straight waves in the grating pattern, one begins to see *three* sets of straight waves (it is difficult to get anything beyond first order): one set propagates in the original direction along the principal axis, and two sets go off on either side of the principal axis in the direction of first-order constructive interference. The fact that the constructive interference gives *straight* waves (not curved ones) moving in a sharply defined direction is the qualitative indication of the sharpening produced by the grating. One receives no signal from any direction slightly different from that in which the straight wave is propagating. With ordinary light, the plane waves emerging from the grating must, of course, be focussed by a lens in order to form an image on a screen. (This focussing is performed for us by the lens of the eye when we look through a grating.)

These are some of the physical insights that get lost in many modern demonstrations relying on the laser. This is said not with the intent of deprecating the use of lasers in optics demonstrations: the device has made important demonstrations more visible and easier to perform by many orders of magnitude. It is tremendously valuable. One need only be careful about those issues that might be glossed over or confused.

9.15 YOUNG'S ELUCIDATION OF THE DARK CENTER IN NEWTON'S RINGS

An illustration of how an element of historical knowledge can enhance physics teaching resides in examining Thomas Young's modification of the well known Newton's rings experiment—a modification that lent powerful support to the wave model of light at a time of vigorous controversy over the wave and particle models. The modification, at the same time, exhibits to the students a fine illustration of a crucial experiment.

One troublesome feature of the attempt to explain Newton's rings with a wave model was the fact that the reflection pattern for a film of air between glass boundaries was observed to be *dark* rather than bright at the center—where the film thickness is essentially zero and where the incident and reflected waves would interfere constructively rather than destructively if the phase difference between them were zero.

It is easy now, in retrospect, to assert phase inversion at one interface, but, at the time, this was a substantial puzzle. The entire problem is a subtle one for students new to the study of physics. First, it takes time and thought for them to begin to appreciate the fact that there is indeed a nontrivial problem and that something needs to be explained. Furthermore, they are still very vague as to what is "waving" in the case of light, and, since the connection must be made to mechanical situations by analogy rather than by direct sense experience, the reasoning is subtle and requires thought and discussion for genuine understanding.

Young, who also had no notion of what was "waving," pursued the analogy to mechanical behavior and surmised that the undulations, whatever they might be, kept the same phase or were inverted in phase depending on whether they were reflected from "less dense" or "more dense" interfaces, something like waves at free and fixed boundaries of strings.

To test this hypothesis, he utilized a lens of crown glass having an index of refraction of about 1.5 and a plate of flint glass having an index of about 1.7. Between the glasses he placed, instead of air, a film of sassafras oil, selected because it had an index of refraction falling *between* those of the two glasses. Young predicted that, if the wave analogy was correct, the reflected ring pattern should now have a bright, rather than dark, center. He reasoned that both reflected waves would either retain the same phase or both be inverted in phase (relative to the incident wave) since both reflections took place at interfaces at which the reflecting medium was optically more dense than the incident medium. Young showed that the center of the ring pattern, under these circumstances, is indeed bright, and this lent powerful support to the wave model.

This story provides the students an opportunity to engage in qualitative physical thinking of great importance and in a very rich context, one linking their still shaky grasp of mechanical wave reflections with growing insight into the nature of light. It also provides an excellent illustration of the design of an experiment to test a hypothesis. Although some writers have denied the existence of really "crucial" experiments, it seems to me that this experiment comes very close to satisfying the classical criteria for a crucial experiment.

9.16 SPECULAR VERSUS DIFFUSE REFLECTION

Although some textbooks present clear diagrams and emphasize the distinction between specular and diffuse reflection, many gloss over these concepts and their attendant physical effects much too quickly and casually. Apparently the ideas are deemed too simple to be worth valuable time and space. Yet many students, even when using texts that do not shortcut the concepts, emerge without having absorbed their significance. This is probably due to the fact that, even when the effects are described, qualitative questions about them rarely arise in homework or on tests. Assigned problems tend to concentrate on numerical or geometrical determination of image positions with mirrors and lenses, and students are rarely impelled to visualize the overall array of physical phenomena, especially those having to do with light radiated or reflected from the object.

Probably the most significant gap that develops in this connection is the failure of students to become explicitly conscious of the fact that *all* nonself-luminous objects that we see are seen by diffuse reflection of ambient light and that each point on the object acts as a point source, reflecting light in *all* directions. As a result, very few students are aware of any connection at all between mirrors on the one hand and ordinary objects (nonmirrors) on the other.

Similarly, many fail to recognize that each point on a luminous source is also radiating light in *all* directions. Many diagrams that students see in texts (especially science texts at pre–high school level) are incorrect or, at best, misleading since they are likely to show only special rays in preferred directions without showing that a special selection is made out of an infinite bundle. Since students are rarely, if ever, asked to sketch diagrams in which they themselves show the multiplicity of rays emerging from each point on the object and then select a preferred ray for further tracing, they end up without firm assimilation of the concept.

The consequence of this widely implanted gap is that few students start a ray diagram involving image formation by mirrors or lenses with explicit recognition that each point on the object is an independent point source of a spherical wave front and an infinite bundle of rays. Eventually, many misconceptions regarding images and image formation can be traced back to this gap as at least one of the root causes.

In the initial stages of dealing with ray diagrams, it is advisable to lead students to sketch the spherically divergent bundle of rays from an object point and only then to select the special ray that will be readily traced through the system. Only after the concept has been firmly registered should the clutter of unneeded rays be dispensed with.

Students should then be led to see the difference between a mirror and an ordinary object, that is, the difference between specular and diffuse reflection.

9.17 IMAGES AND IMAGE FORMATION: PLANE MIRRORS

Goldberg and McDermott (1986) have studied student misconceptions regarding images and image formation by plane mirrors. They show that certain misconceptions are held by large numbers of students and that these misconceptions persist through exposure to conventional instruction, which concentrates on quantitative aspects and fails in leading students to deal with, and interpret, the qualitative observations and phenomena.

Goldberg and McDermott report student performance on four tasks. In Task 1, students were asked to put a finger at the location of the image of a vertical rod they saw in a plane mirror. The majority of preinstruction students (65 to 75%) responded correctly by locating the image behind the mirror while 20 to 30% located it on the mirror. (The latter response is found to be common among individuals encountering the question for the first time.) The great majority of postinstruction students (95%) located the image behind the mirror. It is apparent that this aspect of instruction does register fairly firmly.

In Task 2 the student was asked to keep the finger on the image position while considering the following question: "Suppose you were sitting where I (the interviewer) am now, about two feet to your left, and I asked you to put your finger above the image. Would you put your finger at exactly the same place it is now or at a different place?" About 45% of the preinstruction students responded correctly, namely that the image would remain in the same place, while the majority (about 55%) responded that the image position would change. Most of the latter responses were based on the supposition that the image would lie along the line of sight between the viewer and the object (in these interviews, the object was placed closer to the mirror than either of the two viewers.) Among the postinstruction students, 70% answered correctly while 30% still maintained that the position would change. Many of the latter changed their view, however, as the interviewer asked them to draw their own ray diagrams; those who drew correct ray diagrams were able to revise their prediction.

In Task 3 the mirror is kept covered during the interview. The student is seated in a position well beyond the right edge of the mirror, and the vertical rod is also placed beyond the right edge in a position such that the line of sight from the student to the rod intersects the mirror. (No light from the object, however, can be reflected to the student's eye.) The question is then asked "If we uncover the mirror, would you see an image of the rod?" After this question is answered, the interviewer asks "Would I see an image of the rod?" (The interviewer is seated to the left of the student in a position where light from the object would be reflected to his eye.) Among the postinstruction students, 70% gave the correct response (no/yes) while 5% said "yes/no" and 25% said "yes/yes." Virtually all the students who gave the last response used one kind of reasoning to predict what they themselves would see and another to predict what the interviewer would see. They erroneously decided that they would be able to see the image because it would be on the line of sight to the rod. On the other hand, by correctly applying the law of reflection, they concluded that the interviewer would be able to see the image.

These investigations show that incorrect views and interpretations persist beyond conventional class instruction with large numbers of students—even in the relatively simple situation involving plane mirrors. Goldberg and McDermott also conclude that instruction that explicitly raises phenomenological questions such as those used in the interview tasks is markedly effective in improvement and retention of understanding.

My own experience coincides with that reported by Goldberg and McDermott. I have also observed that one of the difficulties behind the off-the-cuff incorrect responses of the students is the gap discussed in Section 9.16: few students explicitly invoke, in thinking about the phenomena involved, the idea that each point on the object sends out rays of light in all directions and that the image in the plane mirror does exactly the same thing for all those rays (from the object) that strike the mirror. Thus they do not, on exposure to conventional instruction about plane mirrors, begin to form a clear conception of what is meant by "virtual image." This concept offers still greater difficulty to many students when it arises in the more subtle contexts of lenses and curved mirrors. Stronger emphasis on the virtual image concept in the case of plane mirrors greatly facilitates later learning. Just invoking the name, however, is not enough; students must be helped to articulate the ideas in their own words after looking into real mirrors and drawing their own diagrams, and they should be led to sketch entire *bundles* of rays, not just isolated special rays.

9.18 IMAGES AND IMAGE FORMATION: THIN CONVERGING LENSES

Goldberg and McDermott (1987) report an investigation, similar to that described in Section 9.17, of student conceptions regarding image formation by converging lenses. All the tasks involved questions concerning events on an optical bench that the interviewer and student viewed from the side. The apparatus on the bench included (1) an object consisting of the luminous horseshoe-shaped filament of an unfrosted light bulb; (2) a converging lens (diameter 7.5 cm; focal length 17 cm); and (3) a translucent screen. The inverted image of the filament was focussed on the screen, and the student viewed the screen from the side facing the lens.

In Task 1 Goldberg and McDermott framed the following question: "If the lens is removed, leaving the object and the screen where they are, would anything change?" The majority of preinstruction students (60%) responded that the image on the screen would become erect. About 40 to 45% of postinstruction students (who had received conventional instruction in geometrical optics) gave the same response. On further investigation, Goldberg and McDermott found that the response in this task was, to some extent, confounded by the sharp nature of the object (the luminous filament); fewer students gave the incorrect response when an obviously diffuse source (a frosted light bulb) was used as the source. The overall results, however, indicate that large numbers of students, even after instruction, fail to recognize the absolute necessity of the lens for image formation. Even though they are fully aware that images of objects in the room do not form on the walls, they do not invoke such everyday experience when viewing the apparatus on the optical bench.

In Task 2 the interviewer holds a piece of opaque cardboard above the lens but does not cover any part of it. The following question is then asked: "Suppose I were to bring this cardboard down and cover the upper half of the lens, leaving the lower half uncovered, would anything change on the screen?" Between 90 and 95% of preinstruction students predicted that half of the image would vanish, and between 55 and 75% of postinstruction students made the same prediction. Only a minority of the latter recognized that the entire image would remain and become less bright. In other words, even after instruction, only a minority of the students recognized that *any* portion of the lens is capable of forming the image.

The misapprehension arising in this context stems largely from the routine way in which students learn to draw ray diagrams. The texts show only the principal rays in order not to clutter up the diagrams. The students never draw anything *but* the principal rays; they never show the divergent bundle of rays emitted from every object point (cf. Section 9.16.) They are never led to visualize what happens to the bundle that passes through any part of the lens regardless of where the principal rays happen to be. When they look at a diagram showing only the principal rays, they see the cardboard as cutting off the principal rays in the upper half of the diagram and conclude that half the image must disappear.

If one presents students with a problem sketched in such a way that the arrow representing the object is taller than the converging lens, many students find it impossible to draw the principal ray parallel to the principal axis because it never intersects the lens shown in the diagram. They fail to recognize that, once the plane of the lens is given, *all* the principal rays can be drawn regardless of whether or not all the principal rays actually pass through the lens. This reflects a gap in understanding closely related to that in the interview task.

In Task 3, starting with the image focussed on the screen, Goldberg and McDermott invoked the question: "Suppose I were to move the screen toward the lens. Would anything change on the screen?" Among the postinstruction students only 35 to 40% recognized that the image would become fuzzy and disappear; the remainder expected the image to persist, changing in size and perhaps becoming somewhat fuzzy. Goldberg and McDermott report the following:

> "Many of the remarks made by the students indicated that the function of the screen was widely misunderstood. Often they did not think of it as a diffuse reflector or transmitter that, when located at a particular position for a given object distance, makes it possible for an observer not looking along the axis of the lens to see the image. Instead they seemed to believe that an image can be seen on a screen no matter where it is placed along the axis. In some cases this claim seemed to be buttressed by a misinterpretation of the experience of watching someone else use a slide projector. The students may have remembered that, in order to make the image larger, the screen had to be moved further from the projector. However, they did not recall that it was necessary simultaneously to refocus the projector by changing the object distance.

Many students fail to recognize the significance of the ray diagram in the sense that the intersection of the principal rays determines a *unique* image position and that an image is therefore not formed in any location other than the image plane. Furthermore, students have never been led to interpret the role of the screen in terms of diffuse reflection, that is, they have not been led to show each image point on the screen as a source of a divergent bundle of rays (again cf. Section 9.16.)

In Task 4, the student, while looking at the image on the screen, is initially asked if he or she would still be able to see the image from his or her present position (on the side of the screen facing the lens) if the screen were removed. Virtually all the students recognized that the screen was necessary for reflecting the image so that it could be seen, although several of the preinstruction students volunteered that they might be able to see the image on the wall several meters beyond the screen.

The students were then asked whether they would be able to see the image if the screen were removed and they were free to take up any other position they wished. Goldberg and McDermott report:

> *Many of the students especially the prestudents, seemed to have difficulty in understanding the question. Their remarks indicated that they could not conceive of an image as existing in free space, independent of a surface. The majority of the students, both pre and post, said either that they would not be able to see an image or that they might be able to see an image if they could place their eye at the screen position. Several of the students who gave the latter response seemed to think of their eye as simply replacing the screen.*
>
> *At this point in the interview, the investigator actually removed the screen and directed the student to move about two meters beyond the initial screen position and to look along the lens axis toward the lens. With this guidance, almost all of the students were able to see the aerial image. Many appeared surprised that they were able to see anything.*
>
> *When the investigator asked for the location of the image, only a very few students were able to state correctly that the image was located at the same position that the screen had been. The rest of the students gave a variety of answers. The prestudents, especially, tended to say that they thought the image was at or in the lens.*

It is my own experience that, if one asks students to give an overall description in their own words of what happens to the light which emanates from the object, is intercepted by the lens, and ends up forming the image on the screen, a substantial number tell a story that effectively boils down to something like "The image travels from the object to the lens; in the lens it is turned upside down; then it travels to the screen."

Again in my own experience, I have found it helpful to require students, when drawing ray diagrams on homework and tests, to write out full verbal descriptions of how each principal ray is drawn (this in connection with both converging and diverging lenses.) Many of the initial versions tend to be gibberish, indicating that the students have not registered the ideas involved and are just following memorized diagrams or procedures without making explicit connection to the definitions of principal foci and rays. Only after the verbal description has been written out correctly and clearly at least once (preferably in connection with a problem that involves image formation by two lenses in sequence) can this requirement be relaxed.

Another very useful technique for tests and homework is to help the students reverse the direction of reasoning. Instead of asking for the image location given an object position, one can give the image and lens locations and ask where the object must have been. (I know of very few texts that ask for diagrams under these circumstances.) Still another version, of course, is to ask for the lens position given object and image.

Generally speaking, very little genuine understanding of images and image formation is registered under homework that consists principally of numerical calculations utilizing either the Gaussian or Newtonian lens equations. This remark is not meant to advocate elimination of the numerical work. The latter is necessary and important, but it needs to be coupled directly, preferably in the same problems, to the phenomenological aspects that are shown to be far more difficult to assimilate than the numerical routines.

It is clear that homework assignments, tests, and *laboratory work* should contain questions and problems of the type used in the tasks described by Goldberg and McDermott. Many variations on these specific tasks could and should be devised. One cannot avoid the conclusion that most conventional instructional routines fail to lead many students to the kind of understanding we would like them to achieve.

9.19 NOVICE CONCEPTIONS REGARDING THE NATURE OF LIGHT

Just as students come to the study of motion with many deeply rooted, commonsense preconceptions, they also come to the study of light with a variety of preconceptions. Many of these preconceptions go unnoticed by teachers and texts and interfere with the development of understanding of the physics. Watts (1985) summarizes some of the notions held by novices. (Not every individual of course, simultaneously holds all the views described; percentages vary for different groups and different age levels.)

Relatively few students hold a conception of light as a physical effect existing apart from its sources and effects. Light illuminates objects so that they are seen, but the act of seeing is not explicitly associated with the arrival of light at the eye of the observer. (I have observed that a few students even reinvent the ancient idea proposed by Parmenides that something emanates to an object from our own eyes to make seeing possible.) Sources of light (such as lamps or candles) are "seen" at a distance but are not thought of as sending out light.

Reflection is something that happens with mirrors but not with a sheet of paper, walls of a room, or other objects. (See Section 9.16.) Color filters are seen as *adding* color to white light.

It is important for a teacher to be aware of the fairly wide incidence of such preconceptions and to help students unsettle them through appropriate questions leading to encounter of contradiction and inconsistency. Rapid assertions of the "correct" view do little good.

9.20 PHENOMENOLOGICAL QUESTIONS AND PROBLEMS

Preceding sections have indicated the importance of giving students qualitative, phenomenological questions about both mechanical waves and optical phenomena in order to strengthen their intuition and build firmer understanding of both the phenomena and the underlying concepts that we invent. One very powerful type of question is the "What will happen if . . . ?" variety that cultivates hypothetico-deductive reasoning. Most of the researches that were cited above used such questions as revealing probes, and the very questions that have been quoted cry out to be incorporated in day to day instruction.

In addition to the illustrations that have been provided in geometrical optics, it is desirable to add similar questions in various aspects of physical optics. For example, in addition to making calculations of wavelength from an interference pattern (say with parallel fringes formed by a wedge-shaped film between glass plates), one might ask students to sketch how the pattern would change if one of the plates were to be displaced parallel to itself, either thickening or thinning the film; sketch how the pattern would change from the one observed if the two plates did not have the same index of refraction; sketch how the pattern would change if the color of the incident light were changed from red to green. Such changes are rarely demonstrated in connection with discussion of patterns formed by thin films, and even if demonstrated, the ideas do not register unless students are asked to sketch the effects for themselves. (This also applies to changes in patterns formed by slits and gratings. Although these are usually demonstrated, many students fail to produce correct sketches of their own unless led to practice. The tendency is to try to memorize what was shown rather than to reason it through to the underlying principles.)

The brightness of an image formed by a lens invites return to ratio reasoning and scaling without substitution in a formula (see discussion in Chapter 1.) The combined effects of focal length and aperture of the lens offer many students, even those in engineering-physics courses, very severe difficulty, and they endeavor to avoid the ratio reasoning (especially that part associated with area) by trying to substitute in formulas without doing the qualitative reasoning as to what makes for a larger, and what for a smaller, effect in what ratio.

CHAPTER 10

Early Modern Physics

10.1 INTRODUCTION

There is an understandable desire in the physics community for earlier introduction of students to aspects of modern physics. In some quarters these dreams of accelerated learning extend to advocacy of injection of the results of quantum mechanics, nuclear physics, and high-energy physics as early as freshman, or even high school, level. In light of what we have been learning in recent years about cognitive development and concept formation, I doubt that genuine learning and understanding of such material is feasible at such early stages. One would only cultivate blind memorization of end results to be used in artificial homework exercises and to be tested for as what Eric Rogers calls "cheap recall."[1] I doubt that it is wise for us to succumb to subject matter pressure as the chemists have done and force our students to memorize end results without understanding.

What seems to me to be feasible and highly desirable in an introductory course is to get to the insights gained in early twentieth-century physics: electrons, photons, nuclei, atomic structure, and (perhaps) the first qualitative aspects of relativity. To achieve this, it is impossible to include all the conventional topics of introductory physics. One must leave gaps, however painful this may seem. How does one decide what it to be left out? One powerful way, in my experience, is to define what I call a "story line." If one wishes, say, to get to the Bohr atom, one should identify the fundamental concepts and subject matter from mechanics, electricity, and magnetism that will make understandable the experiments and reasoning that defined the electron, the atomic nucleus, and the photon. The selected story line would develop the necessary underpinnings and would leave out those topics not essential to understanding the climax. For students continuing in physics, the gaps would have to be recognized, accepted, kept in mind by the

---

[1]At a recent meeting, I heard a high school teacher, winner of a presidential citation, report on how he teaches students modern elementary particle physics. Without examining any of the "How do we know . . . ? Why do we believe . . . ?" questions (which are, of course, far beyond the students at that stage), and merely asserting the end results, he had his students "learn" about quarks and gluons and hadrons and mesons and leptons and color and charm and top and bottom and so on and so forth. He seemed to be convinced he was doing something modern and fascinating and exciting. It seemed to me, however, that the now much vilified and disdained task of memorizing the capitals of the 50 states would generate far more practical and useful knowledge for the victims.

faculty, and closed in subsequent courses. (If the students were given a chance really to learn, understand, and absorb the most basic concepts, they would subsequently close at least some of the *seeming* gaps on their own.) Some efficiency could be gained by putting certain topics (e.g., elementary dc circuits, geometrical optics) entirely in the laboratory and not devoting them appreciable class time. Such topics are far more effectively developed in a "hands on" context in any case (cf. Sections 7.4–7.9 and Sections 9.17 and 9.18.)

If one has carefully thought out the story line to be developed, it is possible to inject, along the way through earlier material, many questions and exercises that prepare the students for thinking and reasoning that will be encountered toward the end. Quite a few textbooks are now attempting to do this, but most of them include so much material that such preparatory exercises get lost in the clutter. Most textbooks that do deal with the early twentieth century developments tend to go through the material so rapidly that much opportunity for physical insight and reasoning is extruded; it is the end results that are dwelt on and not the reasoning that yielded them.

It must be kept in mind that all of what we call "modern physics" deals with levels of insight not directly accessible to our senses. Students need time to absorb and comprehend the inferences drawn from the classical experiments that led to the deep insights we now assert so quickly and casually. Furthermore, the classical experiments and the reasoning they entail provide an exceedingly rich and valuable opportunity for the kind of spiraling back that has been advocated throughout this book. Many students begin to show their first reasonably firm mastery of basic concepts such as velocity, acceleration, force, mass, momentum, energy, centripetal force, electric charge, electric field strength, and magnetic B-field when they synthesize them in rich contexts such as those provided by the Thomson experiment and the Bohr atom.

In the light of the issues outlined above, this chapter will concentrate on the intellectual growth students can achieve in the study of early twentieth-century physics. This happens to be an instance in which at least parts of the historical development (not all the intimate details) are deeply conducive to learning and understanding. Unfortunately, neglect of some of these historical aspects greatly diminishes the effectiveness of many treatments of this area of subject matter.

10.2 HISTORICAL PRELIMINARIES

The insights we usually associate with the term "modern physics" began with the qualitative study of gaseous discharge and cathode rays in the 1870s and 1880s and rose to something of a crescendo with Roentgen's discovery of x-rays in December 1895, Becquerel's discovery of radioactivity in early 1896, and Thomson's experiment in 1896–97.

Replicas of the tubes that Crookes used in his study of cathode rays are available in most physics preparation rooms, and the demonstrations are invariably of great interest to the students. There is an unfortunate tendency, however, for lecturers to rush through the demonstrations, asserting very quickly what each one implies about the cathode beam. The impact of these demonstrations can be greatly enhanced if time is allowed for contemplation, discussion, and inference. As the demonstration is performed, it is more effective to ask the students what is to be inferred from the observations. For one thing, this helps many students who are still in need of exercise in sharpening their discrimination between observation and inference[2]; for another, it allows articulation of alternative explanations and inferences—which should be tolerated and debated rather than dictatorially suppressed. Such discussion provides a very important underpinning for the study of the Thomson experiment since Thomson was motivated to resolve the debate as to whether the cathode beam consisted of particles, as had been conjectured by Crookes, or of some hitherto unknown radiation, as was being argued by Lenard (see Section 10.3). Students inclined to support a radiative model should be, at least temporarily, encouraged to do so; they would be in good company.

Many textbooks give adequate and sufficient, albeit abbreviated, discussions of the discoveries of x-rays and radioactivity. (Although there is much good physics to be learned in considering these stories in greater detail, there are limits to the time one can devote.) A few aspects, important for subsequent study, are, however, insufficiently emphasized, and students tend to lose sight of their significance. One of these aspects is the fact that both x-rays and radioactive emanations were quickly discovered to ionize air, the conductivity being observed and recognized through the discharge of electroscopes. Becquerel, in fact, initially surmised that the rays from uranium were weak x-rays. Thomson was studying x-ray induced conductivity just before undertaking his classic study of the cathode beam, and his awareness of the ionization played a very important role in the cathode beam experiments.

The conceptual importance of the discovery of ionization of gases is underlined by Millikan (1917):

> *... up to this time the only type of ionization known was that observed in solution, and here it is always some compound molecule like sodium chloride which splits up spontaneously into a positively charged sodium ion and a negatively charged chlorine ion. But the ionization produced in gases by x-rays was of a wholly different sort, for it was observable in pure gases like nitrogen and oxygen, or even in monatomic gases like argon and helium.[3] Plainly, then, the neutral atom even of a monatomic substance must possess minute electrical charges as constituents. Here was the first direct evidence (1) that an atom is a complex structure, and (2) that electrical charges enter into its makeup. With this discovery, due directly to the use of the new agency, x-rays, the atom as an ultimate, indivisible thing was gone, and the era of the constituents of the atom began. ...*

A second aspect, either not mentioned at all or too quickly glossed over, is the discovery by the Curies that a vial of radium compounds maintains itself permanently above room temperature and that, when placed in a calorimeter, "each gram of radium gives off 80 calories per hour ... sufficient heat ... to melt its own weight of ice." Thus a serious question was raised, from the very beginning, about the origin of all this energy and the validity of the energy conservation law.

---

[2] Many students are exceedingly weak on such discrimination. Unless guided by questioning, they do not ask themselves what were the facts and evidence on the one hand and the inferences drawn from the facts on the other. They fail to make similar discriminations in other disciplines (history, for example). Yet such discrimination underlies, together with other processes, the intellectual behavior one would characterize as "critical thinking." (See Chapter 13 for more detailed discussion.)

[3] It was, of course, well known that flames rendered gases conducting, but flames involved chemical reactions, introduced new products into an initially pure gas, caused convection currents and extraneous effects due to temperature differences. A steady, controlled, reproducible electrical process could not be achieved under these circumstances.

Furthermore, both α and β radiations were shown to have mass. How could one account for the continuous emission of material particles in the apparent absence of chemical change (the work of Rutherford and Soddy on the transformation of the elements was still to come) or other alteration in the state of the radioactive material? How, in particular, could one account for material emission from *elements* (pure metallic uranium and radium) without interaction with atoms of other substances? Thus the law of conservation of mass was also being called into question.[4]

As part of awareness of their own intellectual history, it is desirable that students face and appreciate these initial questions and that the story that unfolds eventually show how they were explicitly resolved. It is through such experience that scientific literacy is enhanced, not through glossing over of the questions and rapid assertion of names and end results.

A question, which frequently arises when one elects to use elements of the historical sequence in teaching an introduction to modern physics, concerns what was known about Avogadro's Number N_0 and about the sizes of atoms and molecules prior to Millikan's determination of the quantum of electrical charge and the advent of x-ray diffraction. (This is a question I am asked from time to time by my own physics colleagues.) The facts are as follows:

The orders of magnitude of these quantities were firmly established well before the end of the nineteenth century and were used in guiding both experimental work and theoretical analysis, but the values were far from precise. The sources of information were the kinetic theory of gases on the one hand and experimental data on the transport phenomena (viscosity, thermal conductivity, diffusivity) and on departures from ideal gas behavior on the other. The story began with the theoretical foundations laid by Clausius in 1857 and 1858 and by Maxwell in 1860 [see Brush (1965) for translations and reprints]. These works established the mean free path and its connection to the transport coefficients. In modern notation, for example:

$$\lambda = \frac{1}{\sqrt{2}\pi n \sigma^2} \qquad (10.2.1)$$

and

$$\eta = \frac{1}{3} nm\lambda \bar{v} \qquad (10.2.2)$$

where λ denotes the mean free path, n the number of atoms or molecules per unit volume; σ the atomic or molecular diameter (assuming spherical shape); η the coefficient of viscosity; m the mass of one atom or molecule; and $\bar{v}$ the mean atomic or molecular velocity. From the Maxwellian distribution, $\bar{v}$ is given by

$$\bar{v} = \sqrt{\frac{8RT}{\pi M}} \qquad (10.2.3)$$

---

[4] It might be noted that it was during this period of convulsion in physical science that Henry Adams made (in *The Education*) his oft quoted remark "Chaos is the law of Nature; order is the dream of Man."

where R is the universal gas constant; T the absolute temperature; and M the relative atomic or molecular mass. In this notation

$$m = \frac{M}{N_0} \quad (10.2.4)$$

Combining Eqs. 10.2.1 and 10.2.2 gives

$$\eta = \frac{m\bar{v}}{3\sqrt{2}\pi\sigma^2} \quad (10.2.5)$$

Equation 10.2.5 implicitly relates the experimentally measurable quantity η to the two unknowns N_0 and σ. It also contains the prediction, since n has dropped out, that the viscosity coefficient of an ideal gas is independent of the pressure—an intuitively unanticipated behavior that helped provide powerful reinforcement for the newborn theory.

In 1865, Loschmidt made what appears to be the first calculation of molecular size by taking the intrinsic volume excluded by the molecules in the gas to be equal to the volume occupied by the substance in the solid or liquid state, that is, the very low compressibility of liquids and solids justifies the assumption that the atoms or molecules are exceedingly close together in these states. If ρ denotes the density of the solid or liquid, the volume of one mole of molecules M/ρ is given by

$$\frac{M}{\rho} = \frac{1}{6}N_0\pi\sigma^3 \quad (10.2.6)$$

Combining Eqs. 10.2.5 and 10.2.6, Loschmidt obtained a value of molecular diameter. He could readily have obtained the value of n, which came to called the "Loschmidt Number," but did not actually do so. One can also calculate what came to be called Avogadro's Number, N_0.

After van der Waals, in 1873, put forth his modified equation of state for departure from ideal gas behavior,

$$(p + \frac{a}{v^2})(v - b) = RT$$

recognition that the constant b must be approximately equal to four times the volume excluded by one mole of molecules in the gas phase made possible an improved calculation based on modification of Eq. 10.2.6:

$$\frac{b}{4} = \frac{1}{6}N_0\pi\sigma^3 \quad (10.2.7)$$

If one takes modern values for nitrogen, for example, of $\eta = 178$ micropoise at 27°C and $b = 0.03913$ liters per mole (l/mol), one obtains, from the van der Waals approach, combining Eqs. 10.2.5 and 10.2.7: $\sigma = 3.1$ Å and $N_0 = 5.1 \times 10^{23}$.

The experimental values were considerably less accurate in the nineteenth century, but the preceding calculation illustrates that the orders of magnitude were right and provided reliable guidance. That the estimates were deemed important is indicated by the fact that figures such as Stoney, Lothar Meyer, and Kelvin participated in their development. More accurate determinations did not come until those of Perrin (in 1908), based on Einstein's 1905 paper on Brownian motion, combined with experimental observation of gravitational stratification and Brownian motion of particles in colloidal suspension. Further refinement of the values of N_0 awaited Millikan's (1909, 1911) determination of the corpuscle of charge and the advent of x-ray diffraction.

Unfortunately, this story does not lend itself to use in an introductory physics course, the kinetic theory base being far beyond what is realistic at that level. I include the story here, not to advocate its use in teaching, but because it might enhance the perspective of others, as it did my own when I first explored it. One can tell students about it qualitatively if one wishes to do so and ask them to take the assertions on faith. (Although I see strong objections to asking students to take assertions on faith in early portions of the course, when, because of past inexperience, they only feebly discriminate what is fully substantiated and what is not, I see no objection to doing so occasionally after their ability to discriminate has been strengthened.)

10.3 PRELUDE TO THOMSON'S RESEARCH

Prior to the time at which Thomson embarked on his investigation of the cathode beam, two divergent views had evolved regarding its nature: British scientists adhered to the particle model advocated by Crookes, while Continental Europeans, led by Philipp Lenard (then at Bonn and later at Heidelberg), preferred a wave or radiation model.

The latter view was by no means naive. Lenard had conducted numerous careful experiments on the transmission of cathode rays through very thin metal foil "windows" in the end of the cathode ray tube and on the penetration of the rays through different gases after passing through the foil. Finding that the rays penetrated the foil and continued on for another centimeter or two, still in straight lines, Lenard became convinced that the cathode rays could not be corpuscular or material in character, but must be wave disturbances in the ether. He could not conceive material charged particles penetrating a substance as dense as the foil without deflection, and, although still numerically crude, kinetic theory was far enough advanced to make it convincing that an atomic or molecular beam would not penetrate so far in air at atmospheric pressure.

The opposing points of view of Lenard and Crookes exemplify the sharp distinction between corpuscular and wave phenomena that had emerged in nineteenth-century scientific thought. It was believed that these two manifestations were mutually, absolutely exclusive, that any given phenomenon must be either of one class or the other, that no manifestation could exhibit both corpuscular *and* wavelike aspects. This complete dichotomy is something well worth leading students to think about and examine as a prelude to subsequent introduction of modern views of wave–particle duality. It sets in perspective a significant episode in intellectual history.

Another experimental fact that, at the time, stood in the way of the corpuscular hypothesis was failure to achieve electrostatic deflection of the cathode beam by passing it between capacitor plates built into the tube. It was well known that a magnetic field caused deflection of the beam in the direction that would be expected of moving negatively charged particles, but attempts to produce electrostatic deflection yielded null results. (This was one of the first obstacles that Thomson proceeded to resolve.)

This difficulty led to some questioning of a result that had been obtained by Perrin. Perrin had inserted an electrometer cup at the end of the tube opposite the cathode and had collected negative charge, as exhibited by the electroscope to which the cup was connected. The question was raised as to whether the charge being thus collected was indissolubly connected with the cathode beam or was an independent manifestation.

It was at this juncture, late in 1896, that Thomson embarked on the famous experiments that led to the determination of the charge-to-mass ratio in the cathode beam.

10.4 THOMSON'S EXPERIMENTS

Thomson (1897) set out to resolve the argument concerning the nature of cathode rays. Referring to the conflicting corpuscular and wave hypotheses, he revealed some of the factors that moulded his thought:

The electrified particle theory has, for purposes of research, a great advantage over the aetherial theory, since it is definite and its consequences can be predicted; with the aetherial theory it is impossible to predict what will happen under any given circumstances, as on this theory we are dealing with hitherto unobserved phenomena in the aether, of whose laws we are ignorant. The following experiments were made to test some of the consequences of the electrified particle theory.

Thomson first repeated Perrin's electrometer cup experiment (see Section 10.3), but, instead of placing the cup at the end of the tube opposite the cathode, he sealed it into the side of the tube where it did not directly receive the undeflected cathode beam. When the tube was turned on, the electrometer showed no charge, but when the beam was deflected magnetically so that it entered the cup, the electrometer collected negative charge. Thomson writes:

This experiment shows that however we twist and deflect the cathode rays by magnetic forces, the negative electrification follows the same path as the rays, and that this negative electrification is indissolubly connected with the cathode rays.

Although the logic may seem obvious to us, it turns out that quite a few students have difficulty seeing why Thomson took the trouble to perform this experiment even though he knew of Perrin's results, and they have difficulty articulating its significance. Few have had the opportunity to think through such a sequence in previous study; this is not the nature of conventional end-of-chapter exercises.

Thomson then attacked the problem of electrostatic deflection:

An objection very generally urged against the view that the cathode rays are negatively electrified particles is that hitherto no deflexion of the rays has been observed under a small electrostatic force. . . . Hertz made the rays travel between two parallel plates of metal placed inside the discharge tube, but found that they were not deflected when the plates were connected with a battery of storage cells; on repeating this experiment I at first got the same result, but subsequent experiments showed that the absence of deflexion is due to the conductivity conferred on the rarified gas by the cathode rays. On measuring this conductivity it was found that it diminished very rapidly as the exhaustion increased; it seemed then that on trying Hertz's experiment at very high exhaustions there might be a chance of detecting the deflexion of the cathode rays by an electrostatic force.

As a result of his previous year and a half of experimenting and thinking about conductivity in gases induced by x-rays, Thomson was very sensitive to the possible role of this phenomenon. He realized that cathode rays as well as x-rays induce conductivity, and he was well prepared to visualize the consequences. As it happened, newly developed vacuum techniques made it possible for him to test his ideas by achieving sufficiently high vacuum to suppress the conductivity:

At high exhaustions the rays were deflected when the two aluminium plates were connected with a battery of small storage cells; the rays were depressed when the upper cell was connected with the negative pole of the battery, the lower with the positive, and raised [when the connections were reversed]. The deflexion was proportional to the difference of potential between the plates and I could detect the deflexion when the potential difference was as small as two volts.

It was only when the vacuum was a good one that the deflexion took place, but that the absence of deflexion is due to the conductivity of the medium is shown by what takes place when the vacuum has just arrived at the stage at which the deflexion begins. At this stage there is a deflexion of the rays when the plates are first connected with the terminals of the battery, but if this connection is maintained the patch of fluorescence gradually creeps back to its undeflected position. This is just what would happen if the space between the plates were a conductor, though a very bad one, for then the positive and negative ions between the plates would slowly diffuse until the positive plate became coated with negative ions, the negative plate with positive ones; thus the electric intensity between the plates would vanish and the cathode rays be free from electrostatic force. . . .

As the cathode rays carry a charge of negative electricity, are deflected by an electrostatic force as if they were negatively electrified, and are acted on by a magnetic force in just the way in which this force would act on a negatively electrified body moving along the path of these rays, I can see no escape from the conclusion that they are charges of electricity carried by particles of matter.

Now that vacuum tubes are things of the past, students no longer encounter the concept of space charge or have occasion to visualize the behavior of ions in a rarified gas. I have, on a qualifying examination, given graduate students all the information about the initial failure to achieve electrostatic deflection of the cathode beam, told them that Thomson achieved deflection on sufficient improvement of the vacuum, hinted that the difficulty had to do with ionization of the residual gas, and asked for an explanation, with rough pictures of what had been happening. Very few graduate students were able to give a reasonably competent answer. I suggest that qualitative physical thinking of this kind should *not* be eliminated from introductory physics. A rich context, such as the research being described, is an invaluable opportunity to give such thinking and visualization substance and meaning.

Thomson then went on to conduct the measurements that are very cursorily described in most texts: determining the deflection on application of a single field, eliminating the unknown velocity of the particles by restoring the beam to initial position through application of crossed electric and magnetic fields, and evaluating the charge-to-mass ratio of the hypothetical particles.

Much physics is left out of the majority of text presentations. For example, students are not asked to consider the significance of the fact that the beam remains coherent (the spot on the screen does not smear out) when it is deflected, either electrically or magnetically, from the intial position; it does not occur to them that there is physical information in what does *not* happen as well as in what does, and it is necessary to lead them to think about such matters. The coherence, of course, supports the hypothesis that the entities in the beam are identical in their properties and in their velocity and that the equation for the trajectory of one particle applies to all.[5]

Most text presentations simply eliminate the velocity of the particles from the two available equations, as though it were of no interest, and solve for the charge-to-mass ratio. This not what Thomson does. He calculates the numerical value of the velocity, shows it to be of the order of one tenth the velocity of light, and argues this to be strong evidence against Lenard's hypothesis of electromagnetic disturbance in the ether. This is physics that learners should be led to confront.

An aspect that Thomson does *not* dwell on, since it is trivial to a physicist, is, however, not trivial for students. I have pointed out to students that there is no observable *gravitational* deflection of the cathode beam whereas there is large electric and magnetic deflection, and asked how they account for the unobservability of the gravitational deflection. Many students (even graduate students on qualifying examinations) respond that the gravitational deflection is so small because the mass of the particles is so small. They do not immediately invoke what they supposedly learned about the uniformity of g for all masses, and they do not associate the smallness of the drop with the enormous velocity. This opportunity to spiral back to earlier macroscopic concepts in an entirely new microscopic context is invaluable for assisting in mastery of the earlier concepts.

Using the method of the crossed electric and magnetic fields, Thomson then went on to determine the charge-to-mass ratio (actually he gives the mass-to-charge ratio in the old cgs electrostatic units) in tubes with electrodes of different metals (aluminum, platinum, iron) and with different residual gases (air, hydrogen, carbon dioxide). His results (quite inaccurate by modern standards) fell into a relatively narrow range—given the large room for error and uncertainty in this early undertaking.

10.5 THOMSON'S INFERENCES

In his paper, Thomson remarks on certain systematic errors that he believed made his values of charge-to-mass ratio somewhat low. At this juncture, however, he was not striving for high accuracy; rather, he was interested in orders of magnitude, and he was trying to establish whether or not the charge-to-mass ratio associated with the cathode rays varied over a large range, as it was known to do with different ions in electrolysis and in conducting gases.

The results of all his different measurements fell (converted to modern units) between 0.67 and 0.9×10^{11} C/kg. This being a very much narrower range than that observed for different ions in electrolysis, and also being within the range of uncertainty of his experimental measurements, Thomson was led to conclude that the negatively charged particles have the same charge-to-mass ratio in all cathode

---

[5]It is worth noting that a beam of *positive* ions, formed, by acceleration through suitable electrodes, out of a region of ionized gas, *does* smear out when deflected magnetically since the beam is not homogeneous in velocity of the ions. This is why a velocity selector is necessary in a mass spectrometer. I find many graduate students completely oblivious to matters of this kind since they have never had a chance to think about the phenomena. Surely the groundwork should be laid in the introductory courses.

beams regardless of electrode material and ambient gas. He also conjectured that, when it would become possible to determine the two properties separately, it would very likely be found that the particles all have the same charge and the same mass. On this account he denoted the mass of the particles by m and the charge by the special symbol e (rather than a more conventional symbol for an arbitrary quantity of charge), and the symbol e/m is the one used to this day.

Furthermore, the additional homogeneity of the beam, reflected in the facts (1) that the spot does not smear out under electrostatic deflection (showing that, since e/mv_x^2 must be the same for all particles, they have all fallen through the same accelerating potential difference) and (2) that the velocities of the particles are all the same (as indicated by crossed field determination), implied that the particles must all originate in the immediate neighborhood of the cathode—either by ejection from the cathode or through formation by ionization of gas at the cathode. (We now recognize, of course, that the particles emanate from the cathode through the process called "field emission.")

To establish the physical significance of the observed charge-to-mass ratio, Thomson appealed, for comparison, directly to the well-established electrolytic data. For hydrogen ions, for example

$$\frac{q_H}{m_H} = \frac{96,500}{0.0010} = 9.6 \times 10^7 C/kg$$

whereas the value for oxygen is 1.2×10^7 C/kg. The charge-to-mass ratio of ions produced by irradiation of gases was, at that time, much less precisely established but was known to be of the same order of magnitude.

Although this may sound trivial to some teachers, many students have not acquired, in their course of study, explicit realization of the fact that numerical magnitudes convey almost no information when standing alone and that information and inference come primarily from *comparison* of magnitudes. The present context forms a vivid and valuable illustration. Concerning the charge-to-mass ratios, Thomson remarked:

> *Thus for the carriers of electricity in the cathode rays [e/m is very large] compared to its value in electrolysis. The [size of e/m] may be due to the smallness of m or the largeness of e, or to a combination of these two. That the carriers of the charge in cathode rays are small compared with ordinary molecules is shown, I think, by Lenard's result as to the rate at which the brightness of the [fluorescence] produced by these rays diminishes with the length of path traveled by the ray.*

As was pointed out in Section 10.2, the orders of magnitude of mean free path and of atomic–molecular size were well established at this time, and Thomson was making use of this information in his interpretation of the data. Since mean free paths were known to be of the order of 1000 Å at atmospheric pressure, and since, therefore, an atomic beam would have been scattered beyond detectability after traveling only ten times that distance, Thomson interpreted Lenard's data, showing penetration of the order of a centimeter or two, as indicating the smallness of the mass of particles in the cathode beam. This smallness implied, of course, the existence of a subatomic entity.

Students should be led to think about why one would *not* be inclined to ascribe the difference between the e/m values in the cathode beam and in electrolysis to a difference in e rather than to a difference in m, or even, perhaps, to differences in both properties. (Thomson, in the quotation given above, acknowledges all the possibilities but rejects the latter two without much discussion.) This is, again, a kind of thinking that students have rarely had the opportunity to engage. Everything has been presented to them in the form of polished and seemingly inevitable end results. When they begin to see the role played by faith in simplicity and order in nature, by the plausible reasoning and inductive guesswork based on such faith, they advance to a more realistic grasp of the nature of scientific thought than the one most of them hold in the absence of this intellectual experience. Here is one of the elements of scientific literacy.

The coda to this story is provided by a few episodes that followed in quick succession. Zeeman, working under H. A. Lorentz at the same time Thomson was performing the preceding experiments, discovered the splitting of spectral lines in a magnetic field. Lorentz, with his deep grasp of Maxwell's electromagnetic theory, guided Zeeman to the classical interpretation, and they estimated the e/m associated with the line splitting to be of the order of 10^{11} C/kg. This indicated entities with this charge-to-mass ratio to be bound within the structure of atoms and not just materializing in a free state in the cathode beam.

In 1899 Thomson published a paper in which he showed that the electrical charge ejected from metals on incidence of ultraviolet light (the photoelectric effect) was associated with entities having the same e/m as the cathode rays. His son, G. P. Thomson, remarked in a lecture in 1956:

> He also showed in the same paper that the negative particles emitted from a hot wire had approximately the same e/m. This really completed the proof. Opposition to the idea of particles smaller than atoms did indeed continue, but it was merely the spasmodic dying kicks of the older physics, a matter of muscular contraction rather than brain.

(The published version of the talk [Thomson, G. P. (1956)] gives more detail and uses somewhat different words.)

10.6 HOMEWORK ASSIGNMENT ON THE THOMSON EXPERIMENT

If, before starting study of the Thomson experiment, one asks students to write a one-page note on what they see to be the meaning of the word "electron" and how it is that we come to know about such entities, the resulting documents form a sobering study in their own right. The great majority of students pour forth what amounts to pure gibberish. They have all heard the term from early days of schooling, but the term never acquired anything like the meaning it has for science. There is only an imprecise half-remembered jargon, the residue of names and end results implanted without understanding. Only a very few students have the security and self-confidence to say that they had never understood the jargon and that they have no idea of what the term really means or how knowledge of such an entity originates.

The appendix at the end of this chapter details a written homework assignment based on a Socratic sequence that I have used for many years in connection with the previously outlined study of the Thomson experiment. It leads the students to address the questions of "How do we know about . . . ? Why do we believe in . . . ? What is the evidence for . . . ?" electrons. It also takes full advantage of the opportunity to spiral back to the use of physics concepts developed earlier in the course. As has been pointed out repeatedly, it is such spiraling back in increasingly rich context that helps in attaining mastery of the basic material, a mastery attained by only a very few students on the first encounter.

The assignment, although it follows Thomson's thought and exposition quite faithfully, departs from Thomson's paper in certain details. Thomson does not enlarge on the negligible gravitational effect on the cathode beam, and he shortcuts the derivation of the expression for the deflection of the beam under the influence of a single field with idealizations and approximations that students would find difficult to follow. The Socratic sequence has been structured to evoke explicit consideration of elements that Thomson legitimately assumed his professional readers did not need.

My experience with this assignment has been quite favorable. Some students initially consider it an unwanted burden (they have become accustomed to solutions of end-of-chapter problems without any verbalization whatsoever), but the great majority, as they find themselves progressing and grasping the synthesis, reflect on it as a very valuable and helpful learning experience. For many, the most striking aspect is the interconnection they naively find among the concepts studied throughout the course, especially the ones studied much earlier and then put aside without further use. This naiveté is, in itself, a measure of the importance of such spiraling back, especially when it can be placed in a sufficiently simple, understandable, yet conceptually powerful and important, context.

10.7 THE CORPUSCLE OF ELECTRICAL CHARGE

Many nineteenth-century scientists saw Faraday's law of electrolysis as implying the atomicity of electrical charge. A typical statement is that of Helmholtz in a Faraday lecture at the Royal Institution in 1881:

> *Now the most startling result of Faraday's law [of electrolysis] is perhaps this: if we accept the hypothesis that the elementary substances are composed of atoms, we cannot avoid concluding that electricity also, positive as well as negative, is divided into elementary portions which behave like atoms of electricity.*

Since the order of magnitude of Avogadro's number was known (Section 10.2), it was clear that the order of magnitude of the elementary charge was 10^{-19} C, obtained by dividing the Faraday (10^5 C) by Avogadro's number (10^{24}). Attempts at direct measurement were made by Townsend through study of total charge carried on clouds of water droplets in gases irradiated with x-rays. Thomson refined Townsend's technique and reported values between 1.8 and 2.8 $\times 10^{-19}$ C. As is well known, the definitive measurements were achieved with Millikan's oil drop experiment, the results of which were first published in 1911, with refinements continuing for a number of years.

The Millikan experiment is adequately treated in many textbooks, and there is no need of detailed elaboration here. The principal element lacking in many presentations, however, is the opportunity for the student to see some actual data exhibiting discreteness through the experimental scatter. There is also significance in the identification of the least common multiple among observed quantities of charge and changes of charge. [Holton (1978) gives a pedagogically useful analysis of Millikan's approach to his data.]

One verbal aspect deserves care and attention: It has become conventional in many texts and lectures to announce that "Millikan measured the charge on the electron." The term "electron" has had a long and complex history, and there is little to be gained in exploring this history in an introductory course. It is true that, in early discourse, the term was used in reference to a corpuscle of electrical charge regardless of its carrier, but, in modern terminology (which crystallized in the early 1900s), the word "electron" denotes the particle in the cathode beam and in the structure of atoms. This is the way students understand the term. Thus, given the modern terminology, saying that "Millikan measured the charge on the electron" becomes profoundly misleading; students get the impression that Millikan dealt directly with electrons.

What Millikan did was measure the size of the elementary charge as it was to be observed, accreted on oil droplets, in an ionized gas. Although some of the ions might have been electrons, most were probably not. The connection to the cathode ray particles was, of course, immediate, but it must be remembered that this was a matter of guesswork and plausible reasoning, based on faith in simplicity in nature (how would nature manage to maintain perfect electrical neutrality if the elementary charges were not all identical?) and not on direct work with electrons as such. This is a beautiful illustration of how real scientific thinking works and progresses—another aspect of enhancing scientific literacy. The opportunity is lost, however, in the misstatement that "Millikan measured the charge on the electron."

10.8 FROM THOMSON'S ELECTRON TO THE BOHR ATOM

If one wishes to bring the modern physics story at least to the point of the early quantum picture of atomic structure, there are certain ingredients—in addition to the two discussed above—necessary to the generation of a sequence in which the student can see coherence, plausibility, and intelligibility rather than just disconnected assertions of end results. Here it is not possible to follow the historical sequence in rigorous detail. The time demanded would be excessive, and much of the conceptual and mathematical material lies at a level far beyond that of an introductory course. It is quite possible, however, to form a sequence that addresses the "How do we know . . . ? Why do we believe . . . ?" questions by using appropriate *segments* of the historical development; that maintains plausibility and continuity; that capitalizes on opportunities to spiral back to earlier concepts; and that makes comprehensible the necessity of the departures from classical theory. [As one example of such a sequence based principally on historical material, see Arons (1965). There are many other presentations that depart from the historical base completely but still provide a coherent and intelligible development, e.g., *PSSC Physics*.]

Among the blocks of subject matter essential for such a development are: (1) Bright line spectra of gases; their assumed connection to absorption and emission of light by accelerated charged entities on the microscopic level; and the empirically obtained Balmer–Rydberg formulae for the spectral series of atomic hydrogen. (2) Nature and properties of radioactive emanations. (3) Determination of atomic dimensions and emergence of the nuclear model. (4) The photoelectric effect and the photon concept. (Broader insight and synthesis can, of course, be achieved if items such as the role of radioactive decay in transmutation of elements, positive rays and isotopes, and properties of x-rays were to be included. There is not likely to be time for such inclusion, however, in an introductory course, and the omissions do not destroy the coherence of the story line that has been selected.)

The photoelectric effect will be discussed separately in the following section (Section 10.9). The remaining, essential items listed above are presented well in many sources and will not be discussed in detail here except for the following few peripheral comments.

1. *In connection with bright line spectra*: It is very effective to point out the parallelism between the role of Kepler's empirical laws in the evolution of the theory of gravitation and the role of the empirical Balmer–Rydberg formulae in the evolution of the theory of atomic structure. Many students do not yet have a clear idea of what is meant by "empirical," and these two episodes set the term in clear contrast with "theoretical." Furthermore, these episodes illustrate how modern science sometimes advances through combination of Baconian empiricism with the formation of new concepts and theories and how the two modes fruitfully interact with each other. Here is still another step toward scientific literacy.

2. *In connection with the nature and properties of radioactive emanations*: Since α-particles were crucial to development of the nuclear model of atomic structure, it is important to set them in an intelligible perspective. Rutherford (and others), showed, through measurement of electric and magnetic deflection, that β-rays had the same e/m as the particles in the cathode rays, but Rutherford at first thought both α- and γ-rays to be "undeviable." Then, becoming suspicious of this conclusion for alphas, he turned to a power company for the use of a very much more powerful magnet than was previously available to him and succeeded in deviating the α-rays. After machining better pole pieces for the magnet, he measured both q/m and the velocities of the α-particles from radium and from "radium emanation" (radon) by electric and magnetic deflection. The observations were repeated later with greater accuracy [see Rutherford (1903) and (1906)]. The velocities were observed to lie between 2 and 3×10^7 m/s. Since the observed value of q/m applies to either singly ionized hydrogen molecules or doubly ionized helium atoms, Rutherford faced the problem of making positive identification.

The Curies had reported the heat generated in radium to be about 80 calories per gram per hour (cal/g/h). Rutherford had determined (by observation of scintillations on a fluorescent screen) the number of alphas emitted per unit mass of radium per unit time, and he had the velocities from the electric and magnetic deflections. From these data, he could estimate the total kinetic energy of the alphas on either assumption as to their identity, kinetic energy being twice as great if they were helium atoms than if they were hydrogen molecules. The data were crude, but agreement with the Curies' value was better if the particles were assumed to be helium.

Ramsey and Soddy had previously noted that, as radon gas decays in a sealed tube, the spectrum of helium in the residual gases becomes more intense. This led Rutherford to a definitive experiment [Rutherford and Royds (1909)]. A schematic diagram of the apparatus is shown in Fig. 10.8.1.

Radon gas, obtained by pumping it away as it emanates from a radium compound, is placed in the inner, extremely thin-walled (0.001-cm) glass tube. This tube is sealed within a larger heavy-walled evacuated tube. The α-particles from the decaying radon have enough energy to penetrate the thin-walled tube but are trapped in the thick-walled outer container over a seal of mercury. After a sufficient period of time (about a week), enough gas accumulates in the outer container to allow a spectroscopic test. Additional mercury is let in to compress the small amount of gas into the capillary, which is fitted with electrodes. Electrical discharge between the electrodes causes emission of a line spectrum from the trapped gas. Rutherford and Royds reported positive identification of the helium line spectrum and thereby settled the question as to the identity of α-particles.

(This summary is included here because this beautiful experiment is not usually described in introductory texts, yet it answers a very basic "How do we

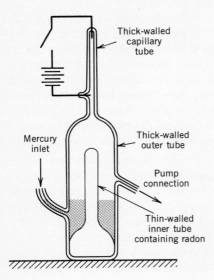

Figure 10.8.1 Schematic diagram of the Rutherford and Royds apparatus for showing spectroscopically that helium gas is formed from α-particles.

know . . . ?" question, exhibits fine, readily understandable, experimental technique, and offers an opportunity for simple, qualitative physical reasoning rarely available to the students.)

A second important aspect of dealing with radioactivity resides in exposure of the students to the arithmetic of exponential decay. Since the arithmetic is similar, this is also a valuable opportunity to make the connection with exponential growth; otherwise few students become explicitly aware of the intimate connection between the two processes.

If one is dealing with students at the calculus-physics level and has access to the exponential function, the discussion can be carried out in such terms. In this context, it is desirable to refer to the basic differential equation, and to point explicitly to other physical situations the students may have encountered (or will encounter) in which the same differential equation applies (e.g., capacitive or inductive circuit elements, decay of gas pressure due to leakage through a small hole in a container, pressure variation with height in an isothermal atmosphere, monomolecular chemical reaction, etc.).

It is not essential, however, to be able to deal with the differential equation and the exponential function. The same insights can be gained through the arithmetic of half-life or doubling time. (In fact, even the calculus-physics students need exercise in this arithmetic just as much as the noncalculus students.) Since such exercises involve ratio reasoning, many students have serious difficulty. The difficulty is greatly reduced for these students (and their security with ratio reasoning is enhanced) if they are shown how to use a simple, concrete graphical aid such as that sketched in Fig. 10.8.2 to keep track of the effect of passage of an integral number of half-lives or doubling times.

Such coupling of growth and decay calculations offers an excellent opportunity to acquaint students with Bartlett's (1976–1979) powerful series of articles on the dangers inherent in unrestrained exponential growth.

3. *In connection with atomic dimensions and the nuclear model*: As soon as the size of the corpuscle of electrical charge e is available from the Millikan experiment, students can be led to calculate Avogadro's number N_0 from the Faraday of electricity: $N_0 = 96{,}500/e$. Having Avogadro's number, they can be led to calculate

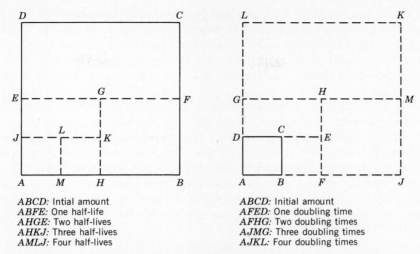

Figure 10.8.2 Keeping track of growth and decay ratios for integral numbers of half-lives or doubling times.

the order of magnitude of atomic–molecular size by turning to the basic assumption first put to such use by Loschmidt (see Section 10.2), namely that the very low compressibility of solids and liquids (compared to that of gases) implies that the constituent particles are essentially contiguous to each other. One obtains a very reasonable order of magnitude, 3 Å for example, if one calculates the volume of one water molecule in liquid water ($18/N_0$) and takes the effective molecular diameter to be the side of a cube having this volume. (Students are usually handed the end results of these simple calculations but are rarely led to carry the calculations through themselves, including the rounding off to one significant figure rather than listing all eight or ten figures emerging on the display of the hand calculator.)

It is important to keep emphasizing to the students that numerical values have little or no meaning standing by themselves. Meaning is generated through *comparison* with other values. A valuable homework question (rarely found in textbooks) would lead students to make such comparisons after completing the above calculations: How does atomic–molecular size compare with wavelengths of ultrasound, visible light, and x-rays; with the smallest distance resolvable in a good microscope; with the "smoothness" of an "optically smooth" surface; with the average separation of molecules in a gas at atmospheric pressure?

The latter question offers an excellent opportunity to revive geometrical scaling in an intellectually significant context: since the density of gases is roughly several thousand times lower than that of liquids and solids, the average spacing in gases must be the cube root of this ratio and therefore of the order of tens or hundreds of times atomic dimensions. Such crude but significant order of magnitude estimates are a powerful antidote to students who have imbibed the (essentially poisonous) idea that "science is exact." Furthermore, such experiences convey to students the insight that "estimating" in science is not wild, unsubstantiated guesswork (as most of them have come to think) but is based on careful, imaginative reasoning, however crude the numerical data may be.

The next coordinated step involves formation of the nuclear model and establishing the order of magnitude of nuclear size. The story of the backward scattering of α-particles, as observed by Geiger and Marsden and interpreted by Rutherford to imply the presence of minute but massive scattering centers, is detailed in one way or another in many texts. In relatively few texts, however, are students given the opportunity to think through the rich phenomenology that is involved and to make the simple calculation that sets an upper bound to nuclear size.

First, it is necessary to get students to go back to laboratory (or demonstration) experience with macroscopic collisions and explicitly recapture the insights: (1) that only forward "scattering" occurs when a more massive object collides with a less massive, stationary one, and (2) that bouncing back ("large-angle scattering") occurs only when the incident object is less massive than the target. These effects may seem obvious to the teacher, but many students never really registered them on first encounter, or have lost sight of them in the interim, and do not spontaneously invoke these ideas in connection with the microscopic phenomena now under consideration.

A second element requires consideration of the fact that, if one tentatively adopts a nuclear model to account for the bouncing back of an α-particle, the positively charged projectile must pass through a "cloud" of negative charge in order to approach the nucleus. Rutherford, in his first paper on this interpretation, makes the full mathematical calculation for an assumed spherical distribution, but this is not really essential for the students. It is sufficient to invoke the idea (that should have been developed in connection with the inverse square laws of both gravitation and electrostatics) that the field is zero anywhere within a uniform spherical shell and that the effect of the negatively charged cloud would therefore decrease very rapidly as the α-particle penetrated it.

The simple numerical calculation the students can then make is that of the closest approach of the α-particle to the center of the target. Given the mass of the α and the velocity of the order of 2×10^7 m/s, one has the kinetic energy of the projectile, the distance of closest approach being determined by the radial separation from the target at which all this kinetic energy is stored as potential energy within the target–projectile system. (In the Geiger–Marsden observations, the target was a gold film, and Rutherford took the number of elementary positive charges on the gold nucleus to be about $100e$, or about half the relative atomic mass of gold. At this time the meaning of the atomic number in the periodic table had not yet been appreciated.) On this calculation, the distance of closest approach is about 3×10^{-4} Å —about 1/10,000 the atomic–molecular dimension. (It is important to lead the students to articulate the perception that this calculation gives an *upper bound* to nuclear size, not the nuclear size itself. Again we have an illustration of the making of a profoundly important *estimate* rather than obtaining an "exact" result.)

The preceding calculation should be tied directly to the students' recapitulation of calculating, given the initial kinetic energy, how high a stone goes when thrown vertically upward, and of calculating escape velocity from the earth (or moon or planet) if the escape velocity idea has been previously invoked. The calculations should be accompanied by descriptions, in the students' own words, of the energy transformations taking place (as previously suggested in Sections 5.3). It is such spiraling back to earlier problems, and making explicit connection among seemingly completely unrelated situations, that help the students master the concepts and reasoning—an effect not achieved, for the majority of students, by didactic exposition, however lucid.

10.9 THE PHOTOELECTRIC EFFECT AND THE PHOTON CONCEPT

It is not possible to tell the story of black-body radiation in an intellectually honest and meaningful way in an introductory course, and hand waving about it only leaves the students mystified. The clear and intelligible way to the quantum concept is through the photoelectric effect, and this is the path almost universally adopted in current texts. Although some texts give presentations that honestly address the "How do we know . . . ?" questions, many, unfortunately, abbreviate the story to the point at which students are presented with end results that are memorized without understanding.

The bulk of the relevant experimental work is that of Philipp Lenard (1902). To understand what transpired it is necessary to know how the experiments were conducted, and it is thus essential to start with at least a schematic diagram of the apparatus (such as Fig. 10.9.1) and an insight into what was being measured. (Quite a few texts give good descriptions of what Lenard did, and only a very concise summary will be given here for the sake of illuminating the pedagogical discussion.)

Referring to Fig. 10.9.1: Electrode P is connected through a sensitive galvanometer (or microammeter) A to the midpoint G of the slide wire resistor CD. (Lenard actually made his measurements with two separate electrometers, connected to electrodes M and P, respectively.) If slide contact S is at point G, the potential difference between M and P is zero. If S is to the left of G, plate M is positive relative to P, and electrons ejected from M would be attracted back toward this electrode and retarded in their motion toward P. In the following, the convention is adopted of denoting this as a "retarding" or negative potential difference. Similarly, with S to the right of point G, ejected electrons would be repelled from M and accelerated toward P. This is denoted as an "accelerating" or positive potential difference. Light from an arc is incident on metallic plate M through window Q. Monochromatic light is obtained by using filter F.

Since students in introductory courses get very little experience in interpreting the functions of a simple electrical circuit, the apparatus of Fig. 10.9.1 offers

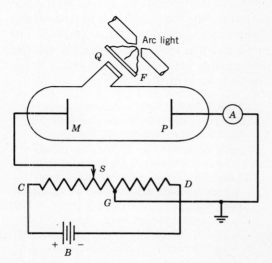

Figure 10.9.1 Schematic diagram of Lenard's apparatus for measuring photoelectric current.

a valuable pedagogical opportunity. One can use this rich context to supplement conventional numerical exercises on the photoelectric effect with the opportunity to think through what the apparatus does and how the ejected electrons are affected at various positions of the contact S. Teachers who have not invoked such situations will be shocked by the inability of many students (even those in engineering-physics courses) to interpret what happens to the potential difference between plates M and P as contact S is moved along the wire. (This is not a matter of complexity of the ideas; it is merely a matter of lack of practice on the part of the students.)

Using three different light sources (an arc light with carbon electrodes, another with zinc electrodes, and a spark discharge between zinc spheres), Lenard made a systematic study of the influence of light intensity and of the potential difference between plates M and P on the photocurrent. The leading observations are summarized as follows:

1. With a steady light source and a fixed, positive (accelerating) potential difference, the photo-current was observed to increase as P was moved toward M. When P came to within about 5 mm of M, the current was observed to level off at a final, maximum value. All the further observations were made at this final spacing between the plates. (Students should be led to interpret the point and purpose of these observations and the inference to be drawn: The experiment showed that electrons must be ejected at all angles relative to M and virtually all of the ejected electrons were being collected on P when the spacing was sufficiently small.)

2. Lenard then varied the light intensity in two ways: by changing the current through the arc and by moving the source to a greater distance from the window Q. (In the latter case, he made use of the inverse square law for variation of intensity with distance from a small source.) He found that the maximum (saturation) photocurrent under accelerating potential difference was directly proportional to the intensity (energy/unit area/second) of the incident light (Fig. 10.9.2).

3. Lenard was especially impressed by the fact that the direct proportionality to incident light intensity extended all the way down to extremely low intensities—one three-millionth of the highest intensity available—with no evidence at all of a threshold level. (A threshold was expected on the supposition that some finite level of energy flux would be required to eject any electrons at all from confinement within the metal. No such intensity threshold has ever been detected for the photoelectric effect; if the incident light is capable of producing the effect at all, some electrons are always ejected, no matter how faint the light.)

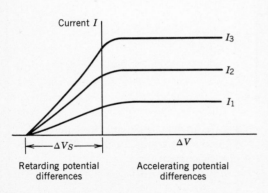

Figure 10.9.2 Idealized curves showing results of Lenard's observations of photoelectric current with apparatus such as that in Fig. 10.9.1. The curves indicated are obtained with three different intensities of the same wavelength of monochromatic light. Values of saturation current (I_1, I_2, I_3) are directly proportional to incident light intensity. Stopping potential difference (ΔV_S) is independent of light intensity; it depends only on the wavelength of the incident light and on the material composing plate M.

10.9 THE PHOTOELECTRIC EFFECT AND THE PHOTON CONCEPT

4. When contact S was moved to the left of point G (Fig. 10.9.1), and the potential difference acquired retarding values, the observed photocurrent did not drop to zero immediately despite the retarding effect. It decreased more or less linearly (Fig. 10.9.2) with increasing retarding potential difference, reaching zero at a potential difference of the order of two volts. The value ΔV_S at which the photocurrent is cut off is called the "stopping potential difference." Lenard observed that the value of the stopping potential difference was completely unaffected by the intensity of the incident light (Lenard varied the intensity of the incident light by a factor of more than 1000, with everything else held constant, without detecting significant change in ΔV_S); it was altered only when a different type of light source was used or when the metal in plate M was changed. (This was, of course, one of the most surprising aspects of the observations. It was anticipated that electrons would be ejected from the metal with a distribution of kinetic energies, but it was also expected that this distribution would be influenced by the intensity of the incident light.)

In the latter observations, one encounters again the reasoning concerning kinetic energies of ejected electrons and the storing of potential energy when a retarding field is imposed between capacitor plates. It is inevitably disappointing to a conscientious teacher, but one must steel oneself to the fact that quite a few students will still have difficulty with these concepts in the new context despite what one might have done on preceding occasions. It takes *several* encounters, in altered context, for many students to master these abstract concepts—not just one or two encounters. Thus the added opportunity is very useful.

Given the summary of Lenard's observations, students should be led to address questions such as the following:

1. Sketch several possible trajectories of electrons ejected from plate M when plate P is a centimeter or two away and when it is within 5 mm. Do this for the cases of both accelerating and retarding potential difference. Describe cases of projectile motion that correspond to sketches you have made for the behavior of the electrons.

2. The quantity $e\Delta V_S$ is said to be equal to the maximum kinetic energy of electrons ejected from plate M. Explain the reasoning behind this statement in your own words. What is the corresponding expression for the case in which a stone is thrown vertically upward? Why is it that the mass of the particle is present in the expression in the gravitational case and absent in the electrical case?

3. Why is it surprising that the maximum kinetic energy of the ejected electrons is unaffected by the intensity of the incident light? Redraw Fig. 10.9.2 so as to show a set of three curves that would *not* have been surprising to Lenard. [Note that this question leads the student to confront, explicitly, what is *not* the case in contrast to what *is*.]

4. It is clear that, although the intensity of the incident light has no effect on the kinetic energy of individual electrons, it does affect the saturation current, the latter being directly proportional to the intensity, with no evidence of a threshold. From the variation of the saturation current, what are you forced to conclude concerning the sole effect of brightness of incident light in the process being observed? Can you, in terms of the physics you have learned so far, account for the observed fact that the intensity of the incident light determines *only* the rate of ejection of electrons without having any effect on

their individual energies? [Lenard explicitly points out that it is very difficult, if not impossible, to explain the observed photoelectric phenomena in terms of classical Maxwellian theory.]

Einstein, in his famous 1905 paper proposing the photon concept [see translation by Arons and Peppard (1965)], deals with the photoelectric effect only as a secondary matter. His main object is to rederive the black-body spectrum, not as Planck had approached it through quantization of the energy of the atomic or molecular oscillators in the walls of the cavity, but through statistical thermodynamics and quantization of the radiation itself. (This material is far beyond the level of an introductory course.) Having generated the concept, however, and having successfully derived the black body distribution, Einstein then turns to the photoelectric observations for further support. He shows that the photon concept is capable of providing a very simple resolution of the photoelectric paradoxes, and these arguments are properly given in most introductory texts. (The essential added ingredient is to give students the opportunity to present the arguments in their own words instead of desperately trying to remember text assertions without ever having restated these for themselves.)

The most important ingredient for future use is, of course, the linear relation between stopping potential difference and the frequency of incident monochromatic light, with its different threshold frequency for each different metal. It is important to note that this linear relation was *not* among the original photoelectric observations, although some texts imply that it was. Einstein *predicted* the linear relation, together with its now familiar interpretation, and this remained to be confirmed. Initial support was provided by Richardson and Compton (1912) and by Hughes (1913). The final and most definitive confirmation was provided by Millikan through the magnificent experiment that involved preparing a clean, uncontaminated metal surface by enclosing what he described as a small "machine shop in a vacuum" [Millikan (1916)].

Millikan subsequently said of his own work [Millikan (1949)]:

[Einstein's explanation of 1905] ignored and indeed seemed to contradict all the manifold facts of interference and thus to be a straight return to the corpuscular theory of light which had been completely abandoned since the time of Young and Fresnel. . . . I spent 10 years of my life testing the 1905 equation of Einstein's, and, contrary to all my expectations, I was compelled in 1915 to assert its unambiguous experimental verification in spite of all its unreasonableness since it seemed to violate everything we knew about the interference of light.

As a step toward enhancing scientific literacy, one might note in this connection that there are other similar episodes of theoretical prediction and *subsequent* experimental confirmation in the history of science. Dalton, for example, *predicted* the Law of Multiple Proportions before he found the regularity already present in the composition data—data that had never been converted from percentages to the form that revealed the small whole number ratios required by a corpuscular model (see Chapter 12 for more detail).

10.10 QUOTATIONS FROM EINSTEIN'S 1905 PAPER ON THE PHOTON CONCEPT

Because of their clarity and eloquence, it is very useful to have at hand some of Einstein's own words concerning the photon concept. A few especially useful passages are quoted in the following, all being taken from Einstein (1905a) [translation by Arons and Peppard (1965)].

It should first be noted that the title of the paper is "Concerning a Heuristic Point of View Toward the Emission and Transformation of Light." Thus Einstein emphasizes the heuristic nature of his proposal. The paper begins in a highly characteristic manner; just as he does in the relativity paper, Einstein points to a certain asymmetry and lack of conceptual consistency in existing theories. The first paragraph of the following quotation points to the fact that Planck's theory of black-body radiation requires quantization of the energies of the material particles in the walls of the cavity while the electromagnetic radiation that is being emitted and absorbed is treated as continuous:

> ... the total energy of a ponderable body must, according to the present conceptions of physicists, be represented as a sum carried over the energies of the atoms and electrons [that make up the body]. The energy of a ponderable body cannot be subdivided into arbitrarily small parts [i.e., Planck's theory], while the energy of a beam of light from a point source (according to Maxwellian theory of light or, more generally, according to any wave theory) is continuously spread over an ever increasing volume.
>
> The wave theory of light, which operates with continuous spatial functions, has worked well in the representation of purely optical phenomena and will probably never be replaced by another theory. It should be kept in mind, however, that the optical observations refer to time averages rather than instantaneous values. In spite of the complete experimental confirmation of the theory as applied to diffraction, reflection, refraction, dispersion, etc., it is still conceivable that the theory of light which operates with continuous spatial functions may lead to contradictions with experience when it is applied to the phenomena of emission and transformation of light [i.e., interactions on the microscopic scale].
>
> It seems to me that the observations associated with black body radiation, fluorescence, the photoelectric effect, and other related phenomena ... are more readily understood if one assumes that the energy of light is discontinuously distributed in space. In accordance with the assumption to be considered here, the energy of a light ray spreading out from a point is not continuously distributed over an increasing space, but consists of a finite number of energy quanta which are localized at points in space, which move without dividing, and which can only be produced and absorbed as complete units. [The term "photon" emerged later.]

In the preceding paragraph, Einstein points to fluorescence as one of the phenomena that Maxwellian theory had not dealt with successfully. He is referring to what is known as "Stokes's Rule," the then well established fact that, when a substance fluoresces under incident radiation such as ultraviolet light, the "transformed" radiation (emitted light) invariably has a longer wavelength or lower frequency than the exciting radiation. Concerning Stokes's Rule, he suggests that perhaps each incident photon, absorbed by the fluorescent material, stimulates the emission of one or more photons, leading to the reemission of the energy that was absorbed. If each absorption and corresponding reemission is an elementary process, independent of other incident photons, conservation of energy requires that the energies of the emitted photons (and therefore their frequencies) be equal to or less than that of the incident photon.

Einstein then continues with regard to Lenard's photoelectric observations:

> The usual conception, that the energy of light is continuously distributed over the space through which it propagates, encounters very serious difficulties when one

attempts to explain the photoelectric phenomena, as has been pointed out by Lenard in his pioneering paper. According to the concept that the incident light consists of energy quanta of magnitude hν, however, one can conceive of the ejection of electrons by light in the following way. Energy quanta penetrate into the surface layer of the body, and their energy is transformed, at least in part, into kinetic energy of electrons. The simplest way to imagine this is that a light quantum delivers its entire energy to a single electron; we shall assume that this is what happens. . . . An electron to which kinetic energy has been imparted within the body will have lost some of this energy by the time it reaches the surface. Furthermore, we shall assume that in leaving the surface of the body each electron must perform an amount of work W_0, characteristic of the substance of which the body is composed. The ejected electrons leaving the body with the largest normal velocity will be those that were directly at the surface. The kinetic energy of such electrons is given by[6]

$$KE_{max} = h\nu - W_0 \qquad [10.10.1]$$

If the emitting body is charged to a positive potential difference relative to a neighboring conductor, and if ΔV_S represents the potential difference which just stops the photoelectric current [it follows that $e\Delta V_S$ must be equal to the maximum kinetic energy of the ejected electrons] and therefore

$$e\Delta V_S = h\nu - W_0 \qquad [10.10.2]$$

where e denotes the electronic charge.

If the deduced formula is correct, a graph of ΔV_S versus the frequency of the incident light must be a straight line with a slope that is independent of the nature of the emitting substance. [This constitutes the prediction of what came to be known as the Einstein equation for the photoelectric effect.]

So far as I can see, there is no contradiction between these conceptions and the properties of the photoelectric effect observed by Lenard. If each energy quantum of the incident light, independently of everything else, delivers all its energy to a single electron, then the velocity distribution of the ejected electrons will be independent of the intensity of the incident light; on the other hand, the number of electrons leaving the body will, if other conditions are kept constant, be proportional to the intensity of the incident light. . . . [This covers Lenard's observed anomalies.]

10.11 BOHR'S FIRST QUANTUM PICTURE OF ATOMIC HYDROGEN

Like Thomson's research on the cathode beam and Lenard's investigation of the photoelectric effect, Bohr's first simple model of the hydrogen atom, from which he obtained the Balmer–Rydberg formula for the spectral series of atomic hydrogen by making tentative departures from Newtonian and Maxwellian theory, carries an intellectual experience accessible to students in introductory physics. If carefully motivated and interpreted, at a pace slow enough to allow comprehension, the effect on many students is dramatic. They recognize that they are putting together in one context a large volume of the most fundamental material they previously encountered in bits and pieces: circular motion, centripetal force, Coulomb's Law,

---

[6]Einstein's symbols have been altered to correspond to the notation adopted in this book.

kinetic and potential energies, absorption and emission of light, conservation of energy, bright line spectra, electrons, the nuclear model, the photon concept. They find the synthesis to be a revelation of "interconnectedness." They find reinforcement in their ability to put it together, and they sense how the opportunity to spiral back gives them an increasingly firm grasp of the basic concepts. At the same time, they take a first step toward new ideas such as the correspondence principle, discrete energy levels, ground states, and excited states, all of which remain basic to the final, correct quantum theory.

There is very great pedagogical value in treating the hydrogen atom at the level at which this is done in Bohr's very first paper of 1913. Unfortunately, many textbook versions eviscerate the treatment, shortcutting it as they shortcut the Thomson experiment, thus greatly reducing the physical content, the impact, and the intelligibility. The following is a condensed outline of how Bohr handled the problem in his first paper. The analysis is confined to circular orbits, and the necessary quantization rule is obtained not through arbitrary quantization of angular momentum (such treatments are nothing but black magic to students at an introductory level) but through application of the correspondence principle, which is far more plausible and intelligible to the students at this stage even though algebraically more complex. (The appendix to this chapter contains an example of a Socratic homework assignment that leads students to put the story together as a unified sequence.)

Bohr begins with explicit recognition of the fact that Rutherford's nuclear model "seems to be necessary in order to account for the results of the experiments on large angle scattering of α-rays." He points to the resulting conflict with classical theory which requires an orbiting electron to radiate continuously:

> ... *the electron will approach the nucleus, describing orbits of smaller dimensions, and with greater and greater frequency, the electron on the average gaining kinetic energy at the same time the whole system loses energy.*[7] *The process will go on until the dimensions of the orbit are of the same order of magnitude as the dimensions of the electron or those of the nucleus. A simple calculation shows that the energy radiated out during the process will be enormously great compared with that radiated out by ordinary molecular processes.*
>
> *It is obvious that the behavior of such a system will be very different from that of an atomic system occurring in nature. In the first place, the actual atoms in their permanent state seem to have absolutely fixed dimension and frequencies. Further, if we consider any molecular process, the results always seem to be that after a certain amount of energy characteristic for the systems in question is*

---

[7]Students should have been exposed to analysis of these kinetic and potential energy changes in connection with satellite motion under gravity, and perhaps orbital motion (without radiation) of a charged particle under Coulomb's Law, earlier in the course, before any mention of the hydrogen atom. Now they spiral back to recover these previously encountered ideas for the new context. Care should have been taken in the earlier treatments to show why it is not possible to take the zero reference level for energy at $r = 0$ and why the most convenient zero reference level is at infinite separation (the kinetic energy of orbital motion tending to zero as $r \to \infty$ and the potential energy of the system being conveniently taken as zero in the same limit). They should then have been led to see that the system must lose energy to the outside when the radial separation decreases, gain energy from the outside when the radial separation increases. They should also be led to see that, on this account, given a finite radial separation, the total energy must be less than that at infinite separation, and therefore negative relative to the reference level adopted. If these aspects have not been slowly and carefully developed earlier, the negative sign that now arises in connection with the total energy of the system presents a serious roadblock; students panic before it, feel they cannot possibly understand the mathematics, and proceed to memorize without comprehension.

radiated out, the systems will again settle down in a stable state of equilibrium, in which the distances apart of the particles are of the same order of magnitude as before the process. . . .

The way of considering a problem of this kind has, however, undergone essential alterations in recent years owing to the development of the [quantum theory of electromagnetic] radiation, and the direct affirmation of the new assumptions introduced in this theory, found by experiments on very different phenomena such as specific heats, photoelectric effect, Roentgen rays, etc. [Here Bohr is referring to the successes of the photon concept, in the years since 1905, in various areas to which Einstein and others applied it.]

The result of the discussion of these questions seems to be the general acknowledgment of the inadequacy of the classical electrodynamics in describing the behavior of systems of atomic size. Whatever alteration in the laws of motion of electrons may be, it seems necessary to introduce in the laws in question a quantity foreign to the classical electrodynamics; i.e., Planck's constant, or as it is often called, the elementary quantum of action. By introduction of this quantity the question of the stable configuration of the electrons in the atoms is essentially changed, as this constant is of such dimensions and magnitude that it, together with the mass and charge of the particles, can determine a length of the order of magnitude required.

In the last sentence Bohr is referring to arguments from dimensional analysis which were widely prevalent at the time. (Although students are usually shown the importance of checking the dimensional consistency of their own work in solving problems, they are rarely shown, in introductory courses, the powerful role that such seemingly crude thinking has had in inductive leaps made in scientific research. The present context provides a valuable opportunity for such exposure.) One line of argument involved angular momentum: It was clearly recognized at the time Bohr was writing that h has the dimensions of angular momentum, and, as a matter of fact, unsuccessful efforts had been made to produce nuclear models in which total angular momentum changed by integral amounts as individual electrons entered or left an orbital ring occupied by several electrons. Another dimensional line of argument (referred to in the last sentence in the quotation above) involved showing that the combination h^2/mke^2 has the dimensions of length and gives an order of magnitude corresponding to atomic dimensions (the expression is for modern SI units, m denoting the mass of the electron, and k the constant in Coulomb's Law).

With this background of motivation, Bohr suggested a direct application of Einstein's photon hypothesis in the following manner:

1. Abandon classical electrodynamics to the extent of assuming that, at radii of the order of atomic dimensions, electrons can revolve in stable circular orbits ("stationary states") without radiating continuously. Retaining the law of conservation of energy, the electron–nucleus system is then assumed to gain or lose energy only when electrons are transferred from one stationary state to another. Thus, if an electron is transferred from an orbit r_1 to an orbit r_2, the system must change energy by the amount $E(r_2) - E(r_1)$, the change corresponding to absorption or emission of energy depending on whether r_2 is greater or less than r_1.

2. Invoking Einstein's photon concept, assume that electromagnetic radiation is absorbed or emitted only in *transfer* of electrons from one orbit to another, and that such absorption or emission of energy by individual electrons is associated with

absorption or emission of *individual* quanta of energy $h\nu$ as suggested in Einstein's heuristic explanation of the photoelectric effect. This gives the relation

$$h\nu = |E(r_2) - E(r_1)| \qquad (10.11.1)$$

3. Turning to the empirical Balmer–Rydberg formula for hydrogen and writing it as an expression for observed bright line frequencies ν (instead of as an expression for $1/\lambda$), one has

$$\nu = cR_H \left(\frac{1}{n_f^2} - \frac{1}{n_i^2} \right) \qquad (10.11.2)$$

Equation 10.11.2 strongly suggests, on comparison with Eq. 10.11.1, that the two terms on the right-hand side of the Balmer–Rydberg formula are referring to stationary states at two different radii. Thus Bohr writes what corresponds to

$$h\nu = |E(r_2) - E(r_1)| = hcR_H \left| \frac{1}{n_f^2} - \frac{1}{n_i^2} \right| \qquad (10.11.3)$$

4. Since only certain discrete frequencies ν are observed in the hydrogen spectrum, Eq. 10.11.3 implies that only certain *discrete* stationary states or energy "levels" (i.e., only certain discrete values of orbital radii r) occur in the structure of the hydrogen atom. This poses the problem of finding, in some way, an additional condition that restricts the allowed energy levels, that is, in modern terminology, a quantization rule.

5. Finally, there is the question of how to visualize the specific atom now under consideration, namely hydrogen. Bohr writes:

> *General evidence indicates that an atom of hydrogen consists simply of a single electron rotating round a positive nucleus of charge e. [This conclusion] is strongly supported by the fact that hydrogen, in the experiments on positive rays of Sir J. J. Thomson, is the only element which never occurs with a positive charge corresponding to the loss of more than one electron.*

At this point Bohr gives a footnote citing Thomson (1912) in which Thomson, describing his work with a crude early version of a mass spectrometer (in which a beam of positive ions was separated by being subjected to *parallel* electric and magnetic fields, causing each species of ion present to fall on a parabolic track on the screen at the end of the tube) remarks that

> *All the elements I have examined give multiply charged atoms with the exception of hydrogen on which I have never observed more than one charge.*

(Giving students this background of physical insight to savor makes the story far more interesting and effective than the bland assertion that hydrogen must consist of one electron and one proton because it is the lightest element known. Bohr saw fit to support his model with much more sophisticated evidence than this statement.)

6. It is now required to say something about the energy of a stationary state, and Bohr takes the classical result

$$E(r) = -\frac{ke^2}{2r} \tag{10.11.4}$$

as the total energy of the electron–proton system. Here he sees it necessary to say something about the mix of assumptions being made:

> [It is assumed] that the dynamical equilibrium of the systems in the stationary states can be discussed by the help of ordinary mechanics, while the passing of the systems between different stationary states cannot be treated on that basis.

(Here, of course, are some of the crucial points on which the early quantum theory broke down. It was unable to say anything about the probability of the transitions, and the mixture with Newtonian mechanics had to be foregone entirely. Bohr, however, was fully conscious of the tentative nature of his exploration, and he clearly put forth the dubious aspects.)

If only certain discrete energy levels are "allowed," it follows that the electron can occupy only certain discrete orbits of radius r_n with energy given by

$$E_n = -\frac{ke^2}{2r_n} \tag{10.11.5}$$

Also, since each of the two terms on the right-hand side of Eq. 10.11.3 must be a quantity of energy, it is implied that any particular energy level E_n might be related to the Rydberg constant R_H by

$$E_n = -\frac{hcR_H}{n^2} \tag{10.11.6}$$

where n is an integer and cannot be equal to zero.

7. Elimination of E_n from Eqs. 10.11.5 and 10.11.6 yields

$$r_n = n^2 \frac{ke^2}{2hcR_H} \tag{10.11.7}$$

and the remaining problem, as indicated in paragraph 4 above, is to find a quantization rule that restricts the allowed radii and makes possible the evaluation of the Rydberg constant in terms of the more fundamental universal constants.

8. In his subsequent papers, Bohr utilized the quantization of angular momentum in terms of $h/2\pi$ as described in most textbooks, but he did so only after he had shown in the first paper that the quantization obtained through application of the correspondence principle boiled down to this angular momentum quantization.

The correspondence principle was invoked by introducing the requirement that, as n and the orbit radii r_n become very large (approaching macroscopic dimensions), the frequency of the photon emitted in transitions between adjacent orbits becomes equal to the frequency of orbital motion, that is, the frequency of the radiation that would be emitted on the basis of classical Maxwellian theory.

Examining the behavior of ν as a function of n for jumps between adjacent orbits in Eq. 10.11.3:

$$\nu = \frac{E_{n+1} - E_n}{h} = cR_H \left[\frac{2n+1}{(n+1)^2 n^2}\right] \tag{10.11.8}$$

and, as n becomes large

$$\nu \to cR_H \frac{2}{n^3} \tag{10.11.9}$$

Applying Newtonian dynamics to the orbital motion of the electron (while neglecting motion of the nucleus, that is, treating this as a one-body rather than as a two-body problem, something for which correction was made subsequently), the frequency f_n of orbital motion is given by

$$f_n^2 = \frac{ke^2}{4\pi^2 m r_n^3} \tag{10.11.10}$$

The correspondence principle, requiring ν to approach f_n as n becomes large, suggests that, in light of Eq. 10.11.9 one should set

$$f_n = \frac{2cR_H}{n^3} \tag{10.11.11}$$

Combining Eqs. 10.11.7, 10.11.10, and 10.11.11 then gives the familiar results

$$R_H = \frac{2\pi^2 m(ke^2)^2}{ch^3} \tag{10.11.12}$$

$$E_n = -\frac{2\pi^2 m(ke^2)^2}{n^2 h^2} \tag{10.11.13}$$

$$r_n = n^2 \frac{h^2}{4\pi^2 mke^2} \tag{10.11.14}$$

which, for angular momentum $L_n = 2\pi m r_n^2 f_n$, reduce to

$$L_n = n\frac{h}{2\pi} \tag{10.11.15}$$

From then on, Bohr adopted Eq. 10.11.15 as the quantization rule.

Although the algebra involved in applying the correspondence principle is more complex than that in applying Eq. 10.11.15, the physical reasoning in the former is much more plausible to the students. It is clear that Bohr himself initially felt the same way; otherwise it is unlikely he would have bothered to publish the correspondence principle approach before going on to quantization of angular momentum.

It is worth noting Rutherford's initial response to these ideas. In a letter to Bohr, dated 20 March 1913, Rutherford says [Birks (1962)]:

> I have received your paper safely and read it with great interest, but I want to look it over again carefully when I have more leisure. Your ideas as to the origin of the spectrum of hydrogen are very ingenious, and seem to work out well; but the mixture of Planck's ideas with the old mechanics makes it very difficult to form a physical idea of what is the basis of it. There appears to me one grave difficulty in your hypothesis, which I have no doubt you fully realize, namely, how does an electron decide what

frequency it is going to vibrate at when it passes from one stationary state to another? It seems to me that you have to assume that the electron knows beforehand where it is going to stop.

Some students begin to wonder about matters of this kind if their wondering is not suppressed by an implication that the questions are foolish. If they begin to wonder, they are in very good company indeed and are preparing to understand why the new quantum mechanics eventually had to replace the old.

There are several questions, in addition to the usual numerical calculations concerned with the diagram of energy levels, that help deepen student insight into the quantum model:

1. Some years ago, on a qualifying examination for graduate students, I inserted the following: Starting with a reproduction of the absorption and emission spectra of sodium, placed one above the other, came the question "Explain **QUALITATIVELY**, in a few words, how the quantum model of the atom accounts for the difference between these two spectra, that is, how does it come about that the emission spectrum has the same lines as the absorption spectrum but also has additional lines that the absorption spectrum does not contain?"

Fourteen students took the exam. One of the 14 gave the straightforward response that, in absorption, electrons are elevated from the ground state to higher states and one would see only those transitions, while, in emission, electrons cascade down through intermediate states, in addition to dropping directly to the ground state, and thus produce additional lines. Five students, despite the emphasis on "qualitatively," launched into irrelevant quantum mechanical formalism with selection rules and reached no conclusion. The remaining students left the question blank. These were not incompetent students; they simply had never had the opportunity to confront basic questions of this kind and talk about them in their own words. This experience underlines the need to expose students to such qualitative questions from the earliest encounter.

2. By the time Bohr proposed the quantum model, it had been noted that the first 10 or 12 lines of the Balmer series could be observed in laboratory discharge tubes, while as many as 33 lines of this series had been detected in stellar spectra and in the corona of the sun. Bohr turned this observation to good account:

> *[This] is just what we should expect from the above theory, [according to which] the diameter of the orbit of the electron in the different stationary states is proportional to n^2. For $n = 12$ the diameter is equal to 160Å, or equal to the mean distance between molecules at a pressure of about 7 mm mercury. For $n = 33$ the diameter is equal to 1200Å, corresponding to the mean distance of molecules of about 0.02 mm mercury. According to the theory, the necessary condition for the appearance of a great number of lines is therefore a very small density of the gas; for simultaneously to obtain an intensity sufficient for observation, the space filled with gas must be very great.*

The 7 mm pressure is approximately that in a laboratory discharge tube; it was difficult to obtain sufficient intensities at lower pressures. One has here an opportunity to use numerical values of the radii to achieve physical insight into a significant set of phenomena. In the discharge tube, the atoms begin to interfere with each other around $n = 12$, whereas such interference does not arise in stellar atmospheres until around $n = 33$. In the latter case, one has the benefit of being able to look at a tremendous volume of gas, and observable intensity is achieved despite the very low concentration. This is valuable physical thinking in which students cannot engage unless explicitly afforded the opportunity.

3. An opportunity to spiral back to basic electromagnetic concepts is afforded by one of the outstanding failures of the Bohr model: the prediction of a nonzero magnetic moment for the ground state of atomic hydrogen, whereas the ground state is known to have zero magnetic moment. Many students have, at this stage of the game, forgotten about current loops and the attendant magnetic behavior. Review of this concept and connecting the macro- and microscopic phenomena help register the ideas more firmly. This encounter, at the same time, paves the way for deeper appreciation of one of the early triumphs of the modern quantum theory—at least for those students who continue to that level.

Final comments: The Bohr hydrogen atom is, of course, an essentially ad hoc model and is not the result of powerful theoretical synthesis. The theory that we call "quantum mechanics" evolved later. This does not, however, make the Bohr story pedagogically useless. Much of its vocabulary and physical insight are retained to this day. Its intelligibility to students provides a rational step to modern insights. Similarly, the photon concept has evolved over time to the point that Einstein's completely particlelike localization of the entity has been abandoned. This does not mean that the story must be abandoned (after all, Einstein's Nobel prize was for this development rather than for relativity). One need not conceal from students the fact that the story continues. [Kidd, Ardini, and Anton (1989) give a discussion of the evolution of the photon concept from Einstein to the present and provide an extensive bibliography.]

10.12 INTRODUCING SPECIAL RELATIVITY

Einstein's revolutionary demolition of the classical notions of absolute space and time took a long time to penetrate the scientific community at large. Quick understanding and acceptance came only to a relatively small number of already prepared minds. The ideas involved in the transition to special relativity are exceedingly subtle and contraintuitive. They are more subtle and abstract, even when reduced to careful operational thought experiments, than the Law of Inertia or the abstractions associated with energy, momentum, and electricity. It is little wonder that students, even physics majors, emerge from their first exposure with virtually no conceptual understanding of what has transpired, regardless of how well they might do end-of-chapter problems manipulating the consequences of the Lorentz transformations.

Because of the subtlety of the ideas, there is no quick and easy way of infallibly capturing all beginners. It is quite possible, however, to provide a qualitative, phenomenological introduction that lays the groundwork for subsequent better understanding of the formalism on the part of science students and also gives those who will not go further some comprehension of what is meant by the relativity of space and time. A rather abbreviated outline of how this might be done is given in the following. Readers seeking greater detail will find such treatments, for example, in Arons (1965) and Huggins (1968).

As a start, before going on to dealing with, or even mentioning, different frames of reference, it is necessary to reexamine the ways in which we measure both space and time in a *single* frame—the one most familiar to us. Many texts give adequate operational descriptions of the measurement of length, and this issue will not be belabored here. The principal difficulty, as far as student understanding is concerned, relates to what Bridgman (1962) called "spreading time over space," that is, synchronizing clocks that are widely separated and giving meaning to "simultaneity." Although such synchronization, the use of signals, and so on, is mentioned in many texts, the missing ingredient is usually that of sufficiently strong emphasis on the fact that spreading time over space involves *definition, convention, invention* on our part and is not already out there independent of us. This may sound trivial to a physicist long acclimated to these ideas, but, in fact, there is very great resistance to accepting this view among most newcomers—young or old. They do not believe that clock synchronization and simultaneity must be defined by an agreed-upon convention. (That Einstein himself regarded the concept of "simultaneity" as crucial to understanding is indicated by the prominent role he gave it in his own popularization of relativity [Einstein (1961)].)

It is necessary to separate two aspects of simultaneity: local and distant. When we speak of simultaneity, most students immediately think only of the former—the sense we have about events taking place together, right in front of us, here and now—and they fail to see that there is a problem concerning establishing simultaneity (or nonsimultaneity) for a remote event with one taking place before us, or for two remote events. Bridgman (1962) points out that we accept "local simultaneity" intuitively as a *primitive*, as mathematicians now accept "point" and "line," without futile attempts at definitions that turn out to be circular. "Distant simultaneity," however, requires careful operational *definition*, and that is where synchronization of remote clocks comes in. Novices find it difficult to accept the idea that our intuitive sense of local simultaneity does not automatically extend to remote simultaneity, and many of them strongly resist the idea that what we do with clocks is a *definition* and not just the quantification of preexisting reality. (Resistance can be shaken to some extent by asking the student how we would ever establish what is happening on α-centauri, four light years away, right *now*.)

Without full awareness of the fact that synchronization is a matter of definition even in a *single* frame of reference, students are unprepared to understand what happens when we start comparing observations from two different frames. On the other hand, having developed such awareness, they are better prepared to comprehend the disagreements between different observers. Thus the introduction to special relativity is more effective if it starts with a slow, careful look at spreading time over space in a single frame and with strong emphasis on the fact that "distant simultaneity" must be operationally *defined* and is not something that already "exists" independently.

There are a variety of valid treatments and gedanken experiments concerning clock synchronization and simultaneity available in the literature, each with its own merits. For the purpose of later use in comparing assessments of simultaneity and length measurement in different frames of reference, however, it seems to me that the simplest and most direct gedanken experiment is that in which time is spread over space (along the x-axis) by synchronizing clocks placed at equal distances on either side of a central point from which a light pulse is emitted. The two clocks are defined as having been started simultaneously by arrival of the spherical wave front from the central point. (Later, it will be easy to see that, from the point of view of another observer moving along the x-axis, the two clocks could not have been started simultaneously.)

While still in our own single frame of reference, however, it is important to emphasize for the students that the use of light (radio signals, actually) is a matter of convenience and precision rather than of logical necessity. To the best of our knowledge, based on what we know of internal consistency, other, cruder procedures (e.g., sound signals in still air, slow transport of accurate chronometers) would yield synchronization in agreement with each other and with the electromagnetic signal. The reason for concentrating on the electromagnetic signal (apart from its high precision and the fact that this is the way world clocks are actually synchronized) is that, in the final analysis, we will find that the velocity of this signal is the *only* velocity that two *different* frames of reference (i.e., frames moving relative to one another) have in common. Hence this velocity eventually provides the only way of linking observations made in the different frames.

A second aspect that merits slower discussion than is afforded in most texts is that of the passionate nineteenth-century search for the "absolute" frame of reference, which came to be identified with the electromagnetic ether. Many students fail to comprehend the motivations for the search. They have not yet fully absorbed the role of frames of reference in physical theory; they are shaky on the meaning of "inertial frame"; and, unless explicitly prompted, they do not perceive why nineteenth-century physicists hoped that the ether might turn out to be the primary frame for *both* mechanics and electrodynamics. Some discussion of this background, and the opportunity it affords for spiraling back to frames of reference, greatly strengthens their grasp of these underlying aspects of physical theory.

Many students, if given the chance to speak up, show themselves to be uneasy about the use of the word "absolute" in this context, not being at all certain what it means. Such students are in good company, as indicated by Bridgman's comment:

> *The sort of tacit idea that we have before us in using the word "absolute" is itself not very definite, and may be one thing for the theologian and philosopher and another for the physicist. I think most physicists have in the back of their heads when using the word "absolute" not something which cannot be specified in terms of physical operations, as did the theologian and philosopher, but something in which . . . the operations can be specified in terms which do not refer to accidental, temporary, local situations. . . . Thus if the existence of an all-pervading ether could have been established in some way, then velocity with respect to the ether would have been the sort of thing that the physicist would have been willing to call absolute. It is curious that there is a uniquely definable velocity, namely velocity with respect to the fixed stars, which is not felt to have the property of absoluteness implicitly wanted. . . .*

(It is interesting to speculate on what Bridgman might have said about the discovery of the 3°K cosmic microwave background radiation and our motion relative to it.)

After providing an understanding of what motivated the search for the earth's motion through the ether, one appropriately goes on to the experimental attempts and the accompanying null results. These, especially the Michelson–Morley experiment, are well discussed in many texts. Concerning the point of departure into special relativity, Einstein's own statement in the second paragraph of the 1905 paper is still one of the best and clearest. After pointing out, in the first paragraph, certain unhappy asymmetries in the prevailing view of electrodynamic phenomena, he continues:

> *Examples of this sort, together with the unsuccessful attempts to discover any motion of the earth relative to the "light medium,"[8] suggest that the phenomena of electrodynamics as well as of mechanics possess no properties corresponding to the*

---

[8]There has been much discussion by historians of science in recent years as to whether or not Einstein was aware of the Michelson–Morley results at the time of writing the 1905 paper. Many minute details, pro and con (mostly con), have been adduced. Einstein himself in interviews, late in his life, with Shankland (1963) made contradictory statements about this. As far as an introduction to the subject and Einstein's thinking are concerned, however, it is immaterial whether he was explicitly aware of Michelson–Morley. It is clear from this introductory paragraph that he was fully aware of at least some of the null results of efforts to detect motion relative to the ether.

idea of absolute rest. They suggest rather that . . . the same laws of electrodynamics and optics will be valid for all frames of reference for which the laws of mechanics hold good. We will raise this conjecture (the purport of which will hereafter be called the "Principle of Relativity") to the status of a postulate, and also introduce another postulate which is only apparently irreconcilable with the former, namely that light is always propagated in empty space with a definite velocity c which is independent of the state of motion of the emitting body. These two postulates suffice for the attainment of a simple and consistent theory of the electrodynamics of moving bodies based on Maxwell's theory for stationary bodies. The introduction of a "luminiferous ether" will prove to be superfluous inasmuch as the view here to be developed will not require an "absolutely stationary space" provided with special properties, nor assign a velocity vector to a point in empty space in which electromagnetic processes take place.

In connection with the second postulate concerning the velocity of light, it is worth noting Bridgman's (1962) remark that, "The postulate that the velocity of light is independent of the velocity of its source is indispensable to relativity theory and is a much more fundamental postulate than that of the equality of velocity in all frames of reference. . . ."

Having synchronized clocks in a single frame by means of light signals from a central point, and having motivated Einstein's postulates, it is now possible to lead students to perceive that observers in a second frame of reference, moving relative to the first, will not agree that clocks in the first have been synchronized by the chosen operation. It is also easy to see which clock the moving observer believes to have been started ahead of the other.

Referring to Fig. 10.12.1, S represents the first frame; M is the midpoint between locations A and B at which, by definition in S, simultaneous events are triggered on arrival of a light or radio pulse originating at M. S' represents the second frame, moving to the right at velocity u relative to S. The points M and M' coincide at the instant of emission of the light pulse.[9] The first step (and this takes time and effort, quick assertion leaves a blank field) is to lead the students to comprehend that the postulate concerning the velocity of light requires that observers in S claim that the spherical pulse is permanently centered on M and that all of S' is moving to the right relative to the center of this sphere. At the same time, they must be led to see that observers in S' claim the light pulse to be permanently centered on M' while all of S is moving to the left relative to the

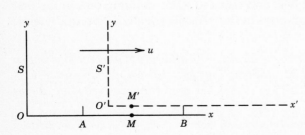

Figure 10.12.1 Light flash is emitted from point M midway between remotely separated points A and B along the x-axis in frame S. By definition in frame S, arrival of the pulse synchronizes clocks or triggers simultaneous events at A and B. Frame S' moves to the right at velocity u relative to S. Points M and M' are coincident when the flash is emitted.

---

[9]In class discussion, each frame of reference can be effectively represented in concrete form by means of a long board with vertical dowels mounted at each of the three indicated positions. Each frame can carry its own color to distinguish it. An instantaneous position of the spreading circle of the wave pulse originating at M can be represented by a hoop placed so that it is centered on M.

center of the sphere. They must visualize the difference in the claims of the respective frames, and they must learn to accept these views as forever irreconcilable. This is what takes time.

Having set up the basic situation as just outlined, one can proceed to inject further detail:

From the point of view of observers in S', with M' permanently centered in the hoop (referring to the suggested demonstration equipment), points A and B have been moving to the *left* within the spreading circle. Thus, with B moving *toward* the advancing wave front and A moving *away* from it, S' claims that the signal would have arrived at B before arriving at A and that the clock at B was therefore started *before* the clock at A. (The reciprocal view, namely that S claims that a clock at A' in frame S' would have been started before a clock at B', follows by the similar line of argument and makes an appropriate homework problem.)

This approach to the relativity of simultaneity is well known and widely used, and I summarize it here not because it is new but because it is the necessary basis for the qualitative description of length contraction that follows. The purely qualitative description of length contraction is also well known, but it appears in very few texts.

The next step in the argument involves another operational concept that students have never confronted and with which they initially have substantial difficulty: that of the measurement of length of a moving object. The problem here is analogous to that of distinguishing between local and remote simultaneity, but, in a sense, even more subtle. Previous experience has been only with objects stationary in our own frame of reference, and one can put a ruler on the object, or mark off the ends against an available scale, at leisure, without considering any time element at all. Thus there is no expectation that time might get inextricably mixed up with length measurement. The latter is a new and very unsettling idea that is not quickly assimilated.

It is necessary to lead students to review the operation of length measurement when they possess the rod, say, in their own frame of reference (defining "proper length" while at it) and then explicitly raise the question of what operations would be necessary to measure the length of a rod which is flying by in another frame. It takes very loaded questioning to extract the unfamiliar and unanticipated notion that one must mark the ends of the flying rod *simultaneously* against the scale in one's own frame of reference.[10]

Once one has arrived at the perception that the ends of the moving rod must be marked simultaneously, length contraction follows directly from the previously established failure of simultaneity. Suppose that, in Fig. 10.12.1, S' is holding a rod parallel to the x'-axis; the proper length is then L_0'. Observers in S measure the length of the moving rod by marking the ends simultaneously, according to their own clock synchronization, along the x-axis. How does S' view this operation? Since, according to S', any clock in S was started ahead of any clock to the left of it, the observers in S must have marked the right end of the rod *before* they marked the left end. Since the rod must have moved over to the right during that time interval, the marks are too close together, and the length L measured in S must be smaller than the proper length L_0'. (The reciprocal argument develops in

---

[10] An alternative, and valid, operation is, of course, to measure the time it takes the rod to pass a fixed point in our frame and then to calculate its length from the time interval and the velocity. Some students come up with this idea and should be reinforced for doing so. It is not possible, however, to discern length contraction in this operation without going to the additional step of discerning time dilation. The two operations are, of course, eventually found to agree with each other.

exactly the same way if S is holding the rod, and the writing out of this part of the story is well left as a homework assignment.)

One has now attained some of the major insights associated with the Special Theory: (1) That clock synchronization and remote simultaneity are a matter of definition in a single frame of reference and not an a priori. (2) That observers in a second frame of reference, moving relative to the first, will not agree with the clock synchronization or simultaneity of events as defined in the first frame, and vice versa. (3) That the operation of measuring the length of an object gives different values in different frames. (4) That the measurement of length is inextricably intertwined with the spreading of time over space. (5) That the only reason we can compare the measurements made in the different frames is that they all still have something in common, namely the velocity of light.

The qualitative development summarized up to this point provides an effective beginning for *any* group of students coming to the concepts *de novo*. It works with high school students as well as with more mature groups. It marks a valid endpoint for a brief introduction to the revolution in point of view toward space and time that is entailed and to the meaning of "relativity" in this context. It is also an appropriate beginning for students who are going to go ahead to the development of the formalism. Very few students who are taken directly to the Lorentz transformations and the subsequent formal derivations of time dilation and length contraction develop the direct insight into the operations and the differences between observers that stem from the qualitative introduction described above. Their understanding of the formalism is significantly sounder if the qualitative introduction (with suitable homework problems) has been provided.

Following the qualitative introduction outlined above, one can start developing the attendant algebraic relations without going directly to the Lorentz transformations. The expression for time dilation follows directly by setting up a "light clock" along the y-axis and comparing the proper time interval between two events for the frame carrying the clock with the longer interval that would be calculated in the other frame. (This is done in just this way in many texts.) Given the expression for time dilation, one can directly derive the expression for length contraction. Given both time dilation and length contraction, one can derive the expression for the failure of synchronization (the difference S' contends exists between clocks separated by a distance x in S—clocks that S contends are synchronized).

Given the three results now listed, one can put them together to obtain the Lorentz transformations. This approach yields the transformations as a consequence of a line of physical argument that assembles the operations and calculations made in each frame. [Derivations following this line can be found in Panofsky and Phillips (1962) and in Arons (1965).] This line of argument turns out to be far more intelligible to many students than the approach of seeking "that linear transformation of coordinates which leaves the velocity of light invariant for the two frames." To the majority of students who have not yet developed the mathematical insights of a born theoretical physicist, the latter approach has virtually no meaning; the end results are memorized, and the underlying arguments are not assimilated.

Once the Lorentz transformations are at hand, one can derive the other usual consequences of Special Relativity. There are many valid sequences and treatments in existing texts, and my own experience does not single out any one approach as pedagogically superior to others. The teacher should use what he or she finds most congenial.

One comment remains to be made concerning the transition from kinematics to dynamics. For many years it was conventional to enter the discussion of dynamics through derivation of the relativistic mass, that is, the mass–velocity relation, and this is probably still the dominant mode in textbooks. More recently, however, it has been increasingly recognized that relativistic mass is a troublesome and dubious concept. [See, for example, Okun (1989).] Not only does it get one into the infelicities associated with longitudinal and transverse masses, but it also tempts one to associate relativistic mass (rather than just rest mass) with gravitational effects. The latter association is basically incorrect. The sound and rigorous approach to relativistic dynamics is through direct development of that expression for *momentum* that ensures conservation of momentum in all frames:

$$p = \frac{m_0 v}{\sqrt{1 - \frac{v^2}{c^2}}} \tag{10.12.1}$$

rather than through relativistic mass. Unfortunately, it is more difficult to derive the momentum expression in a simple way than it is to obtain the mass–velocity relation from the collision gedanken experiments prevalent in the literature. [See Peters (1986) for a recent effort to simplify this derivation.]

In some texts and presentations, the velocity v in the given frame of reference [as in Eq. 10.12.1] is confused with the relative velocity u of one frame with respect to another (as in the Lorentz transformations), and students tend to confuse the two velocities even when the presentation concerning the distinction is clear. It is necessary to call attention to the distinction forcefully and explicitly by extracting a statement of it from the students themselves.

APPENDIX 10A

Written Homework on the Thomson Experiment

[Note to users: The specific wording of such an assignment would have to be adjusted to what background students might have been given in class or lecture.]

This a written homework assignment. All you need do is follow the sequence of questions given below; they constitute the full outline. Make this a continuous story, explaining each step of your reasoning in your own words and interpreting the results.

1. Briefly describe in your own words the goal of Thomson's investigation of the cathode rays.

2. Briefly describe the observations (before Thomson) indicating that cathode rays might somehow be connected to negative electrical charge. What was the point of Thomson's experiment with the charge collector placed on the side of the tube instead of at the end opposite the cathode?

    [As indicated in class, observations of deflection of the cathode beam (prior to Thomson) had revealed deflection by a magnetic field but had failed to reveal deflection by an electrostatic field. Thomson conjectured that the presence of ions in the tube (the ions being charged atoms or molecules formed in the residual gas in the tube) might lead to conditions in the neighborhood of the capacitor plates such that the beam, even if it consisted of charged particles, would be unaffected by the charged capacitor plates. On greatly improving the vacuum in the tube, Thomson found that charged capacitor plates did indeed deflect the beam.]

3. In your own words, and with accompanying pictures or sketches, explain how the behavior of numerous ions in the residual gas would create a condition such that the cathode beam would be unaffected by the charged plates. Would there still be zero electrical field *outside* the capacitor plates? Explain your reasoning.

    [The following questions involving algebraic equations and relationships all have to do with Fig. 1 on the following page. Refer to this figure accordingly. You may redraw the figure for inclusion in your paper or you may simply include in your paper the figure that is given. All the derivations that are carried out have initially to do with the behavior of a *single* negatively charged particle of mass m carrying charge of magnitude e. Be sure to make this aspect clear in your presentation. After the behavior of the single particle has been analyzed, you will consider what would happen if the beam consisted of enormous numbers of particles.]

WRITTEN HOMEWORK ON THE THOMSON EXPERIMENT   265

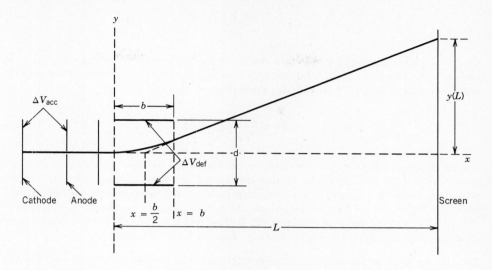

[Suppose that the hypothetical charged particle, starting somewhere between the cathode and anode, is accelerated to a velocity v_x and enters the region between the deflecting capacitor plates with this velocity. (We shall assume that the deflecting electrical field is sharply confined to the region between the plates and that there are no appreciable fringing effects.)]

4. Temporarily ignoring any gravitational effects, argue that the particle will be subjected to a uniform upward acceleration while it is between the plates and will therefore follow a parabolic trajectory from $x = 0$ to $x = b$. (Note that the situation here is *exactly* like the one you studied in projectile motion in introductory mechanics.) What will be the character of the trajectory after the particle gets beyond the edge of the plates? Why? How is this section of the trajectory oriented (or connected) to the parabolic part?

5. Now derive the equations for the trajectories described verbally in part 4: Starting with the basic definitions of electrical field strength and potential difference, argue that, if we denote the potential difference between the plates of the deflecting capacitor by ΔV_{def} and the separation between the plates by d, the electrical field strength $\mathscr{E}$ between the plates is given by

$$\mathscr{E} = \frac{\Delta V_{def}}{d} \qquad (1)$$

Then, with $y(x)$ denoting the vertical deflection as a function of x, show that, while the particle is in the region between the plates, the trajectory is given by

$$y(x) = \frac{1}{2} \frac{\mathscr{E}e}{m} \frac{x^2}{v_x^2} \qquad (2)$$

Then show that the slope of the parabolic trajectory at $x = b$ is equal to

$$\frac{\mathscr{E}e}{m} \frac{b}{v_x^2} \qquad (3)$$

Now go back to the elementary analytic geometry of the equations of straight lines, and, making use of the point–slope form (or any other form you wish), show that the equation of the trajectory in the region between the deflecting plates and the screen is given by

$$y(x) = \frac{\mathcal{E}e}{m}\frac{b}{v_x^2}(x - \frac{b}{2}) \tag{4}$$

Use analytic geometry to prove that the straight line trajectory extrapolates back to intersect the x-axis at $x = b/2$. Letting $y(L)$ denote the vertical deflection of the particle from its initial undeflected position on the screen, show that

$$y(L) = \frac{\mathcal{E}e}{m}\frac{b}{v_x^2}(L - \frac{b}{2}) \tag{5}$$

6. It is an observed fact that the spot produced on the screen by the cathode beam does *not* smear out but remains a sharply defined spot on being deflected from its initial position when one connects a battery to the deflecting plates. If the beam does consist of charged particles, there cannot be just one; there must be enormous numbers. In the light of these observed facts and in the light of Eq. 5, what can we conclude about what the host of charged particles must have in common? Explain your reasoning.

7. Identify the known (measurable and directly observable) quantities in Eq. 5, and identify the unknown quantities. Note that there are three of the latter and that they are all properties of the hypothetical particles.

   [Thomson set out to reduce the number of unknowns. He hit upon an ingenious way of measuring v_x. Starting with the measured deflection $y(L)$ under known deflecting field strength $\mathcal{E}$, he introduced a magnetic field B (in a direction perpendicular to the plane of the figure) by means of Helmholtz coils and increased the current in the coils until the spot was returned to its initial position on the screen. Let us denote the strength of the magnetic field that brings the spot back to zero on the screen by B_0. This is a measurable quantity.]

8. Establish the direction of the magnetic field (in or out of the plane of the paper) that restores the beam to its original undeflected position on the screen. Going back to the basic equations for forces on charged particles in electric and magnetic fields, draw a free-body force diagram of a single particle when under the influence of the crossed fields $\mathcal{E}$ and B_0. Then, starting with basic concepts you studied in electricity and magnetism, show that the velocity of the particle must be given by

$$v_x = \frac{\mathcal{E}}{B_0} \tag{6}$$

Since the spot remains coherent (is not smeared out) under the influence of the magnetic field, what can you infer about the velocities of all the particles? Sketch what you might have seen on the screen if the particles did *not* all have the same velocity. Explain your reasoning. What is the *significance* of the fact that the billions of particles in the beam all have the same velocity? (*Hint:* Would they all have had the same velocity if they had originated at different points in the region between the accelerating plates? Why or why not? Since they all do have the same velocity, what is their most likely point of origin?)

9. Thomson calculated the velocity of the particles and found it to be of the order of one tenth the velocity of light. He then argued that the cathode beam could not be electromagnetic radiation as Lenard had contended. Put his argument in your own words.

10. Combining the relevant expressions that have been developed, show that, although e and m of the particles cannot be obtained separately, one can now obtain a numerical

value for the ratio of charge to mass because it is connected with measurable quantities by the relation

$$\frac{e}{m} = \frac{\mathcal{E}}{bB_0^2} \frac{y(L)}{\left(y - \frac{b}{2}\right)} \tag{7}$$

11. Suppose that in a given tube b and L are 4.00 cm and 20.0 cm, respectively, and that the spacing between the deflecting plates is 1.50 cm. Under a potential difference of 150 V on the deflecting plates, the deflection of the spot on the screen is observed to be 2.6 cm. The magnetic field that restores the spot to the center of the screen has a strength of 4.5×10^{-4} tesla (or webers/m²). Calculate the charge-to-mass ratio and the velocity of the particles in the beam. (Do not report more than the legitimate number of significant figures.)

12. Note the velocity of the particles calculated in part 11. Calculate the vertical deflection the particles sustain under the influence of gravity as they traverse the tube. Would this deflection be observable? (In answering the last question, compare the expected gravitational deflection with the order of magnitude of wavelengths of light and with the order of magnitude of the size of atoms or molecules.) Now explain in your own words why the gravitational deflection is so small and why it was *necessary* to neglect it in any calculations made in connection with this experiment. (Be sure not to include any more significant figures in your results than are justified by the data.)

13. Let us denote the accelerating potential between cathode and anode by ΔV_{acc}. Argue that *if* the negative particle originated essentially at rest at the cathode, its velocity v_x on passing through the anode would be given by

$$v_x^2 = \frac{2e\Delta V_{acc}}{m} \tag{8}$$

[Because of complications with fringing fields, configuration of the cathode and anode, and because of the difficulty of measuring the high accelerating potential difference (tens of kilovolts), Thomson could not make use of this relation to obtain a reliable value of v_x even after he had evidence as to where the particles originated (see part 8). Equation 8 does, however, indicate, in a general way, what happens to v_x as the accelerating potential changes.]

Suppose the accelerating potential is increased while the deflecting potential remains unchanged. What would happen to $y(L)$, that is, would the deflection on the screen increase, decrease, or remain unchanged? What would happen to the required value of B_0? Explain your reasoning clearly in each instance.

[Thomson made many determinations of e/m while changing conditions in the tube. He tried several different metals as cathode material (aluminum, platinum, iron). In addition to air, he made observations with other residual gases in the tube (hydrogen, carbon dioxide). (Note that even though the vacuum was very high, there was still a significant amount of gas present.)]

14. Why did Thomson conduct these experiments? What inferences are to be drawn from the fact that he kept getting the same value of e/m (within a fairly large experimental scatter)?

[When water is decomposed by electrolysis, it is found that the passage of 96,500 coul of charge liberates 1.0 g of hydrogen at the cathode and 8.0 g of oxygen at the anode. It is also well known that a molecule of water consists of two atoms of hydrogen and one of oxygen.]

15. Calculate the charge-to-mass ratio of the hydrogen and oxygen ions in electrolysis, explaining your reasoning. Compare these values with that obtained for the particle in the cathode beam.

    [In connection with the values you have examined in part 15, Thomson writes: "Thus for the carriers of electricity in the cathode rays e/m is very large compared to its value in electrolysis. The size of e/m may be due to the smallness of m or the largeness of e, or a combination of these two."]

16. State in your own words the considerations that make it seem plausible that the smallness of m best accounts for the large charge-to-mass ratio of the cathode particle rather than either of the other two possibilities cited by Thomson. (The principal observations involved are the seemingly perfect electrical neutrality of ordinary matter and the large penetration through air of the cathode beam as noted by Lenard. The latter aspect was discussed in lecture.)

17. In light of the entire story you have now put together, define the term "electron," that is, what does this word mean and what does it apply to?

APPENDIX 10B

Written Homework on the Bohr Atom

This is a written homework assignment similar to the one we had earlier on the Thomson Experiment. Be sure to think through everything carefully for thorough understanding. Explain lines of reasoning in your own words. Show intermediate steps of algebraic derivations. Show the numerical setup (i.e., the numerical substitution you have made in an algebraic expression) as well as the final numerical result of calculations. Pay careful attention to valid numbers of significant figures.

[Much profoundly significant thinking in science is done by a process called "dimensional analysis" in which one examines combinations of physical quantities in terms of the *dimensions* to which the combinations reduce. Complex combinations that end up with very fundamental dimensions may (or may not) point the way to deep underlying scientific connections. Bohr starts off his epoch-making paper of 1913 with a dimensional analysis that does turn out to be profoundly significant.]

1. Bohr introduces his paper by pointing to the fact that the combination

$$\frac{h^2}{mke^2} \tag{1}$$

has the dimensions of length and that the numerical value of the combination is about 20 Å. Confirm both of the preceding statements about the combination, showing your work in detail.

[Note, as Bohr did, that this numerical value of length is of the general order of magnitude of atomic size (what eventually turns out to be missing is simply the numerical factor $4\pi^2$ in the denominator, see Eq. 17). Bohr argued that this is evidence of the fact that Planck's constant h probably has some deep connection with atomic structure and should appear in any equations derived for this structure.]

2. Other investigators, not only Bohr, had also noted that h itself has the same dimensions as angular momentum. Verify this fact. [This encouraged the view that h might have something very directly to do with angular momentum within the atom itself. We shall return to this possibility later in our analysis.]

[Bohr attacked the problem of putting together a model of the hydrogen atom that would account for the observed discrete (bright line) spectra. He first had to settle on the *constituents* of the hydrogen atom, and he took it to consist of one proton (forming the nucleus) and one electron in circular orbit around the nucleus. To justify this choice of constituents, Bohr cites experimental work reported by Thomson. Thomson had investigated positive ions formed in various ionized gases (such as hydrogen, oxygen, nitrogen, carbon dioxide, ammonia, etc.) and reported that he had observed singly ionized atoms and molecules of all the substances. He also reported that, when he elevated the accelerating potential of the cathode beam that ionized the gases in the first place, he began to observe doubly ionized species. He then remarked that he had been able to obtain doubly ionized *atoms* of every species except hydrogen.]

3. Interpret this story in your own words: How do you account for the fact that doubly ionized atoms and molecules were not produced until the accelerating potential of the ionizing beam was elevated? What was highly suggestive about the observation concerning hydrogen, that is, how does this observation support Bohr's choice of constituents of the hydrogen atom? What other information concerning atoms also supports this choice?

[Bohr then *postulated* that the electron occupied a circular orbit around the proton nucleus and did *not* radiate continuously at its orbital frequency (as required in classical electromagnetic theory) when its orbit was of *microscopic* scale, that is, of atomic size. He postulated that an electron could occupy an orbit of given radius indefinitely in what he called a "stationary state." He further postulated that the electron gained or lost energy by absorbing or emitting a photon of frequency ν, and, in doing so, "jumped" from one stationary state (orbit of radius r_1) to another stationary state (orbit of radius r_2). Since, in each stationary state, the electron–proton system has a specific total energy $E(r)$ [we shall derive the expression for $E(r)$ shortly], Bohr was saying he was assuming that

$$h\nu = |E(r_2) - E(r_1)| \qquad (2)$$

4. Explain in your own words what lies behind Eq. 2: What motivates its introduction (i.e., what is the connection to Einstein's heuristic picture of the photoelectric effect)? Would the energy emitted or absorbed in the jumps have any direct relation to the orbital frequency f of the electron's motion? Why have the absolute magnitude signs been introduced in Eq. 2? Under what circumstances is the right-hand side of Eq. 2 positive and under what circumstances is it negative?

5. Now go back to fundamental classical physics with respect to energy quantities in orbital situations such as the one under consideration:

   Show that the potential energy (P.E.) of the electron–proton system, when the electron is in a stationary state of radius r, is given by

$$P.E. = -\frac{ke^2}{r} \qquad (3)$$

This involves going back to the *definition* of potential energy, setting up the relevant integral, choosing and justifying the choice of a *zero level* of potential energy, and carrying out the integration over appropriate limits with careful and correct treatment of all algebraic signs. Be sure to include an appropriate picture and force diagram. Be sure to explain your reasoning, especially in the choice of the zero level of energy, including a clear statement of why it is not possible to take $r = 0$ as the reference level.

Then find the expression for the kinetic energy (K.E.) of the electron in a stationary state of radius r in terms of the same quantities that occur on the right-hand side of Eq. 3.

Finally, show that the *total* energy $E(r)$ of the electron–proton system, relative to a reference level at infinite separation of the particles, is given by

$$E(r) = -\frac{1}{2}\frac{ke^2}{r} \qquad (4)$$

Why is the right-hand side of Eq. 4 negative? How can a total energy possibly be negative? Interpret Eq. 4: Does the total energy increase or decrease when an electron is moved to a "higher" orbit (larger r)? If a photon were emitted (in accordance with Bohr's picture), would the electron end up in a higher or a lower orbit? Explain your reasoning clearly and carefully.

[Now consider the Balmer–Rydberg formula, which gives the wavelengths of lines actually observed in the various series of atomic hydrogen spectra:

$$\frac{1}{\lambda} = R\left(\frac{1}{n_2^2} - \frac{1}{n_1^2}\right) \tag{5}$$

where R stands for the number 10,973,731.2 m^{-1}. (The value of R is obtained *empirically*, not theoretically, i.e., it is calculated from the *measured* wavelengths. Note the extremely high precision attained in modern spectroscopic measurements!)]

6. Show that Eq. 5 can be revised to yield the relation

$$h\nu = hcR\left(\frac{1}{n_2^2} - \frac{1}{n_1^2}\right) \tag{6}$$

[Bohr pointed out that the right-hand side of Eq. 6 contains two separate terms and that, if one adopts the idea behind Eq. 2, each one of these terms might be interpreted as related to the total energy of a stationary state of the electron.]

Putting together the observed facts of the existence of bright line spectra and the postulates of the model being developed, one arrives at the conclusion that only certain discrete stationary states (i.e., only certain special orbital radii) are allowed to exist in the electron–proton system. Present in your own words the argument that leads to this conclusion. (What would be the nature of observed spectra if *all* values of r were allowed?)

Argue that, if only certain special values r_n of r are allowed, only certain energy levels E_n of the system will be possible, and show that these energy levels can be expressed in either of the following two ways:

$$E_n = -\frac{ke^2}{2r_n} \tag{7}$$

$$E_n = -\frac{hcR}{n^2} \tag{8}$$

7. Combining Eqs. 7 and 8, show that they imply that the radius r_n of the orbit associated with the integer n is given by:

$$r_n = n^2 \frac{ke^2}{2hcR} \tag{9}$$

Argue that the radius r_1 associated with the integer $n = 1$ ought to be the lowest allowed value of the radius of the electron orbit, that this should be the "normal" or "unexcited" state of the hydrogen atom, and that higher values of n and of r_n should be associated with larger orbits and "excited" states. In this context, what is meant by the term "excited"? (The $n = 1$ state is now called the "ground state" of the atom.)

[The crucial problem now becomes that of finding out how nature selects or defines the "permitted" or "allowed" values of r_n from among the infinity of continuous values of r. Bohr, in his first paper, attacked this question via what he called the "correspondence principle." This is the idea that requires any strange, new numerical behavior, on a new level of experience, to merge smoothly into what has previously been established as correct in well explored levels of experience. (In relativity for example, one applies the correspondence principle when it is required that the Lorentz transformations for position, time, and velocity reduce to the ordinary classical relations at low velocity and when it is required that the momentum and energy formulas to do the same.) Bohr applied the correspondence requirement in the following way:

We know that, when oscillating on macroscopic scale (e.g., in radio antennae or in macroscopic circular orbits in magnetic fields), electrons radiate electromagnetic waves having the same frequency as the frequency of their periodic motion. Bohr therefore argues that, although at small values of r the electron does not radiate at all while in a fixed orbit (that is the real meaning of "stationary state") and although at such radii the frequencies of the emitted or absorbed photons have no direct relation to the frequencies of orbital motion, nevertheless, as the orbits become very, very large (i.e., approach macroscopic scale), the frequency of a photon emitted in a jump between adjacent orbits should become more and more nearly equal to the frequency of the orbital motion. If that were to be the behavior, the correspondence principle would be obeyed.]

8. Going back to the classical dynamics of circular motion, show that the frequency f_n of *orbital* motion of an electron in an orbit with radius r_n would be given by

$$f_n^2 = \frac{ke^2}{4\pi^2 m r_n^3} \tag{10}$$

where m denotes the mass of the electron.

We want the frequency ν of the photon emitted in jumps between adjacent orbits to approach f_n at large n. To achieve this, we need to look at what happens to the photon frequency ν as n becomes very large. Develop the following argument in detail:

Show that the frequency ν of the photon emitted in jumps between adjacent orbits is given by

$$\nu = \frac{E_{n+1} - E_n}{h} \tag{11}$$

and that (making use of Eq. 8)

$$\nu = cR\left[-\frac{1}{(n+1)^2} + \frac{1}{n^2}\right] = cR\left[\frac{(2n+1)}{(n+1)^2 n^2}\right] \tag{12}$$

Now argue that, as n becomes very large

$$\nu \rightarrow \frac{2cR}{n^3} \tag{13}$$

where the right arrow symbol means that the value on the left-hand side keeps getting closer and closer to the value on the right as n increases.

Argue in your own words that we can satisfy the correspondence principle if we introduce the requirement that

$$f_n = \frac{2cR}{n^3} \tag{14}$$

9. Now assemble the algebraic consequences of what has been done up to this point: You have Eqs. 8, 9, 10, and 14. Use them to solve for the quantities R, E_n, and r_n in terms of the fundamental constants, that is, obtain the following relations:

$$R = \frac{2\pi^2 m(ke^2)^2}{ch^3} \qquad (15)$$

$$E_n = -\frac{2\pi^2 m(ke^2)^2}{n^2 h^2} \qquad (16)$$

$$r_n = n^2 \frac{h^2}{4\pi^2 mke^2} \qquad (17)$$

Finally show that these results combine to give the Balmer–Rydberg formula for lines in the atomic hydrogen spectra, with the Rydberg constant no longer simply an empirical value but fully accounted for in its relation to fundamental constants.

Discuss and interpret these results: How did the energy levels come out to be discrete? Calculate the size of the normal or unexcited hydrogen atom. Would there be any meaning to setting n equal to zero? Why or why not? Calculate the energy and wavelength of a photon that would just ionize a normal hydrogen atom. Explain your reasoning carefully.

Sketch the essence of Eq. 17 for the allowed orbits by sketching a set of at least five orbits to *scale*. Account for the various spectral series (Lymna, Balmer, etc.) by showing what transitions correspond to various observed lines.

10. Look up the meaning of "angular momentum" and write the expression for the orbital angular momentum L_n of the electron in terms of r_n and f_n. Then, making use of the expressions you have derived, show that everything reduces to

$$L_n = n\frac{h}{2\pi} \qquad (18)$$

We said earlier that investigators had noted that h had dimensions of angular momentum and suspected that it might have something to do with angular momentum within the structure of the atom. Eq. 18 shows that angular momentum is "quantized" on the microscopic scale. What does this mean?

[In his subsequent papers, Bohr no longer used the correspondence principle to derive the results as you derived them above. He used the approach given in many textbooks and introduced "quantization of angular momentum" as one of the basic postulates instead of using Eq. 14. One obtains, of course, exactly the same final results.]

11. Show this last statement to be correct by rederiving Eqs. 15, 16, and 17 by using Eq. 18 as the quantization rule without invoking the correspondence principle.

CHAPTER 11

Miscellaneous Topics

11.1 INTRODUCING KINETIC THEORY

A few textbooks plunge directly into derivation of the pressure formula in kinetic theory without saying anything at all about the underlying assumptions; others make a few cryptic assertions concerning the model being adopted but do not try to justify the assumptions through appeal to prior experience available to the student. Only a very few texts discuss the assumptions in detail and provide justification.

Very few students become conscious of such gaps on their own. If no mention is made of assumptions, few realize that assumptions are being made. If the assumptions are asserted rapidly and cryptically, few students pay them any serious attention unless something along such lines is called for on tests, and then the assumptions are simply memorized with little or no consideration of how they are motivated or justified. Failure to lead students through selection of the assumptions and articulation of justifications deprives them of a rich intellectual experience with phenomenology since the underlying assumptions of kinetic theory tie to many of our everyday experiences with behavior and properties of material substances. Students need help in articulating these connections in order to understand and appreciate the structure being generated.

Furthermore, the kinetic theory of the ideal gas is an essential step in the formulation of the microscopic model underlying macroscopic properties and behavior. It is generatecd in the Galilean tradition of idealization and simplification. The idealizations are an essential feature. They must be fully understood, and even the existence of the computer does not obviate this necessity.

The primary feature in approaching kinetic theory is, of course, the acceptance of discreteness rather than continuity in the architecture of matter. Students are so used to having heard the terms "atoms" and "molecules" from early schooling that they are unaware that they have not examined any of the "How do we know . . . ? Why do we believe . . . ?" questions and have been exposed only to a string of names and unsubstantiated assertions. They have no idea that other views might have been legitimately held and that, over a long period in Western thought, atomism was considered atheistic, evil, and heretical and had to be subtly defended by its relatively few courageous proponents. It would seem that cultivation of genuine scientific literacy requires at least some attention to this background. Very few texts, even in chemistry, any longer deal with it, however, and an individual teacher must decide for him/herself whether or not to spend time examining the story.

Although, after having accepted discreteness either through examining evidence or by assertion, one does not wish to expend a great deal of valuable time discussing alternative models, there is something to be said for making students aware that alternative atomistic models were indeed entertained by major figures in the history of science. (A few students, in fact, think of these other possibilities but hesitate to bring them forth because of fear of being considered "stupid" if they do not immediately see the inevitability of the canonical model. Such students find significant reinforcement in knowing that they were in good company even if these initial thoughts did not survive subsequent tests.) The more important alternative models considered at various stages were the following:

1. *The Static Model.* Some atomists, Newton and Dalton among them, held the view that the corpuscles of a stationary, nonflowing gas occupied fixed positions and filled the entire space available to them, expanding and contracting and remaining in contact with each other as the volume occupied by the gas was increased or decreased. (Quite a few students think of this picture initially and do not see, without discussion and consideration, why the modern kinetic model must be regarded as superior.) In the *Principia*, Newton shows that if the corpuscles repelled each other with a force inversely proportional to the distance between their centers, the pressure of the gas would vary inversely as the volume, just as Boyle had demonstrated experimentally. The model therefore had the sanction of very high authority, and it persisted for a long time.

2. *The Boscovich Model.* In 1758 the Serbian scientist Roger Boscovich suggested a model in which matter was to be viewed as composed of indivisible point centers of force. The point centers possess inertia and interact with each other to infinite distances as do gravitating bodies. However, the force between two point centers is repulsive when they are very close together, alternates between attraction and repulsion as the points are moved farther apart, and becomes an inverse square attractive force when the points are widely separated. Thus, in a sense, Boscovichean atoms are infinite in extent. The whole conception involves an attempt to describe material substances only in terms of centers of force and to dispense with naïve notions of "stuff" and "matter."[1] Although Boscovich's model was not fruitful enough to achieve wide acceptance, it did influence the thinking of major figures such as William Hamilton, Michael Faraday, and Joseph Henry.

3. *The Vortex Model.* Early in the nineteenth century, Humphrey Davy proposed a qualitative dynamical theory of heat suggesting that in solids the vibration of atoms or molecules increased as the material was heated, whereas in gases the atoms rotated about about their axes or possessed rotating "atmospheres." For a brief time Joule and other investigators turned to this model and attempted to account for the tendency of gases to expand

---

[1] During the nineteenth century two opposing philosophies were influential in science, and each had its prominent adherents. Proponents of "Naturphilosophie," Oersted and Faraday among them, looked for the "interconvertability" of all forces in nature and accorded "force" a leading role in the conceptual structure. The positivists, on the other hand, aimed at removing from physics what they saw to be the mystique and vagueness associated with the concept of "force."

by visualizing (in an essentially Cartesian tradition) the atoms as spinning, fluid vortices, tending to expand centrifugally when external confining forces were relaxed.

4. *The Kinetic Model.* In his treatise on the mechanics of fluids, published in 1738, Daniel Bernoulli suggested an atomistic model based on visualizing gases to consist of minute corpuscles, moving freely and eternally at high velocities in the volume in which they are confined, exerting a steady average pressure on any boundary by virtue of extremely high frequency of bombardment. This model was neglected for about a century, one of the principal impediments being the reluctance of the scientific community to accept a model that required perfect elasticity in the microscopic interactions. The model was revived quite independently by nineteenth-century scientists (in particular Waterston,[2] Maxwell, and Clausius) who were now strongly motivated to construct a dynamical theory by the advent of the concept of conservation of energy. The kinetic theory, simple and immediately enormously successful in a wide range of applications (see, for example, Section 10.2), became the basis of our modern view.

11.2 ASSUMPTIONS OF THE KINETIC THEORY OF THE IDEAL GAS

Time spent in leading students to understanding and acceptance of the basic assumptions of the kinetic model is well invested because of the range and richness of the phenomena and experiences that must be invoked. It is such hitching together of seemingly disparate, unrelated physical manifestations that leads to mastery of concepts and understanding of the nature of scientific thought. Following is an outline of the thinking involved. The best and most effective way of conveying it to the students is to lead them to articulate the insights through Socratic group discussion rather than through didactic assertion in text or lecture.

1. Having accepted the atomic–molecular picture, a next step is to examine some of the immediate consequences in highly personal terms: If all matter, including our own flesh and bones, consists of discrete particles, what keeps individual particles hanging together to form liquids and solids? Our finger resists being pulled apart under tension; it also resists being compressed. If the structure is discrete, there is no alternative but to accept the existence of interactive forces between the particles. Furthermore, in liquids and solids the particles must be in "equilibrium" locations and spacings (potential wells, if the way has been prepared for the jargon) such that they attract each other very strongly if pulled farther apart and repel each other very strongly if pushed closer together.

   That the interaction is probably electrical is the insight that began to develop during the nineteenth century with the acceptance of atomism and with simultaneous perception of the dominant role of electricity on the microscopic scale (electrolysis, ionization, dielectric breakdown). Very few students see such connections for themselves, but they begin to articulate them under questioning, and they readily appreciate their significance. Given this crude, initial insight, many students naturally want to know more about the mechanism and details of the interaction, and they expect

[2] See Brush (1961).

pat and simple answers. It is a healthy experience for them to recognize that immediate jargon conveys no understanding; that at least four generations of sophisticated scientists asked the question and lived and died without arriving at an answer, that they (the students) might have to defer seeing the answers until they have learned more of the intervening physics.

2. Having established this initial unsophisticated insight, one can turn to gases and what might make them so different from liquids and solids. One prominent, key property is that of compressibility: very high compressibility of gases and exceedingly low (albeit not zero) compressibility of liquids and solids.[3]

Given such explicit consideration of macroscopic phenomena, students begin to perceive that the difference between liquids and solids on the one hand and gases on the other can be readily accounted for on the microscopic level by visualizing the discrete particles to be very close together in the former and far apart, relative to their own size, in the latter. Furthermore, they see that such a picture is consistent with the vastly lower density of gases (the order of magnitude of 1000 for the ratio of densities of solids and liquids on the one hand to gases on the other should be kept in mind) and the fluidity of gases. Having formed this picture, they can now see that, as a first approximation, it would be reasonable to take the long-range interactions of the particles to be negligible in the gaseous state and to expect strong repulsive interaction only during the short interval of direct collision. Furthermore, it becomes reasonable to expect a next approximation in which the longer range interaction would manifest itself as an attractive one (van der Waals forces) leading to condensation into liquid at lower temperature and higher pressure.

With this background, students begin to see the volume of liquids and solids as representing, to a first crude approximation, the volume occupied by the atoms or molecules themselves since the low compressibility implies that the particles are virtually continuous (see Section 10.2 for early use of this idea in arriving at an estimate of Avogadro's number). They can be led to interpret the density ratio of 1000 or more as an indication that the average separation of particles in a gas must be of the order of at least the cube root of 1000 or 10 times the size of the particles. (This is an opportunity to return, in a rich and conceptually important context, to ratio reasoning and scaling, the difficulty of which has been emphasized in earlier chapters.)

3. Now one can turn to evidence that the particles of the gas are in translational motion. One appropriate bit of evidence, commonly cited, is that of diffusion of odors or of colored gases (e.g., bromine) through air. A caveat, however, is posted by the confounding phenomenon of convection under small temperature differences, and more evidence is desirable. Another commonly cited

---

[3]This requires some thought and discussion. Many students, in the absence of previous instruction, initially believe liquids and solids to be completely incompressible since they do not experience visual evidence of compressibility. Here one must return to phenomena that transcend direct sense experience (see Section 3.12). Two commonly performed lecture demonstrations help in this context. One is the breaking under tension of a metal rod (that has been expanded on heating) as it cools after having had its ends pinned in a massive frame. The other is the breaking of a closed container of water on freezing. Students should be led to visualize that the breaking in each case occurs not because the materials are intrinsically inextensible or incompressible but because the surrounding structure is being asked to extend or compress the objects to their initial length or volume and that, although not impossible, such extension or compression requires very large forces.

aspect is the tendency of the gas to fill the entire space available to it. This is highly relevant, but it needs reinforcement by consideration of a concomitant aspect few students think of spontaneously yet perceive with just a hint. (This involves something that does *not* happen. The importance of being aware of what does *not* happen as well as of what does has been pointed to repeatedly; here is still another instance.) If one calls attention to the picture formed up to this point (discrete particles in gases, spaced much farther apart than their own diameters) and asks what would happen in the room if the gas molecules were stationary, students will reply that the molecules would fall out onto the floor in a very thin layer, but very few perceive this without the hint. The *combination* of the observations listed above presents a fairly compelling basis for the initial assumption of perpetual translational motion. The assumption is subsequently reinforced by feedback from the success of the picture of pressure as stemming from collisions with the wall.

4. Perpetual motion of the particles immediately implies collisions with each other and with the walls of the container. Here one confronts what was the greatest impediment to early acceptance of the kinetic theory: the question of perfect elasticity of the collisions and hence of perpetual motion on the microscopic scale. Students are not as sophisticated about this problem as were mature scientists of an earlier day, but, if given the chance, they do express concern about perfect elasticity. They know that macroscopic collisions are inelastic, that motion would run down in a macroscopic system. They have been told repeatedly that "perpetual motion is impossible," and they do not immediately perceive the possible differences at the microscopic level. First they must be led to perceive that, if one accepts the picture developed so far, there is no choice but to accept elasticity since the pressure on the walls does not diminish and the gas molecules never do fall out to the bottom of the container, implying that the motion persists. Furthermore, elasticity of interaction on the microscopic level is reasonable if it is essentially electric or electromagnetic. (At this point it is effective to appeal to "noncontact" collisions of charged pith balls or magnets that might have been demonstrated during study of electricity and magnetism.)

5. Acceptance of perpetual motion on the microscopic scale leads to visualization of the intrinsic randomness of the system and the concept of distribution of speeds and directions of motion. At some point in the sequence, either here or earlier, it is, of course, highly effective to put into use one of the many available demonstrations of the kinetic model that utilize a multiplicity of small beads kept in continual random motion, or a comparable computer simulation. (In my own experience, the material beads present far more conviction to the students than do the computer simulations—at least in the early, unsophisticated stages of development. The computer simulations become useful and more impressive as one gets further along into more sophisticated aspects.) The demonstrations make vivid the fact that the particles keep changing their speeds and directions of motion, that collisions with the walls and among particles take place at all possible angles, and that instantaneous velocities range from zero to very high values and that one cannot think of a single velocity but must think in terms of distributions and averages.

6. It is now appropriate to visualize, without any formalism and in a purely qualitative way, the generation of pressure on the wall of the container.

Here one invokes, of course, the situation that should have been examined in single particle dynamics—the force associated with change of momentum on elastic bouncing from a wall—and sets it in the context of force per unit area and an enormously high rate of collision. Discussion should elicit the perception that the force per unit area should depend both on the frequency of collisions (and thus on the density of the gas) and on the velocity of the incident particles. (This combined dependence may seem obvious to us after long familiarity with the model, but, in fact, it eluded many of the early thinkers and constitutes a sophisticated penetration on the part of the students.) Ultimately one finds that the collision frequency also depends on the velocity, and the pressure thus depends on the *square* of the velocity, but this must be allowed to emerge from the more formal derivation.

7. The model now merits further refinement through examination of the implications of the isotropy of the gas and the steadiness of pressure on the wall. How is it that the pressure remains so steady despite being caused by individual collisions? How is it that isotropy is maintained? Students respond with relatively little difficulty to those aspects that are associated with high frequency and large numbers, that is, they see that the constancy of pressure can stem from the enormous numbers of collisions and they can visualize the fluctuations that would set in as the numbers were decreased. The moving-bead demonstrations help a great deal in this context.

   What is more subtle and difficult to perceive (and what merits slower and more careful discussion) is the "principle of detailed balancing"—the notion that constancy and isotropy stem not simply from large numbers as such but from the fact that the large numbers lead to a situation in which any change in speed and direction of motion of a particle somewhere in the gas is invariably compensated by exactly the reverse change of another particle, always maintaining the steady state once it has been achieved from some initial, transient situation (such as the crowding of the particles toward one end of the box).

   Another aspect depending on the principle of detailed balancing is the justification for treating the bouncing of molecules from the wall as, on the average, specular reflections. Students should be led to sketch a "molecule's eye" view of the wall and realize that the wall must be "rough," full of bumps and hills and valleys on the molecular scale. It is the averaging out of the departures from "angle of incidence equals angle of reflection" over enormous numbers of collisions that justifies the idealization of specular reflection at the walls.

8. Having gone this far with the qualitative picture, it is possible to anticipate that temperature will be somehow related to molecular velocity. Most students are aware that the pressure of a gas increases when the temperature is increased without change in volume (i.e., at constant density). With constant density, given the model developed so far, the only way that pressure can increase is through an increase in velocity of the molecules. Thus, temperature must be intimately connected to molecular velocity. Now the uniformity of temperature in a gas at equilibrium can also be related to the large numbers of particles and the principle of detailed balancing.

9. Finally, one can invoke the energy concepts and connect them with the evolving qualitative picture. The assumption of negligible forces of interaction (except during extremely short intervals of collision, "short" meaning short

relative to the average interval between collisions) among the widely separated molecules, implies negligible storage of potential energy in the ideal gas. (Students should not lose sight of the fact that storage of potential energy in molecular interactions does play a significant role in real gases and that large amounts of potential energy are associated with the latent heat of condensation to liquid state.) The transfer of heat to a gas at constant volume is then visualized as increasing the internal thermal energy of the gas by increasing the random kinetic evergy of the molecules. If the molecules have no internal degrees of freedom, their kinetic energy is purely translational. If they can be excited into rotation, then rotational kinetic energy would be involved. If internal vibration is possible, then the molecules would possess both internal potential and kinetic energies of vibration that could be exchanged in collisions, etc.

On the one hand these qualitative insights pave the way for understanding one of the principle failures of classical theory and initial successes of quantum theory (namely, the accounting for the observed anomalies in the specific heats of gases), and, on the other hand, to an appreciation of why one starts the most elementary kinetic theory with the consideration of point–mass particles, thus avoiding, at least at the beginning, the complexity imposed by molecular internal energy other than the purely translational. Without such qualitative background, the assumption of point–mass particles is unmotivated and carries no meaning to the students. They memorize it, if necessary, among the various inexplicable assumptions, but they regard it as still more unintelligible black magic.

10. A final valuable homework exercise, not strictly within the realm of kinetic theory of the ideal gas, nevertheless helps students register many essential ideas. This assignment might run as follows:

a. Describe in terms of motion and migration of molecules what it is that happens in evaporation of liquid water from a container that is open to the air. Supplement your verbal description with sketches, however crude the latter may be. Keep in mind the fact that some water molecules that have escaped from the liquid may return to it in their random motion and collisions. In terms of the kinetic model, explain how it is that all the liquid eventually evaporates.

b. Describe in similar terms what happens when water, with an air space above it, is put into a tightly closed container. In terms of the kinetic model, why is it that, after a relatively short time, no further *net* evaporation of water takes place under these circumstances? Do molecules cease leaving the water when the net evaporation ceases? The space above the liquid water is said, under these conditions, to be "saturated" with water vapor, and the contribution of water molecules to the total pressure of the gas phase on the walls of the container remains constant. How does this contribution manage to remain constant despite the continual random motion of the molecules and the fact that some of them must return to the liquid phase during their wanderings?

c. Consider the situation in which a lump of sugar is placed in water. Although the process is slow in the absence of stirring, all the sugar eventually dissolves if the amount of water is fairly large. If the amount of water is relatively small, however, the dissolving ceases, and the solution is said to be "saturated." Describe in terms of the kinetic-molecular model, with the aid of sketches, what happens in these two cases, indicating how it is that equilibrium is achieved in one case and not in the other. Compare this situation with the evaporation discussed in parts (a) and (b).

11.3 HYDROSTATIC PRESSURE

Understanding the concept of "hydrostatic pressure" involves a good bit more than acquiring the definition "force per unit area." A major step toward understanding resides in appreciation of the full significance of Pascal's Law: that the pressure at any point in a fluid is the same in *all* directions. The usual formal "proof" of this idea is given by examining the static equilibrium of a small volume of fluid, for example, one having a parallelopiped shape and a triangular cross section. This treatment is so abstract that, even if it is presented, very few students assimilate its physical implications. Many introductory courses now eschew such treatments entirely in order to save time for other subjects. This is a legitimate choice, but teachers must then remember that something very subtle and fundamental has been left out and must be prepared to help close the gap at some subsequent point.

The subtlety of the insight, and the fact that many individuals (including many active physicists) have not really acquired it, are indicated by the responses given to the following question: A container of the shape in Fig. 11.3.1 initially contains a uniformly dispersed mixture (or colloidal suspension) of two immiscible liquids of different densities (e.g., oil in water; cream in milk). As time goes by, the lower density fluid separates and collects in the throat of the container. How does the final pressure on the bottom of the container (after the separation is complete) compare with the initial pressure (when the fluids were uniformly dispersed)? Is it the same, greater, or smaller than the initial pressure?

Novices tend to say they do not know, but the majority of those who have had some physics but have never thought about such manifestations of fluid pressure tend to say that the pressure must remain the same. Those who suggest a line of reasoning rather than just making an intuitive guess say that, since the weight of the fluid is unchanged, the total force on the bottom, and therefore the pressure on the bottom, must remain unchanged.

This reasoning is incorrect because the pressure on the bottom of a container of any shape other than purely cylindrical is not equal to the weight of fluid divided by the area of the bottom. Since the fluid pressure is the same in all

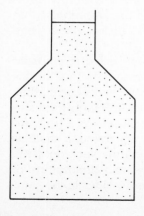

Figure 11.3.1 Initially uniform mixture of two immiscible fluids separates, and lower density fluid collects in throat of container. Does pressure on the bottom of container change with the separation of the liquids?

directions at any point, the fluid at a wall exerts a force on the wall, and the wall, in turn, exerts a force on the fluid. The sloping walls of the container exert a *downward* component of force that influences the pressure on the bottom of the container. Since, on separation, the average density of the fluid in the central column is less than it was initially, the pressure at the bottom of the central column has decreased. The pressure at the level of any point along the sloping walls has also decreased; the downward force exerted by the wall has decreased; the pressure has decreased all over the bottom.

Students can be helped to understand the physics of the phenomena through the drawing of simple force diagrams, for example, force diagrams of a central column of fluid and of columns under the sloping walls. The drawing of such diagrams helps develop understanding of the physics and of the real meaning of fluid pressure, and the spiraling back to force diagrams helps strengthen the usually still uncertain grasp of the force concept (as discussed in Chapter 3).

An illustration of how an alteration in context can help deepen understanding: If, after discussion of the phenomena of fluid pressure, one turns students' attention to the pressure at the base of a vertically standing solid metal cone or cylinder, and asks what will happen to the pressure at the base of the cylinder if it expands under a uniform increase in temperature, many students (especially those who have memorized the formula $p = \rho g h$ without real understanding of where and when it applies) will say that the pressure at the base increases because the height increases. In the case of the solid, the pressure at the base *is*, of course, equal to the weight of the object divided by the area of the base. In the case of thermal expansion, the change in height has no effect, and the pressure at the base decreases (very slightly) because of the increase in area.

Such contrasts in phenomenology play a powerful role in enhancing understanding, but few students can raise such questions on their own. They begin to do so only if they are led into the effort, gradually, and through many examples.

11.4 VISUALIZING THERMAL EXPANSION

Thermal expansion is another subject that is frequently skipped nowadays in introductory physics for the sake of other topics. This is just as legitimate as the skipping of fluid phenomena, but, again, teachers should be cognizant of certain widely prevalent gaps and misconceptions, the most important of which is the failure to visualize the overall effect of linear expansion of solids.

Many students fail to see that linear expansion in all directions means that all lengths, all areas, and all elements of volume in an object expand simultaneously. Thus they fail to see that the *circumference*, not just the diameter, of any circle drawn in the object increases in length. As a result, they fail to comprehend that a hole in a metal plate will increase in size on thermal expansion rather than decrease.

If thermal expansion is to be understood, students must be led to confront the qualitative phenomena in addition to the standard numerical exercises. In this instance, one must visualize the details of effects that cannot be seen directly; they transcend direct sense experience. Such exercises are essential in building up students' capacity for abstract logical reasoning and for using concepts as a basis for understanding more complex phenomena. Simple, everyday phenomena, such as thermal expansion, can provide very valuable contexts for building such capacity. It pays to invoke them, at least as minor digressions, without the expenditure of excessive time on numerical exercises such as those common in older texts.

11.5 ESTIMATING

A widely prevalent faculty complaint concerns students' unwillingness and resistance to doing simple calculations that involve estimating magnitudes of any kind. The thinking that goes into such calculations is, nevertheless, very sophisticated, however simple it may seem to experienced physicists. Students can be helped to develop the capacity, but only through being given the chance to practice in meaningful and interesting contexts. Since scaling and ratio reasoning are inextricably involved in almost all estimating, and, given the difficulty that the majority of students have with such reasoning (see Sections 1.6–1.12), teachers must recognize the underlying difficulty and provide the necessary guidance and simpler exercises where necessary. Only then will the capacity begin to develop.

A great deal of thought, effort, and imagination has gone into invention of fruitful problems of this variety—problems that can be readily incorporated into homework and tests in existing courses even though very few occur in standard texts. Teachers interested in drawing on such resources will find a rich harvest in various references in the bibliography, for example, Bartlett (1976–1979), Crane (1960, 1969a,b, 1970), Hobbie (1973), Kunz (1971), Lin (1982), Memory and Jenkins (1977), and Morrison (1963).

11.6 EXAMPLES OF MATHEMATICAL PHYSICS FOR GIFTED STUDENTS

A profoundly important aspect of intellectual development in students who are likely to become physicists (or engineers who use physics at a sophisticated level) resides in being able to extract subtle physical interpretation from analytical results. This is, again, a matter of practice and experience, but very little such experience is generated in standard problems in the texts until later, more advanced, levels of study. It is highly desirable to expose gifted students to such practice at the earliest possible moment, but, at introductory physics level, it is necessary to invoke situations that are sufficiently simple to be analyzed with the conceptual and mathematical tools then available.

Such situations must be sufficiently rich phenomenologically to offer nontrivial problems of interpretation, that is, the situation must not be intuitively obvious, and the mathematical solution must be essential for penetration of the physics. Such problems are not easy to find; a few that I have found valuable are illustrated in the following. It must be emphasized that these problems are not original and are quite well known. They do not, however, generally occur in textbooks, and the key to their use is not so much in the problems themselves as in how they are presented to the student. The basic approach in each instance is to have the student set up the problem, obtain the formal solution, and then interpret the solution by extracting its physical implications. Playing with actual apparatus (where this is feasible) can follow the experience of doing the abstract mathematical physics. Reversing the procedure is perfectly possible, but it destroys the mathematical-physics experience.

Problem A.
A bob of mass m is attached directly to a spring of finite length having spring constant k, and the system is whirled around in a horizontal circle at angular velocity ω as shown in Fig. 11.6.1. (The axis of rotation is the vertical line through the end of the spring.) At the initial, relaxed condition the distance of the center of

284 MISCELLANEOUS TOPICS

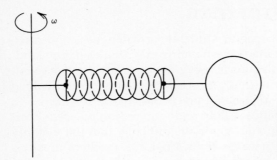

Figure 11.6.1 Bob on spring of finite length is whirled around in a horizontal plane. Axis of rotation is vertical line through end of spring.

mass of the bob from the axis of rotation is denoted by r_0, and, at angular velocity ω, the radius of the circular motion is denoted by r. Examine the dynamics of this system from the point of view of determining the way in which the equilibrium value of r varies with imposed values of ω. Interpret the mathematical result by describing the physical effects predicted. Do not try to treat this situation as one in which r varies with time; just find how values of r depend on steady values of ω.

Solution: Since, at any final radius r the spring has been stretched by an amount $r - r_0$, the centripetal force exerted on the bob is given by $k(r - r_0)$. The dynamical equation is therefore

$$k(r - r_0) = mr\omega^2 \tag{11.6.1}$$

Solving for r, we obtain

$$r = \frac{r_0}{1 - \frac{m\omega^2}{k}} \tag{11.6.2}$$

Interpretation: Equation 11.6.2 shows that, starting from zero angular velocity, r increases as ω increases as would be expected. What is unexpected, however, is that the system blows up as ω approaches the angular frequency of the natural oscillation of the mass on the spring.

Problem B.

A pendulum bob of mass m is fixed at the end of a thin rigid rod of length L. The mass of the rod is negligible compared to that of the bob. The bob and rod form a conical pendulum that can be rotated at angular velocity ω around a vertical axis through the pivot at the end of the rod (Fig. 11.6.2). Obtain the analytical expression for the angle θ assumed by the rod at various steady angular velocities ω and interpret what it predicts about the physical behavior of the system.

Solution: From the force diagram in Fig. 11.6.2, we have

$$T \cos \theta - mg = 0 \tag{11.6.3}$$

and

$$T \sin \theta = mL \sin \theta \, \omega^2 \tag{11.6.4}$$

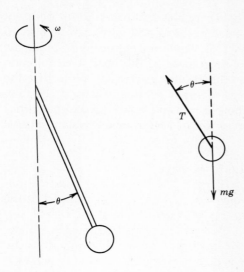

Figure 11.6.2 Conical pendulum (bob on rigid rod) is rotated at angular velocity ω around a vertical axis through the pivot at the upper end of the rod. At any steady angular velocity, the rod takes up an angle θ from the vertical.

Combining Eqs. 11.6.3 and 11.6.4 gives

$$\cos \theta = \frac{g}{L\omega^2} \qquad (11.6.5)$$

Interpretation: Equation 11.6.5 shows that, for values of ω less than $\sqrt{g/L}$, $\cos \theta$ would be greater than 1. Thus, θ would be imaginary and consequently meaningless. The solution implies that the pendulum does not begin to swing out from its zero position until the angular velocity ω exceeds the natural angular frequency of the pendulum. This is indeed the case, as can be readily confirmed by an experimental observation.

Problem C

A thin wire is bent into the shape given by the power function $y = ax^n$ where n takes on integer values starting with unity (Fig. 11.6.3). A bead of mass

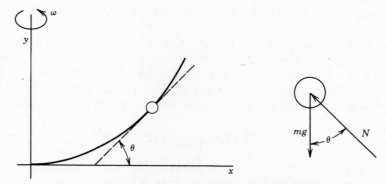

Figure 11.6.3 A rigid wire is bent into the form of the power function $y = ax^n$ and is rotated around the vertical axis. A frictionless bead of mass m can slide along the wire and take up equilibrium positions at various steady values of angular velocity ω.

m is free to slide along the wire with negligible friction and can take up different equilibrium positions along the wire. The system (bead and wire) can be rotated at angular velocity ω around the y-axis, which is oriented vertically. Obtain the analytical expression for the equilibrium x-coordinate of the bead as a function of steady angular velocity ω and interpret the results physically. Note that the results are quite different for different values of n; pay especially careful attention to interpretation of results for the cases where $n = 1$ and $n = 2$. (Note that, since both x and y are to have the physical dimension of length, the constant a must have dimensions of $(\text{length})^{-n+1}$).

Solution: From the force diagram in Fig. 11.6.3, we have

$$N\cos\theta - mg = 0 \tag{11.6.6}$$

and

$$N\sin\theta = mx\omega^2 \tag{11.6.7}$$

Combining Eqs. 11.6.6 and 11.6.7 gives

$$\tan\theta = \frac{x\omega^2}{g} \tag{11.6.8}$$

Since, from the shape of the wire

$$\tan\theta = nx^{n-1} \tag{11.6.9}$$

we obtain

$$x^{n-2} = \frac{\omega^2}{nag} \tag{11.6.10}$$

Interpretation: For values of n greater than 2, the equilibrium value of x is stable and increases, as would be expected, with increasing ω. For the case of the straight wire ($n = 1$), the equilibrium position is given by $x = ag/\omega^2$, but this is an unstable equilibrium in which the bead would slide inward on one side and outward on the other.

The most interesting case is that for $n = 2$. Here, when the angular velocity is equal to $\sqrt{2ag}$, the bead will stay wherever it is placed along the wire. This corresponds exactly to the situation in the rotating basin of water in which, at a given steady angular velocity, the water surface adjusts itself to the appropriate parabolic shape, and every particle of water then occupies an equilibrium position.

11.7 CHAOS

One of the deepest scientific insights attained in recent years, owing largely to the possibilities opened by the computer, has been the realization that classical mechanics is intrinsically indeterminate because of ever-present nonlinearity. This indeterminacy is fundamentally different from that of quantum mechanics, but it is universal and just as deeply significant. What this insight reveals is that Laplace was wrong: Given the values of position and velocity of every particle in the universe, it is not in fact possible (even in a classical Newtonian universe) to predict the subsequent history of that universe. Regardless of how closely together initial conditions are taken, the solutions diverge exponentially when the governing equations are nonlinear.

This insight is one to which we should try to expose our students, especially those in engineering-physics courses, and the ubiquity of adequate computers helps make this possible. The mapping of solutions in phase space also enhances grasp of the meaning of phase space and prepares the way for subsequent, more sophisticated levels of study.

How far one then goes with examination of still deeper aspects (e.g., strange attractors, "dimensionality," connection to fractals, etc.) is a matter of available time and of judgement on the part of the teacher. I believe that we should at least aim for the first level of insight concerning the divergence of solutions regardless of the "exactness" of initial conditions.

CHAPTER 12

Achieving Wider Scientific Literacy[1]

12.1 INTRODUCTION

That public understanding of science, or scientific literacy, is in a lamentable state is an old story. Magazine articles, educational literature, and pronouncements of scientific associations all through this century and well back into the last bear testimony to this. Scientists of several decades ago, returning to academic work after time spent in research or serving in World War II, were especially determined to help college students acquire a better sense of the nature, power, and limitations of scientific thought, as well as a better understanding of the interactions between science and society. Since then, the escalating impact of science and technology on moral, ethical, political, and societal problems has only continued to enhance the urgency of the problem of education and to heighten the pertinence of the liberal education objective.

For years, meetings of scientific societies reverberated, and pages of educational journals were filled, with descriptions of new courses that had been designed to lead nonscience majors to greater scientific literacy. Almost every report presented or published was accompanied by the results of "evaluations" of student answers to tendentious questionnaires, the answers invariably demonstrating how much the students loved the course, valued the learning experience, and appreciated the instructor's efforts on their behalf. With but a few exceptions, these course vanished and were rapidly succeeded by "more up-to-date" but essentially identical—and equally evanescent—versions, also accompanied by enthusiastic student testimonials.[2]

These numerous attempts have had very little impact on scientific literacy. Those who were students in such courses, and responded so favorably on the questionnaires, show little or no understanding of science and of its interactions with society. In retrospect, most say that they enjoyed their course very much, but recall nothing of what they were supposed to have learned. (It has been

---

[1] This chapter is based on a paper originally published in *Daedalus* [Arons (1983a)].

[2] I wish to make it clear that I am not deprecating student opinion as such. I do question, however, the specious use of such opinion by faculty who have failed to provide the students with an adequate frame of reference from which to judge what has and has not been learned.

sardonically suggested that, this being the almost universal outcome, perhaps we should direct our efforts to devising courses still more enjoyable and still easier to forget.) Yet the clamor for making science a more effective component of liberal education continues unabated, and with an urgency indicative of how little past efforts have achieved.

The notion that understanding of science can be achieved by purely verbal inculcation seems to me to be a principal source of failure. Experience makes it increasingly clear that exclusively verbal presentations—lecturing to large groups of intellectually passive students and having them read text material—leave virtually nothing in the students' minds that is permanent or significant. Much less do such presentations help the student attain what I would consider the marks of a scientifically literate person. Since such marks, however, underlie the contentions and recommendations in this chapter, it is well to stop at this point in order to enumerate some of them.

12.2 MARKS OF SCIENTIFIC LITERACY

I suggest that an individual who has acquired some degree of scientific literacy will possess the ability to:

1. Recognize that scientific concepts (e.g., velocity, acceleration, force, energy, electrical charge, gravitational and inertial mass) are invented (or created) by acts of human imagination and intelligence and are not tangible objects or substances accidentally discovered, like a fossil, or a new plant or mineral.

2. Recognize that to be understood and correctly used, such terms require careful operational definition, rooted in shared experience and in simpler words previously defined; to comprehend, in other words, that a scientific concept involves an idea *first* and a name *afterwards*, and that understanding does not reside in the technical terms themselves.

3. Comprehend the distinction between observation and inference and to discriminate between the two processes in any context under consideration.

4. Distinguish between the occasional role of accidental discovery in scientific investigation and the deliberate strategy of forming and testing hypotheses.

5. Understand the meaning of the word "theory" in the scientific domain, and to have some sense, through specific examples, of how theories are formed, tested, validated, and accorded provisional acceptance; to recognize, in consequence, that the term does *not* refer to any and every personal opinion, unsubstantiated notion, or received article of faith and thus, for example, to see through the creationist locution that describes evolution as "merely a theory."

6. Discriminate, on the one hand, between acceptance of asserted and unverified end results, models, or conclusions, and, on the other, understand their basis and origin; that is, to recognize when questions such as "How do we know . . . ? Why do we believe . . . ? What is the evidence for . . . ?" have been addressed, answered, and understood, and when something is being taken on faith.

7. Understand, again through specific examples, the sense in which scientific concepts and theories are mutable and provisional rather than final and unalterable, and to perceive the way in which such structures are continually refined and sharpened by processes of successive approximation.

8. Comprehend the limitations inherent in scientific inquiry and be aware of the kinds of questions that are neither asked nor answered; be aware of the endless regression of unanswered questions that resides behind the answered ones.
9. Develop enough basic knowledge in some area (or areas) of interest to allow intelligent reading and subsequent learning without formal instruction.
10. Be aware of at least a few specific instances in which scientific knowledge has had direct impact on intellectual history and on one's own view of the nature of the universe and of the human condition within it.
11. Be aware of at least a few specific instances of interaction between science and society on moral, ethical, and sociological planes.
12. Be aware of very close analogies between certain modes of thought in natural science and in other disciplines such as history, economics, sociology, and political science; for example, forming concepts, testing hypotheses, discriminating between observation and inference (i.e., between information from a primary source and the interpretations placed on this information), constructing models, and doing hypothetico-deductive reasoning.

I hasten to indicate that this list is neither exhaustive nor prescriptive. It illustrates some of the insights that I believe characterize scientific literacy and that I find most college undergraduates, given time and opportunity, and having the willingness to exert some intellectual effort, can encompass. Readers will have valid modifications, preferences, and priorities of their own. These can be interpolated and examined in light of the following discussion, which I will confine to efforts being made in schools, colleges, and universities to upgrade scientific literacy.

12.3 OPERATIVE KNOWLEDGE

Researchers in cognitive development describe two principal classes of knowledge: figurative (or declarative) and operative (or procedural) [Anderson (1980); Lawson (1982)]. Declarative knowledge consists of knowing "facts"; for example, that the moon shines by reflected sunlight, that the earth and planets revolve around the sun, that matter is composed of discrete atoms and molecules, that animals breathe in oxygen and expel carbon dioxide. Operative knowledge, on the other hand, involves understanding the *source* of such declarative knowledge (How do we know the moon shines by reflected sunlight? Why do we believe the earth and planets revolve around the sun when appearances suggest that everything revolves around the earth? What is the evidence that the structure of matter is discrete rather than continuous? What do we mean by the names "oxygen" and "carbon dioxide" and how do we recognize these as different substances?) and the capacity to use, apply, transform, or recognize the relevance of the declarative knowledge in new or unfamiliar situations.

To develop the genuine understanding of concepts and theories that underlies operative knowledge, the college student, no less than the elementary school child, must engage in deductive and inductive mental activity coupled with interpretation of personal observation and experience. Unfortunately, such activity is rarely induced in passive listeners, but it can be nurtured, developed, and enhanced in the majority of students providing it is experientially rooted and not too rapidly paced, and providing the mind of the learner is actively engaged.

There is increasing evidence that our secondary schools and colleges are not doing a very good job of cultivating operative knowledge in *any* of the formal disciplines, and that the teaching of science is not unique in this respect—although the failures in science are more immediately obvious [Arons (1976); Chiappetta (1976); McKinnon and Renner (1971)]. Consider some specific illustrations:

1. Almost any individual (child, student, or adult) who is asked about the origin of the light coming to us from the moon will respond with the assertion that the moon shines by reflected sunlight. When one asks, however, for the evidence for this conclusion, one very rarely obtains a meaningful or logical response. The knowledge is purely declarative and has been received from authority without accompanying evidence or support. It is interesting to note a deeply related misconception: Most people, including nonscience college faculty, if asked how they account for the unilluminated portion when they see a bright crescent moon, respond that the dark portion is the shadow of the earth. Very few people have ever watched the moon in its changing phases and taken the intellectual step of noting the simultaneous location of the sun. It is perfectly possible to lead young children to full understanding of what is going on and why we conclude that "the moonlight is the sunlight" (see Tennyson's "Locksley Hall Sixty Years After"), but this is very rarely done.

2. In a more subtle and sophisticated context, virtually any individual will tell you that the earth and planets of our solar system revolve around the sun. Most people do not even see anything paradoxical about this because, unlike the ancients, few of us now have occasion to sleep out under the sky and watch the procession of the celestial bodies. If asked for the evidence, for the reasons why we accept a helio- rather than a geocentric model, the vast majority, including college science majors, react only with dismay or embarrassment. A few might mutter something memorized and unintelligible about "stellar parallax," but even these have no realization that the Newtonian picture was firmly accepted long before stellar parallax was actually observed and that the observation was simply a confidently expected confirmation. Thus most individuals have "learned" what Whitehead describes as an inert "end result." They possess only declarative knowledge received from authority, and they have no understanding of the first grand synthesis provided by modern science. They are probably even less sophisticated than their medieval counterparts, who would have put forth the geocentric model but would have qualified it as only "saving the appearances" rather than representing a final Truth. Such modern-day reactions of purely declarative knowledge are typical with respect to many other aspects of science, and, I submit, are not what we have in mind when we speak of "an understanding of science."

3. An example drawn from experience with both pre- and in-service elementary school teachers in undergraduate science courses (the two groups turn out to be indistinguishable in their levels of understanding of science subject matter): Somewhere in their general science courses in the schools, or in other circumstances, they had all heard expositions about "electrical circuits," had seen diagrams in books or on chalkboards, and listened to assertions of the facts and concepts of current electricity. When they are given a dry cell, a length of wire, and a flashlight bulb and are asked to get the bulb lighted, they almost invariably do one of the following things: they either hold one

end of the wire to one terminal of the battery and touch the bottom of the bulb to the other end of the wire, or they connect the wire across the terminals (i.e., short the battery) and hold the bulb on one battery terminal. They have no sense of the two-endedness of either the battery or the bulb; few notice that the wire gets hot when connected across the battery terminals, and fewer still infer anything from the latter effect. It takes 20 to 30 minutes before someone in the class discovers, by trial and error, a configuration that lights the bulb. (Seven-year-old children, when confronted with the same situation, go through exactly the same initial steps, and 20 to 30 minutes elapse before someone gets the bulb lighted.) Lacking the synthesis of actual *experience* into the concept of "electrical circuit," the college students, despite the words they "know" and the assertions and descriptions they have received as passive listeners, have no more understanding of the ideas involved than the seven-year-old approaching the phenomenon *de novo*. Purely verbal inculcation has left no trace of genuine knowledge or understanding. Such is the outcome of the majority of our present modes of science instruction.

12.4 GENERAL EDUCATION SCIENCE COURSES

The majority of college courses that purport to cultivate scientific literacy in the nonscience major tend to fall into two principal classes: courses that in one quarter, one semester, or even one year attempt to give students an insight into the major achievements of a science (e.g., in physics, everything from Galileo and Newton through the laws of thermodynamics, relativity, quantum mechanics, and current particle physics); and courses that focus on some narrower topical area such as the energy crisis, spoliation of the environment, the application of science to military problems, ethical and moral questions lying behind modern advances in molecular biology, philosophical questions posed by relativity and quantum mechanics, and so on.

Courses in the first category have been invented and reinvented in essentially the same form countless times ever since general education curricula sought to provide courses addressed to nonscience majors. Despite pretensions to being substantive and not merely "surveys" and despite the always glowing student "evaluations," these courses have had so short a half-life and so little effect on the generations of students subjected to them, that they are still being reinvented to fill the persisting vacuum. Young scientists, completely unaware of past experience, seem to think the vacuum is there because this mode has never been tried before and that the solution lies in presenting the material in their own specially enthusiastic and impeccably lucid way. The truth is that the vacuum is there because this mode has no prospect whatsoever of educational success, yet its proponents continue to justify it on the ground that students are given a "feeling" for the content of science and the nature of modern scientific thought and that they now "know" something about current scientific progress. Meanwhile, complaints about the lack of scientific literacy continue to escalate.

Such efforts founder—as their replications will continue to founder—first, because they invariably subject students to an incomprehensible stream of technical jargon that is not rooted in experience accessible to the learner; second, because the subject matter is poured forth much too rapidly and in far too great a volume for significant understanding of ideas, concepts, or theories to be generated and assimilated. The pace makes difficult, if not impossible, the development of a sense of how concepts and theories originate, how they come to be validated and accepted, and how they connect with experience and reveal relations among seemingly disparate phenomena. Both the pace and the volume preclude any meaningful reflection on the scope and limitations of scientific knowledge or of its impact on our intellectual heritage and view of man's place in the universe. The "stream of words" courses have not solved, and will not solve, our educational problem, however handsomely illustrated the texts and however liberally salted they may be with allusions to pollution, ethics, energy crises, black holes, or Kafka.

Courses in the second category, although supposedly narrower in scope, nevertheless suffer from related difficulties. It seems to me that intellectual integrity would demand that students acquire some genuine comprehension of the scientific concepts, theories, and insights underlying the great topical problem being examined, and that students should not be encouraged to discourse vacuously on matters they essentially do not understand. With students who already have the requisite conceptual background, one can, of course, enter these discussions directly. But with students who have no notion of what "energy" means (many regard it as some kind of material substance) and no comprehension, however qualitative, of the restrictions imposed on us by the laws of thermodynamics; with students who have no basis for belief in the discreteness of the structure of matter (knowing only a string of names such as "atom," "molecule," "nucleus," "electron" that have been thrown at them by assertion without examination of any of the experiential evidence and reasoning underlying the names); with students who have no idea what is meant (and what is *not* meant) by "electrical charge"; with students who have no understanding of the grounds on which we accept the proposition that the earth and planets revolve around the sun; and with students who are still Aristotelian in their use of teleological locutions and in their unawareness of the law of inertia—with such students it is intellectually specious and dishonest to pursue the initial discussion without first helping them form and understand the essential prior concepts. Indeed, once they see where intellectual integrity lies and what they must understand to talk meaningfully and intelligently about the original problem, few students object to the digression necessary for understanding the underpinnings.

Such backtracking to the necessary scientific understanding, however, drastically reduces the amount of coverage. To the best of my knowledge, very few courses have made the sacrifice, and the students emerge with no more understanding of the scientific concepts or of the nature and limitations of scientific thought than do the victims of courses in category one.

What is the alternative? It seems to me that it is essential to back off, to slow down, to cover less, and to give students a chance to follow and absorb the development of a small number of scientific ideas at a volume and pace that make their knowledge operative rather than declarative. Depending on the time available, they might, for example, be led through one or more of the following questions:

1. Why do we believe the earth and planets revolve around the sun? In what context of concept and theory is this picture "correct"?
2. Why do we believe that matter is discrete rather than continuous in structure? What is the *evidence* behind our belief in atoms and molecules?
3. What do we mean by the term "electrical charge"? How does the concept originate? Is "charge" some kind of substance? What is meant by "like" charges; in what sense is the word "like" being used? On what grounds

do we believe that there are only two varieties of electrical charge? What (hypothetical) evidence would force us to conclude we had discovered a third variety? What is the evidence that electrical charge plays a fundamental role in the structure of matter?

4. Why do we believe that atoms have discrete structure on a subatomic scale? What is some of the evidence? (Ex cathedra assertion of terms such as "electron" and "nucleus" is not evidence at all and cultivates only declarative knowledge; yet much teaching of "science" is done in this fashion.) What experiments and observations lead to (and reinforce) creation of the concept "electron"? What is the evidence that such an entity is a universal constituent of matter? What is the evidence that it is subatomic? What role does the electron play in atomic structure?

5. Since we can take thermal energy out of the atmosphere or the ocean without violating the conservation principle, why is it that we cannot solve energy shortages by using the atmosphere or ocean as energy sources?

6. In what way does Einstein's Special Theory of Relativity alter our fundamental conceptions regarding space and time?

Any one of these questions can be dealt with in an honest way under restricted coverage of subject matter and can be used to cultivate and enhance aspects of scientific literacy such as those defined in Section 12.2.

By contrast, I suggest that enterprises such as the following (however popular they may be) are at best useless, and at worst damaging, since there is insufficient time to attack the "How do we know . . . ? Why do we believe . . . ?" questions or the necessary background is far too advanced:

1. Students being told about the "fascinating" particles of high-energy physics (with unintelligible jargon about interactions, angular momentum, mass–energy relations, quantum transitions, quarks, gluons, color, strangeness, the uncertainty principle) when they have inadequate understanding of concepts such as velocity, acceleration, force, mass, energy, and electrical charge, much less any understanding of how we obtain evidence regarding the structure of matter on a scale that transcends our direct sense perceptions.

2. Students who are still essentially Aristotelian, with no significant understanding of the law of inertia, satisfying distribution requirements by taking courses in meteorology or oceanography and hearing incomprehensible assertions about the role of the Coriolis effect.

3. Students who have no notion how to define local noon or the north–south direction, who have no idea of the origin of the seasons or of the phases of the moon (believing the unilluminated portion of the crescent moon to be the shadow of the earth), who are unaware that the stars have a diurnal motion, who do not understand why we believe the earth and planets revolve around the sun, taking "general education" courses in astronomy and hearing lectures on stellar nucleosynthesis, pulsars, quasars, and black holes.

4. Students who do not know how substances are defined and recognized, who have no idea what is meant operationally by words such as "oxygen," "nitrogen," or "carbon," who do not understand why we believe in discreteness in the structure of matter, and who have no idea what is meant by "electrical

charge" or "potential difference" being lectured to about DNA, molecular biology, and the structure of genes, or about nerve and muscle action.

The stream of unintelligible words cannot possibly generate scientific literacy; it simply aggravates the problem we are trying to solve.

12.5 ILLUSTRATING THE NATURE OF SCIENTIFIC THOUGHT

By addressing ideas such as those suggested in the preceding section—at a pace that allows formation and understanding of underlying concepts as well as consideration of the "How do we know . . . ? Why do we believe . . . ?" questions—illustrations of the character and limitations of scientific thought, rather than being injected artificially by assertion, will arise naturally and abundantly.

When, in the *Two New Sciences*, Galileo confronts the problem of describing the change in velocity of a moving body (the idea to which we give the name "acceleration"), he points out that there are at least two alternatives: (1) we observe that the object changes its velocity from one value to another while it traverses a distance of so many cubits, and we might elect to describe this motion by means of the number that indicates how much the velocity changed for each successive cubit of displacement; (2) the same velocity change, however, occurs over a measurable interval of time, and one might also characterize the process by the number indicating how much the velocity changes in each successive second. Which mode of description should be adopted? The choice is not trivial.

Galileo selects the second concept: change of velocity per unit time. His objective is to create a description of "naturally accelerated" motion (free fall), and he has a deep-seated intuitive conviction that free fall is *uniformly* accelerated in this sense but not in the sense of change of velocity per unit displacement. On the basis of a hypothesis, an inductive guess, he selects the concept that will yield the simplest and most elegant description of free fall and proceeds to test the hypothesis by deducing consequences that can be tested by experiment.

Here are, lucidly displayed, several significant facets of the scientific enterprise: the roles of inductive and deductive reasoning; the fact that scientific concepts are created by deliberate acts of human intelligence and imagination and are not "objects" discovered accidentally; that choice may be necessary and that there might be room for aesthetic criteria such as elegance and simplicity.

Here, too, is the new and revolutionary idea of forming an a priori hypothesis and testing its mathematical predictions by experiment. The Greeks had appealed on many occasions to both observation and experiment (they adduced, for example, the resistance of an inflated animal bladder to compression as experimental evidence for the corporeality of air), but they did not test the models or hypotheses they invented for the explanation of natural phenomena.

I outline this well-known story not to lay claim to new or profound insights into the philosophy or history of science, but only to point explicitly to a set of significant ideas that college students, with little prior scientific knowledge, can comprehend and appreciate—ideas that lie just below the surface in any introductory study of physical science, but that students are rarely given the time and opportunity to discover, articulate, and savor. To acquire these insights, students must have the opportunity to stand back and examine what happened, to relive some of the intellectual experience (the doubts as well as the successes), to analyze and assess the line of thought, recognizing the elements of its logic, its strength, and its limitations (what questions were *not* asked and *not* answered).

In only a few texts and courses, however, are students afforded this opportunity. The more common procedure is to throw the standard definitions of velocity and acceleration at the students in two or three cryptic pages (note the difficulty most students have in forming these concepts clearly and correctly in the first place [Trowbridge and McDermott (1980), (1981)]). The concepts are asserted as though they were inevitable, rocklike formations that have existed for all time, while deference is paid to "history" by mention of the name of Galileo and by a few unsubstantiated clichés concerning his invention of the "experimental method" and his paternity with respect to "modern science."

Students can indeed acquire mature and intelligent intellectual perspectives toward the methods, processes, successes, and limitations of science. Such perspectives, however, are *not* automatically conveyed by training students to calculate how high a stone rises when it is thrown vertically upward, or how much a given electrical field between capacitor plates will deflect an electron beam. Such intellectual perspectives can be developed only by coupling understanding of the scientific subject matter itself with the insights gained by standing back and examining not just the end results but what happened in terms of "How do we know . . . ? Why do we believe . . . ? What is the evidence for . . . ?" (See Section 13.2 in the discussion of critical thinking.)

Opportunities to make explicit such facets of the cultural phenomenon that is science arise at almost every turn without the necessity of invoking esoteric fringes of modern science that are completely unintelligible to students at their existing level of concept formation and grasp of subject matter. Consider a few additional examples.

In view of the rapid, assertive way in which scientific concepts are forced on students in school science as well as in introductory college courses, it is understandable that they acquire the notion that scientific terms are rigid, unchanging entities with only one absolute significance that the initiated automatically "know" and that the breathless student must acquire in one brain-twisting gulp. It comes as a revelation and a profound relief to many students to learn that scientific terms go through an evolutionary sequence of redefinition, sharpening, and refinement as one starts at a crude, initial, intuitive level, and, profiting from insights gained in successive applications, develops the concept to its subsequent level of sophistication.

For example, the concept of "force" is legitimately introduced by connection with the primitive, animate, intuitive, muscular sense of push or pull, but, in the law of inertia, we redefine it to apply to *any* interaction that imparts acceleration to a material object (for example, the action on bits of paper of a glass rod rubbed with silk.) We endow completely inanimate objects with the capacity to exert forces on other objects (the charged rod exerting a force on bits of paper, the table exerting an upward force on the book resting on it, the earth exerting a downward force on us—our "weight"—and the ground exerting an upward force at our feet). Following Newton, we then extend the concept even further and create the idea that, when the table exerts an upward force on the book, the book simultaneously exerts a downward force on the table, and, even more subtly, as the earth exerts a downward gravitational force on us, we exert an oppositely directed gravitational force on the earth. By this time we have come a very long way indeed from the original use of the word "force" in connection with an animate, muscularly sensed

12.5 ILLUSTRATING THE NATURE OF SCIENTIFIC THOUGHT

push or pull on another object. And since the concept is very subtle indeed, even very able students have great difficulty assimilating it in the far too short time made available in introductory courses, while many less able students never assimilate it at all (see discussion of these matters in Chapter 3).

Similarly, starting with the crude idea of "speed" as a measure of how fast (as an average over a substantial time interval) an object travels along a straight line, we endow the concept with direction one way or the other along the straight line, and, with the introduction of plus and minus signs, we begin to call it "velocity." We then refine this primitive notion into a conception of "instantaneous velocity" (something even the extremely able Greeks never achieved). We then extend the concept to cover direction in two- and three-dimensional space; and we finally talk about the rate of change of this quantity in both direction and magnitude.

At each stage in the sequence of evolution and redefinition, the original word (whether it be "force" or "velocity")—its meaning having been changed in significant, intrinsic ways—no longer denotes only the first intuitive idea to which it was applied; it denotes a new and more sophisticated concept. (Strictly speaking, we ought to give it a new name in order to emphasize the change, but this, of course, would only enhance the verbal chaos.) Modest self-consciousness about the process of definition and redefinition enormously increases the confidence of students in their own grasp of the new sequence of thought, opens their eyes to similar shifts and extensions in the subsequent generation of concepts such as "energy," "electrical charge," "electron," and carries over into other areas of study by alerting them to similar semantic shifts in courses in other disciplines—shifts, I might add, that are rarely explicitly pointed out or emphasized but that are crucial to genuine understanding. (In this connection, see discussion of the importance of the process of operational definition in Section 13.2.)

During the 1830s and 1840s, Michael Faraday's beautiful and elegant experimental investigations of electricity and magnetism caused him to raise some very deep questions about these phenomena, and students can be led to articulate a few of these themselves if helped Socratically in group discussion: Is there a process by which one electrically charged particle exerts a force on another? If one of the particles is suddenly displaced, does a finite time interval elapse before the force experienced by the other changes? Does a finite time interval elapse between the instant a wire is connected to an electric battery and the instant a neighboring compass needle begins to swing in response to the magnetic effect of the electric current in the wire? If a finite time interval does elapse in each case, what, if anything, happens in the intervening space between the interacting objects? What happens to Newton's Third Law and the whole concept of "action-at-a-distance" if the objects do not exert equal and opposite forces on each other during the interval of change—however short it may be?

To try to address these questions, Faraday invented a "model," a heuristic device completely transcending any direct sense experience—the famous concept of "lines of force," "lines" that stretched, contracted, spread apart, and pulled together, propagating electric and magnetic effects through empty space—the concept that James Clerk Maxwell subsequently elaborated into the sophisticated modern, mathematical notion of "field." Faraday wrote almost apologetically about his highly speculative model [Faraday (1965b)]:

> *It is not to be supposed for a moment that speculations of this kind are useless or necessarily hurtful in natural philosophy. They should ever be held as doubtful*

and liable to error and to change, but they are wonderful aids in the hands of the experimentalist and mathematician; for not only are they useful in rendering the vague idea more clear for the time, giving it something like a definite shape, that it may be submitted to experiment and calculation; but they lead on by deduction and correction, to the discovery of new phenomena, and so cause an increase and advance of real physical truth, which unlike the hypothesis that led to it, becomes fundamental knowledge not subject to change.

This fine description of the point and function of a heuristic model simultaneously reveals a characteristic facet of the thinking of many nineteenth-century scientists: they were indeed convinced that they were stockpiling "real physical truth" and "knowledge not subject to change. " After students learn something of the conceptual revolution accompanying the failures of classical theory at the turn of the twentieth century, it is interesting to ask them to contrast Faraday's statement with the sadder and wiser one of J. Robert Oppenheimer (1963):

We come to our new problems full of old ideas and old words, not only the inevitable words of daily life, but those which experience has shown to be fruitful over the years . . . . We love the old words, the old imagery, and the old analogies, and we keep them for more and more unfamiliar and more and more unrecognizable things.

In light of such perspectives, students spontaneously begin to articulate some sense of why most scientists now view scientific knowledge as mutable and provisional rather than permanent and final. They anticipate limited ranges of validity to successful theories and are prepared to find a regression of unanswered questions behind every answered question.

In still another sequence involving models transcending direct sense experience, one can have students follow and examine the evidence that led to our belief in atoms and molecules—discreteness in the structure of matter—as well as to belief in the structure of atoms themselves. They must be allowed to doubt with the early participants, to articulate uneasiness about interpretation of some of the evidence, and not just be stuffed with a few disconnected and, in themselves, unconvincing arguments, followed by assertion of the end results. (The original doubters were, after all, a far from foolish company.)

Many illustrative gems line the way through such a sequence. Dalton, for example, in his original attempts to develop a quantitative atomic–molecular theory with a unique and internally consistent set of relative atomic masses, confronted chemical data in the form of percentage composition by weight of various known compounds. The only regularity that had been discerned in these data was the so-called law of definite proportions—the recognition of a fixed percentage composition of any definite chemical substance regardless of its place or manner of origin—and even this aspect was a matter of some controversy and uncertainty.

Dalton's preconceptions concerning the corpuscular constitution of matter, however, led him to give particular consideration to cases in which a given pair of elements (say carbon and oxygen) form more than one compound. It occurred to him that if 1.0 g of carbon combined with 1.3 g of oxygen in one compound, then, for the same 1.0 g of carbon in the other compound, one should find perhaps 2.6 or 0.65 or 3.9 g of oxygen—or some other quantity that bore a whole-number ratio to 1.3. One would expect just such simple numerical connections if compounds did indeed consist of molecules made up of small numbers of atoms of the combining elements. The data had never been examined in this way; this particular orderliness lay hidden behind the unrevealing percentage compositions.

Dalton looked and found that order was there; in other words he *predicted* what is now called the "law of multiple proportions." (In many modern courses, this law, if it is considered at all, is presented as though it had been established empirically along with the "law of definite proportions" and had provided a priori evidence for discreteness.) The point here, of course, is that very frequently "facts" do *not* speak for themselves. In this instance, the facts had been available for a long time, but they were not even seen until looked for through the lenses of a theory; then suddenly their uncovered presence fed back as a dramatic confirmation of the theoretical conception that had revealed them.

The story of the famous "Piltdown Man" fraud that was exposed in the 1950s is an inverse illustration of this idea. Many paleontologists accepted as genuine the fossil with the manlike skull and the apelike jaw because their theoretical preconceptions led them to expect an evolutionary sequence in which brain development led the way for changes in other parts of the body. They accepted the forgery for almost 40 years, even though it was well known that it did not fit any reasonable niche in the humanoid fossil sequence. Here again, facts did not speak for themselves; they were viewed through the lenses of a theory, and the theory led many astray.

In a genuinely liberal (i.e., "liberating") educational enterprise, it is essential that the broad, general characteristics of any intellectual enterprise be brought to light, but this cannot be accomplished through vague generalizations disconnected from the visceral effort that goes with forming concepts and understanding the subject matter. To develop scientific literacy, it is necessary to master at least *some* reasonable amount of subject matter to give meaning to the generalizations.

12.6 ILLUSTRATING CONNECTIONS TO INTELLECTUAL HISTORY

In reasonably paced introductory courses, there are many ways of making students aware of the role of science in their own intellectual history.

For example, in working in the field of elementary school science instruction, I have more than once encountered the following sequence: A child asks (or perhaps is asked by the teacher) "Why do objects fall?" and the "correct" response given or solicited by the teacher turns out to be "because of gravity." The impression is instilled in the child that a *reason* has been given that explains both cause *and* effect. There is no inkling on the part of either the giver or the receiver of the "information" that the technical name conveys neither knowledge nor understanding and merely conceals ignorance of the nature of the phenomenon. If one asks a class of college students "Why do objects fall?" the great majority respond "because of gravity," and few, if any, have the courage to say "we have no idea."

Few students, teachers, or citizens have any awareness of the history of this term—that "gravity" started as the designation of a teleological effect, a "drive" or "desire" on the part of the "heavy" elements earth and water (and their mixtures) to seek the center of the earth; that the opposite "desire" on the part of air and fire to rise was called "levity;" that seventeenth-century natural philosophy banished both the teleological view and the word "levity"; that Newton, explicitly eschewing knowledge of mechanism or process of interaction, made the grand surmise that, however it might work, the same effect that makes the apple fall binds the moon to the earth and the earth and planets to the sun; that, despite the beauty and elegance of the General Theory of Relativity, we have, to this day, no idea of how gravity "works." In light of the answer most students give to the question "Why do objects fall?" a small step toward scientific literacy might reside in disabusing them of the foolish answer.

Very few students have any conception of the revolutionary thrust of seventeenth-century science in discarding the notion that the heavenly bodies were made of a "perfect" substance (the "quintessence") intrinsically different from the mundane materials of earth and in discarding, also, the notion that the laws governing the celestial domain are entirely different from those of the terrestrial. The discarding of these notions together with the subsumption of the entire universe under one system of humanly comprehensible natural law, the extension of the universe to infinity, and the concomitant removal of the literal sheltering heaven from close overhead, marked a profound turning point in our intellectual history. The outlook of every individual toward himself and his place in the universe is deeply conditioned by this heritage from Galileo, Descartes, Newton and other seventeenth-century natural philosophers. An educated person should be aware of such a heritage in conceptual, historical, and intellectual terms, not in the mere assertion of end results. Here is another significant step toward greater scientific literacy. Beautiful presentations of this story, with both its scientific and intellectual aspects, are available at undergraduate levels [Holton (1973); Rutherford, Holton, and Watson (1981)], but they are not widely used even in general education courses, and they make virtually no appearance at all in curricula for scientists and engineers.

Yet if these historical and intellectual aspects were brought out explicitly in introductory science courses, they would enormously enhance and enrich any study of the Enlightenment in courses in Western Civilization and would stimulate a healthy diffusion of ideas through the artificial membranes that now divide one course from another. Students are, for example, astonished to find how many aspects of the rhetoric used by the Founding Fathers in our American historical documents are traceable through the Deists back to Locke and Newton [Arons (1975)].

In an entirely different domain, consider some examples from literature. In Shakespeare's *Henry VI*, there is the passage:

> Glory is like a circle in the water
> Which never ceaseth to enlarge itself
> Till by broad spreading it disperse to naught.

This figure is much appreciated by students who encounter it while studying and observing common wave phenomena.

In the "Morning Song of Senlin" by Conrad Aiken (1953), there occur the lines:

> The earth revolves with me, yet makes no motion,
> The stars pale silently in a coral sky.
> In a whistling void I stand before my mirror,
> Unconcerned, and tie my tie.

The first line invokes the law of inertia and Galilean relativity; the "whistling void" refers to the Newtonian cosmology. And Aiken embellishes all of this with a final touch of irony and paradox. This is clearly a modern view; nothing like it could possibly have been written in the sixteenth century. Here students can see a direct influence of science on literature, an episode of intellectual history, and they can do so, not in terms of unintelligible jargon, but in terms of scientific concepts developable and understandable at introductory level. (As a further step for those interested, it is illuminating to compare Aiken's matter-of-course acceptance of the Newtonian world with Milton's equivocation between the Ptolemaic and Copernican systems in *Paradise Lost*.)

Another tacit acceptance of the Newtonian universe emerges in Tennyson's lines in "Locksley Hall Sixty Years After":

> While the silent heavens roll
>   and suns upon their fiery way,
> All their planets whirling round them,
>   flash a million miles a day.

Another profoundly influential sequence, emanating from technology, overarching and eventually unifying all the sciences, and having a pervasive impact in intellectual history, is the story of the "principles of impotence." Beginning in early times with a growing recognition that Nature conspires against our "getting something for nothing," against our achieving endless supplies of matter, motion, change, or warmth without effort and cost, the eighteenth century began to quantify the relevant restrictions. The "nothing can be created by divine power out of nothing" of Lucretius became, through the beautiful experiments of Lavoisier, the Law of Conservation of Mass. The concepts of mechanical energy began to emerge, but their connection to thermal effects was not yet apparent.

The nineteenth century saw the grand synthesis of the First and Second Laws of Thermodynamics: the unification of mechanical energy and heat; the conservation of energy overall; the possible and impossible directions of spontaneous change and the trend toward equilibrium; the *possibility* of converting a given amount of work entirely into thermal energy and the *impossibility* of converting a given amount of thermal energy entirely into work; the subsumption of all changes in state—mechanical, chemical, thermal, electric, magnetic—under just those two laws. The quantified forms of the principles of impotence, emerging out of the technology of the Industrial Revolution, brought impressive and comprehensible order out of the chaos of disparate forms of change, profoundly influencing the thought, for example, of social philosophers such as France's George Sorel [Humphrey (1951)] and of the American historian Henry Adams (1918).

Another sequence can follow a somewhat different direction. The nineteenth century had seen mass and energy as two separate, closed systems, each maintaining its own integrity of conservation. The twentieth century, through the conceptual revolution precipitated by Einstein, began to comprehend mass and energy as a unity rather than a dichotomy, as interconvertible as work and heat. This led to comprehension of how a radioactive material could maintain itself, apparently indefinitely, at a temperature higher than that of its surroundings; how the sun could keep shining for the enormous extent of geological time; and finally, how the prodigious energy confined in the atomic nucleus could be unleashed.

This sequence leads directly to all the moral, ethical, and societal problems that eventuate from the application of the resulting technology to peaceful industry on the one hand, and to warfare on the other. But the story and its influence on human society began long before Fermi and his associates constructed the first nuclear reactor in Chicago. I would hope to have students see some of this, with its ever-growing intensity, in the rich perspective of intellectual history, not just in a narrow view of the immediate scientific, technological, and societal end results.

12.7 VARIATIONS ON THE THEME

The preceding illustrations of epistemological, philosophical, and historical aspects of science, which, I submit, do cultivate some of the aspects of scientific literacy listed in Section 12.2 and do infuse liberal and humanistic perspectives into any curriculum, happen to be personal favorites of mine. I present them in order to be specific rather than vague in my illustrations and not to advocate them above the host of other possibilities. Each teacher must select subject matter and episodes that appeal to him or her and that he or she can articulate for the students in the most stimulating and compelling way.

There is the whole array of aspects to which James Conant referred as the "tactics and strategy" of science and which are effectively illustrated in the *Harvard Case Histories in Experimental Science* (1957). There is the fascinating, partly scientific, partly sociological problem of validation and acceptance of scientific theories. There are the philosophical problems of positivism and the questions concerning the "reality" of entities that transcend our senses: atoms, molecules, electrons, nuclei.

There are also, of course, the pressing social problems that stem from the release of nuclear energy; the application of science to warfare; the possibility of controlling human genetic development and even of synthesizing living matter; the problems of controlling and limiting abuse of our terrestrial environment. I have no intention of minimizing the significance of any of these vital questions, but I do have reservations about launching into analyses or discussions of them prior to the point at which students have some reasonable understanding of the underlying scientific and technical subject matter. If they have engaged in the necessary prior study, or if the necessary underpinnings are conscientiously developed as the need arises, such discussions can be educationally fruitful and significant and can enhance scientific literacy. If the questions are engaged *without* adequate understanding of the underlying science, as is unfortunately frequently done, the enterprise becomes specious. Students are then misled into thinking that they have pursued inquiry and possess understanding when they have, in fact, only used technical terms the meaning of which they did not comprehend and only dealt in vacuous generalizations devoid of genuine thought and substance. In such circumstances, they have been subtly encouraged to embrace the all too widespread rationalization that "any opinion is just as good as any other opinion."

It is clearly impossible to do all the things one would like to do with sequences such as those illustrated specifically in Sections 12.5 and 12.6 and mentioned briefly in this section. It is impossible to consider every important problem, open every significant perspective. The time is long past when we could teach our students all the things they need to know. It is hardly a new or original assertion that the only viable and realistic function of higher education is to put the students on their own intellectual feet: to give them conceptual starting points and an awareness of what it means to learn and understand something so that they may then continue to read, study, and learn, as need and opportunity arise, without perpetual formal instruction. (Such intellectual development is, of necessity, inextricably connected with the capacities for critical thinking discussed in the following chapter.)

An essential criterion is that students must not end up regurgitating secondhand pronouncements about the nature and processes of science without ever having articulated any such insights out of their own intellectual experience with subject matter they can encompass. Without at least some participation in comprehension and interpretation of scientific concepts, theories, and philosophy, students learn no more from secondhand statements about science than they learn from a commentary on poetry without having read the poetry, or a discussion on the methods and philosophy of history if they lack knowledge of the history of anything.

Rather than plunging into levels of material utilizing concepts and relationships well beyond the students' present comprehension, if we were to allow them the opportunity to confront science through some of the more modest insights and experiences I have tried to illustrate above, I am convinced that they would develop lasting insights far more consonant with our postulated objectives of liberal education. Such insights would contribute to the development of better educated individuals and more thinking–reasoning citizens in exactly the same way as do awareness of history and sensibility to literature.

12.8 ASPECTS OF IMPLEMENTATION

Although I would encourage, especially at upper division levels, occasional courses devoted to specific important problems (such as those mentioned in Section 12.7), using team teaching—despite its relatively high cost—to assure interdisciplinary perspectives, I nevertheless believe that a great deal can be accomplished in the lower division level in existing curricular structure.

The walls separating courses in the sciences, the humanities, and the social sciences need to be made more permeable. In the illustrations I have given, I have tried to show how a science teacher might enhance the liberal content of courses by becoming, through personal effort, more literate and more articulate about historical, literary, philosophical, epistemological, humanistic, or societal overtones than the teacher was prepared to be in his or her own education. Materials to this end are readily available for both teachers and students, and only a little ingenuity is required to devise questions, homework problems, and paper assignments that lead students to confront ideas that transcend the mere end results constituting almost the entire burden of the great majority of our textbooks. Understanding what Galileo contributed to the development of modern science does *not* reside in being able to calculate how high the stone goes when we throw it upward; other ingredients must be included in *addition* to the calculation.

It is true that, in our present science courses, infusion of such perspectives would force us to back away from the already excessive volume and pace of "coverage" that are imposed on our willing and uncritical students. It has for long been my contention that we are crushing our students into the flatness of equation-grinding automatons and forcing them into blind memorization of problem-solving procedures. We do not even give them a chance to begin to understand what "understanding" really *means*. Were we to require that they understand concepts and lines of reasoning within the perspectives I have outlined, we would be setting far higher intellectual standards and demanding far higher performance than we now attain, with all our obsession with "coverage. " Some of our colleagues complain that this would mean a "watering down" of our science teaching, but they misunderstand what is being advocated. What I am suggesting would be far more a "raising up" than a "watering down." An essential ingredient would be conscientious attention to the problems of teaching, learning, reasoning, concept formation, and understanding discussed in the earlier chapters of this book. To the development of secure command of concepts and reasoning, we must add some of the intellectual perspectives defined in the last few sections.

The intellectual perspectives, incidentally, should *not* be confined to the general education courses for nonscience majors. Future scientists and engineers are just as much in need of capacity for scientific literacy as are the nonscientists, and such literacy is not cultivated in our existing science and engineering courses. Solving the end-of-chapter problems, however correctly, does not open historical, philosophical, or humanistic perspectives and does not automatically cultivate the insights underpinning genuine scientific literacy. The most compelling evidence of our failure in enhancing the scientific literacy of our scientists and engineers resides in the widespread reluctance of college and university faculty members to inject and articulate humanistic ideas in their own courses. Most of them ran away from humanities courses because they were "not very good in humanities," and they never saw any parallel ideas made relevant in science. The fear of ideas about science (as opposed to the hard content itself) is, in many instances, more a matter of insecurity because of a faulty education rather than just stubborn insistence on the necessity of coverage.

I know at first hand that students can and do respond to the demand both for thorough understanding of the science and for penetration of the intellectual perspectives. It is futile, however, to exhort the students regarding the importance of liberal education and to give them nothing but ringing platitudes. We must be sufficiently knowledgeable and concerned to bring before them compelling and significant instances of intellectual, moral, or aesthetic experience. To this end, we, as instructors in science or engineering, must exhibit our own capacity to deal with such dimensions, else the exhortation will be hypocritical. How can we demand that the students do something we will not do ourselves?

But the burden is not on the technical instructors alone. The walls must be made permeable in the other direction also. Instructors in the humanities and social sciences must cease running from "hard science," as most of them did during their undergraduate careers; they must bring themselves to master at least a *few* of the "How do we know . . . ? Why do we believe . . . ?" questions I mentioned earlier, and they must treat scientific subject matter and its consequences knowledgeably in their own courses—for example, Newton's influence should be *seriously* discussed (*not* casually and superficially) in any history course dealing with the eighteenth century. The humanist and the social scientist must set an example for the students in the same way I demand of my technical and scientific colleagues. Again, how can we expect the students to cease running from anything that makes them feel slightly insecure if we persist in running ourselves?

As indicated repeatedly, to achieve these objectives we must cut back on the volume and pace of coverage that have escalated in all of our courses, not just in the sciences. Students must have time to form concepts, think, reason, and perceive relationships. They must discuss ideas, and they must write about them. Examples of reduced coverage and selection of reasonable story lines do exist in our textbook literature, although the examples are relatively few in number. To illustrate what I mean, rather than advocating these texts specifically, I point to *The Project Physics Course* as an example of how to weave a story line that brings one to the beginnings of modern atomic and subatomic physics, leaving out unnecessary subjects along the way and introducing many of the intellectual perspectives advocated in this essay; I also point to Casper and Noer's (1972) *Revolutions in Physics* as a sequence that deals with Newton and Einstein without trying to cover everything else in between. Building on existing good examples, it is quite possible for an instructor to generate a reasonable sequence that best fits his or her own skills and predilections.

12.9 THE PROBLEM OF COGNITIVE DEVELOPMENT

If we accept the premise that our liberal education objectives entail thinking, reasoning, and understanding on the part of the students, it is essential that we take into account the empirical results of recent studies of cognitive development. Such studies [Arons (1976); Chiappetta (1976); Karplus et al. (1979); McKinnon and Renner (1971); Perry (1970)], usually based on administration of two or more of the now classic Piagetian tasks, are showing, with remarkable reproducibility, that only about 25% of the cross section of college students execute these tasks successfully, indicating that they have developed the capacity for abstract logical reasoning at this rudimentary level; that up to 50% of the students are unsuccessful in the tasks and still use predominantly concrete patterns of reasoning, while the remaining 25% are in transition, exhibiting only partial success.

Students *do* have intellects capable of development beyond this point. Such development, however, is necessarily based on practice and experience with important modes of reasoning—*repetitive* practice leading, in Piagetian terminology, to accommodation and equilibration. Most college students are very much in need of practice in the various modes of thinking and reasoning listed and discussed in Chapter 13.

As has been pointed out earlier, the volume and pace of material thrust on students in the majority of liberal education science courses preclude the exercise of the time-consuming operations of thinking, reasoning, and understanding. The majority of students are thus forced into blind memorization, and they eventually come to see all "knowledge" and "understanding" as the juxtaposition of memorized names and phrases. They are not held to, or tested on, the reasoning since they find it "too hard." They are tested almost exclusively on the memorized end results. In such circumstances, there is essentially no hope of developing insights and understanding that characterize genuine scientific literacy—or, for that matter, literacy in *any* field requiring abstract logical reasoning.

If we really wish to succeed in our liberal education objectives, it is essential to start from the premise that the students *do* have intellects capable of development and use. Then it is necessary to give them the opportunity to think, to reason, and to develop intellectually by providing material at a pace and level that make assimilation possible. Genuine understanding of a limited range of significant subject matter would testify to far higher intellectual standards and requirements than regurgitation of memorized jargon from advanced or topical subjects, however tantalizing the latter might be.

12.10 THE PROBLEM OF TEACHER EDUCATION

It is not difficult to see that far and away the greatest leverage and highest potential for improving public understanding of the subject matter, methods, limitations, and social impact of science reside in the elementary and secondary schools. This is not to say that sophisticated, college-level insights should be developed in early schooling—pupils at that stage are not ready for such discussion; the point is that the groundwork can be laid for the synthesis of adult insight when maturity has been attained.

With genuine understanding of basic concepts that can perfectly well be developed in elementary and secondary school, with concomitant enhancement of capacities for abstract logical reasoning, college and university levels of instruction would be enormously facilitated. Students would enter with levels of knowledge and understanding that would make discussion of philosophical, historical, ethical, and societal questions fruitful and meaningful. They would even be able to penetrate aspects of modern science that are now hopelessly unintelligible. With such improved background, even our mass production system of lecturing to large classes might be substantially more effective, if still not ideal.

Such progress is impeded at the present time, not by lack of adequate curricular materials at elementary and secondary levels, but by inadequate teacher education in the sciences. It is unlikely that *any* curricular materials, however high their potential for cultivating thinking, reasoning, learning, and understanding, will ever be "teacher proof." A teacher can always negate the intent of the materials by attitude, invidious comment, and, most significantly, by what he or she chooses to test for. The real intellectual values of a course are established by tests rather than by texts. If improved curricular materials are to be successfully implemented, we must have teachers who are secure in the use of the instructional materials. "Security" in this context means both thorough understanding of the subject matter and thorough understanding of the underlying pedagogical intent and design of the materials. The latter understanding is impossible without the former.

For example, intrinsic weakness of the instructional materials is *not* the cause of the too modest success in implementing existing fine, hands-on elementary school curricula (e.g., *Elementary Science Study*, Webster Division, McGraw-Hill Book Co., St. Louis, MO; *Science Curriculum Improvement Study*, Delta Education, Nashua, NH). Rather, it is that college science preparation fails to provide elementary teachers with the knowledge and understanding necessary for effective use and implementation. Our work at the University of Washington with both pre- and in-service elementary teachers shows that the two groups are initially indistinguishable: the majority use concrete rather than formal patterns of reasoning; the majority cannot do arithmetical reasoning involving ratios and division (i.e., very few can solve word problems such as those invoked in fifth and sixth grade arithmetic); the majority fail on tasks involving control of variables; the majority cannot visualize, in the abstract, possible or plausible outcomes of changes imposed on a system. Their "knowledge" of science resides exclusively in memorized names, phrases, and technical terms, and, because they lack operational understanding of these terms, they are unable to reason with them in any specific instance.

In other words, our elementary teachers, if they had any physical science at all, have had courses of the variety criticized in Section 12.4. It is quite apparent that such courses do *not* help the future teacher exercise patterns of abstract logical reasoning, and they do *not* engender understanding of the subject matter. The future teacher is left with a half-remembered string of words such as "nuclei," "laser," "strange particles," "angular momentum," "black hole." The teacher is left, furthermore, with no understanding of the concept of density and its relation to floating and sinking, of the law of inertia and its relation to why we believe the earth and planets revolve around the sun, of the distinction between heat and temperature, of the origin of the phases of the moon, of the observed apparent annual motion of the sun, of the definition of "north–south" or of "noon–midnight," of the concept of electric current and its behavior in the very simplest resistive circuit, of the most elementary aspects of wave motion, of the most prevalent everyday phenomena of light.

The better elementary science curricula wisely pursue the realms of everyday experience rather than esoteric vocabularies of modern physics. The teachers find themselves at exactly the same level of initial knowledge as the children, and, because of consequent lack of security, they cannot handle the instructional materials. I have heard honest and perceptive teachers volunteer this assessment of their present condition in exactly these terms.

Time and again in our investigations, we have confirmed the empirical observation that adults (pre- and in-service elementary teachers), when they have not had the prior learning experiences as children, acquire the abstract reasoning patterns and the conceptual understanding fostered in the elementary science curricula only by encountering and overcoming the same obstacles, hurdles, and difficulties that are experienced by the children. And the fact of being adults is of virtually no help; the pace of learning is usually slower than it is among the children.

A major part of the effort expended in short workshops intended to help schoolteachers implement the new elementary science curricula was completely wasted because of the failure to recognize this necessary time element. Their college background has left the vast majority of teachers at the same point as the pupils they are expected to teach. It is illusory to hope that a brief indoctrination about the "philosophy of the program," followed by rapid examination of a few sample units of the materials, will induce understanding of the content; it is also illusory to hope that, once they get "started," the teachers will learn from the beautifully transparent materials along with the children. We have found that only slow, patient traversal of both the subject matter and the patterns of abstract logical reasoning gives the teacher the level of security that leads to effective implementation [Arons (1977)]. In the meantime, unless and until we change our college courses and curricula, we will continue to produce elementary, intermediate, and secondary school teachers who are in need of remediation the instant they graduate. This, however, is not inevitable; investigation in our group has shown that, given the opportunity to start at the beginning with hands-on materials and proceeding at a pace that allows learning and understanding, the great majority of pre- and in-service teachers do develop the requisite reasoning capacities and subject matter knowledge, and take great satisfaction in having done so. What they have lacked is only the opportunity.

A very common manifestation of the current inadequacy of the subject matter background of elementary school teachers is their reaction to the better hands-on curricular materials that are available. In some instances where the materials have been purchased by the schools, they stay locked up in closets, while the teachers, if they are obliged to teach science, proceed to lecture out of the standard older texts. In other instances, teachers, insecure in the face of the new materials, finding them "too difficult" for the children without being aware that the trouble really lies in their own lack of adequate understanding, band together and direct their energies and good intentions to writing materials of their own. The result is invariably trash that is full of errors, misconceptions, and misstatements, and that probably has negative educational value. (It must be reemphasized that the existing excellent elementary school materials were developed by mature and highly competent scientists who had the necessary perspectives and understanding.)

Research on teaching and learning may give us better instructional materials, but these materials will never have a really widespread impact in the schools unless we produce teachers who can implement them. This is a major problem, and it remains rooted at the college–university level. Accusations and recriminations leveled against elementary and secondary school teachers are, for the most part, unwarranted and unjust: "The fault, dear Brutus, is not in our stars . . . ."

Besides understanding of subject matter and enhancement of capacity for abstract logical reasoning, one other aspect of teacher education that deserves attention is behavior in the classroom. The best new science curricula are organized not around lecturing and inculcation by the teacher, but around exploration, trying and erring, talking and listening, arguing and explaining, among the children themselves. (There is evidence that, when competently implemented, such materials simultaneously enhance reading comprehension and arithmetical reasoning capacities in the children [Bredderman (1982); Shymansky et al. (1983); Wellman (1978)]. In many instances, when children are following an erroneous course, it is important that a teacher direct attention to inconsistencies or contradictions and guide them to revision or correction without flat assertion of "the right answer." Such teaching behavior takes confidence, security, and firm command of concepts and lines of reasoning. Lecturing and asserting the "right answer" are *much* easier and are the refuge of the insecure.

But even with adequate subject matter background, it is asking too much to expect such sophisticated behavior of a young person thrown into a new classroom. The preservice teacher needs both prior example and instruction. Many researchers in this country and abroad consistently show that teachers tend to teach as they have been taught. If we do nothing but lecture at our preservice teachers, they will lecture in their own classrooms, regardless of the indoctrination that may have been given (also by lecturing) in their "methods" courses in departments of education. It is essential that we set aside courses for future teachers taught in small classes in exactly the way we wish to have the elementary science curricula implemented. The science course then becomes a methods course as well—as it should. With teacher education of this kind, we might begin to hope for a shift in performance in the schools, a shift that has *not* taken place through the provision of improved curricular materials alone. Such a shift would tremendously enhance the possibility of cultivating adult scientific literacy at college–university level.

12.11 A ROLE FOR THE COMPUTER

One outcome of research and observation over a wide range of students and introductory courses is that many students do not break through to full command of a particular concept or line of reasoning unless they can be reached in one-on-one Socratic dialogue. The breakthrough is not made in homework or exercises or tests or by passive listening to explanations, however lucid. A gradual breakthrough begins only after the learner has begun to articulate ideas, inferences, and lines of reasoning in his or her *own words* [Arons (1976), (1977), (1982)]. This is especially true of pre- and in-service elementary and secondary teachers.

Examples of such areas of difficulty are arithmetical reasoning using ratios and division; translating symbols into words and words into symbols (i.e., interpreting graphs of any kind, e.g., interpreting position–time and velocity–time graphs in kinematics both in words and by executing the indicated motion with one's hand; translating word problems into arithmetical form; translating an arithmetical or algebraic expression into words, etc.); distinguishing operationally between "heat" and "temperature"; forming and using the model for the origin of phases of the moon (i.e., being able to deal successfully with questions such as "Would you expect to see a full moon rising at midnight? Why or why not?"); forming and using the model of current and the concept of electrical resistance to predict brightness changes that would result from alterations (shorts, removal or insertion of other bulbs) imposed on simple circuits composed of batteries and flashlight bulbs [Arons (1982)].

The necessary one-on-one dialog with a single student can easily take as long as 20 to 30 minutes or more. It is barely possible for an instructor to do this in small classes, and it is quite out of the question in large ones. One might adopt the view that students who cannot master such simple and basic concepts or reasoning modes do not belong in science courses. Enlightened self-interest, however, if not a broader objective, dictates a less callous view. Among the students who fail or who simply disappear from our college–university courses (or who never enroll in the first place because of deep-seated fear and insecurity), and who could be saved by one-on-one help at strategic points, are many potentially promising minority students as well as most of our future elementary school teachers, not to speak of many others.

Here is a point at which modern technology can be of assistance. The coming of age of the personal computer with graphic capability offers the prospect of making one-on-one dialogues practicable in spite of numbers [Arons (1984d), (1986)]. The problem becomes one of writing effective dialogues that pull students over the early, most severe obstacles, and help them on the way to further learning, with decreasing dependence on Socratic assistance.

The writing of effective dialogues will of necessity lean heavily on the results of research in teaching and learning and in cognitive development. Not only must a skillful author have a broad awareness of what is being learned in general about cognitive development and utilization of patterns of abstract logical reasoning, but he or she must also draw heavily on the detailed concept-by-concept student protocols, that is, researches such as those discussed in earlier chapters of this book and listed in the bibliography. Many of us incline to make conjectures concerning what students were thinking and misapprehending in various circumstances. In making these conjectures, we are invariably extrapolating from our own personal experience and preconceptions, and such conjectures are, in fact, very rarely correct. The actual difficulty is almost always something plausible to the student that we have glossed over and have not perceived.

Yet for a really effective computer dialogue, the most important (and most difficult) provisions an author must make are the ones that lead a student to rectify *incorrect* responses; it is easy to take care of the correct ones. Socratic rectification of misconceptions and incorrect reasoning can be achieved only if the author has prior knowledge as to the actual incorrect responses likely to be made. This is why authors must be well versed in the research results if they are to write good material.

12.12 LEARNING FROM PAST EXPERIENCE

With present resurgence of long-standing concern about education of scientists and engineers and about public understanding of science and technology, some individuals are pointing to what they regard as the "failure" of the curriculum development efforts of the '50s and '60s, and are advocating a new round of preparation of "up-to-date" text materials, hoping that *this* time the materials will be "better" and "more effective." Without incorporation, however, of the additional ingredients and insights discussed in this book, this hope is fallacious and forlorn. Although existing curricular materials can certainly be improved in pedagogical quality, this alone will not overcome the real obstacles and underlying problems. Curricular materials, however lucid, skillful, and imaginative, cannot "teach themselves." [cf Arons (1981a)]

I do not consider the science curriculum development of the '50s and '60s to have been a failure. At the time these efforts were undertaken, principally under stimulus provided by the National Science Foundation, new text materials, especially ones rooted in laboratory and observational experience, were very badly needed. Existing texts were obsolete, not "up-to-date," full of errors and misstatements, and intellectually sterile. They had been copied and recopied from each other for several generations by authors who themselves did not have adequate understanding of the subject matter.

In the heady, evangelistic atmosphere of 30 years ago, leaders in the various scientific and technical fields brought imaginative insight, current perspectives, and correct logical underpinnings to the new materials. It is true that the materials were not of uniformly high instructional quality. Some were too rapidly paced for the age groups for which they were intended. Some started by asserting end results of sophisticated lines of inquiry, required students to memorize these end results without examination of any underlying "How do we know . . . ? Why do we believe . . . ?" questions, and proceeded to drown the students in incomprehensible consequences of the initial, unsupported assertions. Some of the curriculum developers, not having the benefit of more recent insights into the actual levels of capacity of different age groups for abstract logical reasoning [Arons (1976); Chiappetta (1976); Karplus et al. (1979); McKinnon and Renner (1971)], went off into abstractions completely beyond the grasp of the pupils they were ostensibly addressing (and of many teachers as well). Mathematics, in particular, suffered from such misdirection.

However, despite some misplaced effort and mismatch to intended audience, a significant number of excellent curricula were developed. In my opinion, the best are among those designed for elementary school [e.g., *Elementary Science Study* (1968); *Science Curriculum Improvement Study* (1968)]. In these materials children are started from scratch, with no presuppositions concerning prior "knowledge" of technical terminology. Ideas come first and names afterwards. Concepts are synthesized out of observational experience rather than received through lecture and assertion. Reasoning starts at concrete levels and, provided it is guided by a competent teacher, gradually proceeds toward the abstract. Repeated intellectual experiences of this kind stimulate the emergence of abstract reasoning capacity and enhance its subsequent development. Some sound and potentially effective materials were also developed for junior high school and high school levels. In many instances, however, these materials are well matched only to the upper levels of student development and are mismatched to a large proportion of the audience the developers had in mind. Although these materials are, in some instances, well suited to students one, two, or even three years older, they are rarely used at the higher age levels because of the inhibition imposed by the nominal lower age label.

The point is that, regardless of some deficiencies and mismatch, a substantial body of pedagogically sound and useful material *was* developed. Few of these materials are obsolete or "out of date" even after 20 or more years. Their adoption and implementation has indeed been disappointing, and some critics who wish to energize a new burst of curriculum development refer to them as a failure. The "failure" that is being adduced, however, does not stem from deficiencies in the quality of the materials so much as it stems from external causes that have, so far, not been remedied. Although the better of the existing curricula are surely not the last word in educational sophistication, and although improvements are certainly possible in light of improving insights into aspects of teaching and learning, I am convinced that a new generation of materials would suffer exactly the same "failure" in adoption and implementation from the same extrinsic causes.

There were two principal causes of such failure. One was inadequacy of logistic support, especially in elementary and junior high school, for teachers venturesome enough to try the new curricula. Hands-on, laboratory-oriented materials require continual maintenance and resupply. Busy and overloaded teachers need sustained support in using such materials; they cannot take care of the logistics in addition to all the other duties they are expected to discharge. Although they might make the extra effort in the initial wave of enthusiasm (and many of them did), such effort is impossible to sustain over long periods of time. The necessary logistic support was rarely strong, even at the very beginning, and it rapidly melted away altogether in the seventies with growing financial pressures on the schools.

Thus, one formidable obstacle that must be overcome at all levels in the schools is that of logistic support. It is illusory to suppose that widespread scientific literacy will ever be successfully cultivated through instructional materials based on purely verbal inculcation. The necessary understanding, reasoning, and mastery of basic concepts and ideas will evolve, in the great majority of ordinary individuals, *only* from concrete observational experience. It is true that some gifted individuals do break through to scientific understanding, and to abstract reasoning in general, without such concrete help. But this small fraction of the population has *always* broken through, perhaps with some delay, simply because of the availability of books, classes, homework, and explanation; they have probably done so in spite of, rather than because of, the instructional system. We are indeed fortunate that this is the case, for otherwise literacy of every variety would be far lower than it actually is. Our reiterated goal, however, is a large increase in the literate fraction. This goal will not be attained without providing the majority of our young students with hands-on experience and with sustained guidance in carrying concept formation and reasoning from the concrete to the abstract.

The second, and probably most significant, cause of failure in adoption and implementation of the better curricular materials (including those in social science as well as natural science) lies in what happened in the retraining of teachers. Huge amounts of money and effort were poured into summer institutes and academic year programs, most of them under the direction of college and university faculty. A few of these programs were beneficial because the staff recognized that the teachers had developed little genuine understanding of scientific concepts and subject matter in their previous school and college courses and were very nearly at the same level of conceptual development as the children they were supposed to teach. Accordingly, it was realized that the teachers must be guided slowly and carefully through the same intellectual experiences they were subsequently to convey in their own classrooms. In the majority of teacher training programs, however, this necessity was not understood or appreciated. It was falsely assumed that the teachers, elementary through high school level, really understood the very elementary science they were to be teaching, or that, if understanding was deficient, they would quickly develop it along with the pupils, simply by working with the perfectly lucid new materials.

In the summer institutes, and in other programs, the teachers were thus given lectures about the "educational philosophy" of materials the substance of which they did not comprehend. Or they were given more advanced subject matter, also totally incomprehensible, in order to extend their "perspectives" through awareness of up-to-date knowledge and progress in various fields. Or in the largest number of cases, they were simply given another run through the same excessively rapid, irrelevant, and unintelligible college courses that had had no visible intellectual effect in the past. In such instances, the teachers were no better able to handle the new curricular materials competently than they were before coming into the retraining programs. It was not *their* fault that they had no awareness of this deficiency and that, in subsequent, disappointing classroom experiences with the new materials with which they were still insecure, they ascribed their troubles to the difficulty of the materials rather than to their own inadequate understanding. (Let me hasten to say that I am not tarring *all* teachers with this brush. I know personally, and I continue to meet and hear about, superlative, competent teachers at every level. These are, however, far too small a fraction of the total teacher population to make for a solution to the national problem.)

If we wish to produce widespread improvement in public understanding of science, we will have to take significant steps toward mitigating the two principal causes of our previous failure. The problems are very difficult, but they are not intrinsically insoluble. The remedies are necessarily fairly costly, but by far the most difficult part of the problem is conveying comprehension of it to our college and university faculty colleagues, most of whom still operate on the premise that instruction of future teachers (as well as all other students) can be effectively conducted by sufficiently lucid verbal inculcation and through the range of subject matter "coverage" that has become conventional in introductory science courses. If they continue in this practice, we shall make no progress regardless of expenditure of effort on curriculum improvement.

Curricular materials, however fine, will never achieve our goals entirely by themselves. They may improve matters for the especially gifted, but they will not solve the problem for the vast and perfectly respectable majority. Unless the essential underpinnings I have been trying to define are provided, tinkering with new curricula will again absorb huge amounts of time and money and will leave us exactly where we are at the present time. We will have a new crop of complaints about failure and inadequacy of the materials and a revived demand for still another needless wave of curriculum development. Attainment of wider scientific literacy will keep receding into the infinite educational perspective.

CHAPTER 13

Critical Thinking[1]

The simple but difficult arts of paying attention, copying accurately, following an argument, detecting an ambiguity or a false inference, testing guesses by summoning up contrary instances, organizing one's time and one's thought for study—all these arts . . . cannot be taught in the air but only through the difficulties of a defined subject; they cannot be taught in one course in one year, but must be acquired gradually in dozens of connections.

JACQUES BARZUN

13.1 INTRODUCTION

No curricular recommendation, reform, or proposed structure has ever been made without some obeisance to the generic term "critical thinking" or one of its synonyms. The flood of reports on education in our schools and colleges that has been unleashed in recent years is no exception; every report, at every level of education, calls for attention to the enhancement of thinking–reasoning capacities in the young. A currently favorite cliché is "higher order thinking skills." Few of the documents that come to us, however, attempt to supply some degree of specificity—some operational definition of the concept, with illustrations of what might be done in day-to-day teaching to move toward the enunciated goals.

It is the object of this chapter to try to "unpack" the term "critical thinking"—to list a few simpler, underlying processes of abstract logical reasoning that are common to many disciplines and that can be cultivated and exercised separately in limited contexts accessible to the student. Subsequently, the individual's conscious weaving together of these various modes results in the larger synthesis we might characterize as "critical thought." As Barzun points out in the quotation cited above, this can be done only through practice in, preferably, more than one field of subject matter.

---

[1] This chapter is based on an article originally published in *Liberal Education* [Arons (1985)]. At the time of writing, Prof. Peter A. Facione (Dept. of Philosophy, California State University, Fullerton) is engaged in coordinating the submissions of a selected group of scholars in an effort to achieve consensus on an operational definition of "critical thinking."

13.2 A LIST OF PROCESSES

To glimpse some of the ways in which effective schooling might enhance students' reasoning capacities, it is instructive to examine a few of the thinking and reasoning *processes* that underlie analysis and inquiry. These are processes that teachers rarely articulate or point out to students; yet these processes are implicit in many different studies. The following listing is meant to be illustrative; it is neither exhaustive nor prescriptive. Readers are invited to add or elaborate items they have identified for themselves or sense to be more immediately relevant in their own disciplines.

1. *Consciously raising the questions "What do we know...? How do we know...? Why do we accept or believe...? What is the evidence for...?" when studying some body of material or approaching a problem.* Consider the assertion, which virtually every student and adult will make, that the moon shines by reflected sunlight. How many people are able to describe the simple evidence, available to anyone who can see, that leads to this conclusion (which was, incidentally, perfectly clear to the ancients)? This does not require esoteric intellectual skills; young children can follow and understand; all one need do is lead them to watch the locations of *both* the sun and moon, not just the moon alone, as a few days go by. Yet for the majority of our population the "fact" that the moon shines by reflected sunlight is received knowledge, not sustained by understanding.

Exactly the same must be said about the contention that the earth and planets revolve around the sun. The validation and acceptance of this view marked a major turning point in our intellectual history and in our collective view of man's place in the universe. Although the basis on which this view is held is more subtle and complex than that for the illumination of the moon, the "How do we know...?" should be an intrinsic part of general education; it is, for most people, however, received knowledge—as is also the view that matter is discrete in its structure rather than continuous.

Similar questions should be asked and addressed in other disciplines: How does the historian come to know how the Egyptians, or Babylonians, or Athenians lived? On what basis does the text make these assertions concerning consequences of the revocation of the Edict of Nantes? What is the evidence for the claim that such and such tax and monetary policies promote economic stability? What was the basis for acceptance of the doctrine of separation of church and state in our political system?

Researchers in cognitive development [e.g., Anderson (1980); Lawson (1982)] describe two principal classes of knowledge: figurative or declarative on the one hand, and operative or procedural on the other. Declarative knowledge consists of knowing "facts" (matter is composed of atoms and molecules; animals breathe oxygen and expel carbon dioxide; the United States entered the Second World War after the Japanese attack on Pearl Harbor in December 1941). Operative knowledge involves understanding where the declarative knowledge comes from or what underlies it (What is the evidence that the structure of matter is discrete rather than continuous? What do we mean by the terms "oxygen" and "carbon dioxide" and how do we recognize these as different substances? What worldwide political and economic events underlay the American declaration of war?). And operative knowledge also involves the capacity to use, apply, transform, or recognize the relevance of declarative knowledge in new situations.

"Above all things," says Alfred North Whitehead in a well-known passage on the first page of *The Aims of Education*, "we must beware of what I will call 'inert ideas'—that is to say, ideas that are merely received into the mind without being utilized, or tested, or thrown into fresh combinations." And John Gardner once deplored our tendency to "to hand our students the cut flowers while forbidding them to see the growing plants."

Preschool children almost always ask "How do we know? Why do we believe?" questions until formal education teaches them not to. Most high school and college students then have to be pushed, pulled, and cajoled into posing and examining such questions; they do not do so spontaneously. Rather, our usual pace of assignments and methods of testing all too frequently drive students into memorizing end results, rendering each development inert. Yet given time and encouragement, the habit of inquiry can be cultivated, the skill enhanced, and the satisfaction of understanding conveyed. The effect would be far more pronounced and development far more rapid if this demand were made deliberately and simultaneously in science, humanities, history, and social science courses rather than being left to occur sporadically, if at all, in one course or discipline.

2. *Being clearly and explicitly aware of gaps in available information. Recognizing when a conclusion is reached or a decision made in absence of complete information and being able to tolerate the ambiguity and uncertainty. Recognizing when one is taking something on faith without having examined the "How do we know . . . ? Why do we believe . . . ?" questions.* Interesting investigations of cognitive skill and maturity are conducted by administering test questions or problems in which some necessary datum or bit of information has been deliberately omitted, and the question cannot be answered without securing the added information or making some plausible assumption that closes the gap. Most students and many mature adults perform very feebly on these tests. They have had little practice in such analytical thinking and fail to recognize, on their own, that information is missing. If they are told that this is the case, some will identify the gap on reexamining the problem, but many will still fail to make the specific identification.

In our subject matter courses, regardless of how carefully we try to examine evidence and validate our models and concepts, it will occasionally be necessary to ask students to take something on faith. This is a perfectly reasonable thing to do, but it should never be done without making students aware of what evidence is lacking and exactly what they are taking on faith. Without such care, they do not establish a frame of reference from which to judge their level of knowledge, and they fail to discriminate clearly those instances in which evidence has been provided from those in which it has not.

3. *Discriminating between observation and inference, between established fact and subsequent conjecture.* Many students have great trouble making such discriminations even when the situation seems patently obvious to the teacher. They are unused to keeping track of the logical sequence, and they are frequently confused by technical jargon they have previously been exposed to but never clearly understood.

In the case of the source of illumination of the moon cited earlier, for example, students must be made explicitly conscious of the fact that they *see* the extent of illumination increasing steadily as the angular separation between moon and sun increases, up to full illumination at a separation of 180°. This direct observation leads, in turn, to the *inference* that what we are seeing is reflected sunlight.

In working up to the concept of "oxygen" (without any prior mention of this term at all) with a group of elementary school teachers some years ago, I had them do an experiment in which they heated red, metallic copper in an open crucible and weighed the crucible periodically. What they saw happening, of course, was the copper turning black and the weight of crucible and contents steadily increasing. When I walked around the laboratory and asked what they had *observed* so far, many answered, "We observed oxygen combining with the copper." When I quizzically inquired whether that was what they had actually *seen* happening, their reaction was one of puzzlement. It took a sequence of Socratic questioning to lead them to state what they had actually seen and to discern the *inference* that something from the air must be joining the copper to make the increasing amount of black material in the crucible. It had to be brought out explicitly that this "something from the air" was the substance to which we would eventually give the name "oxygen." What they wanted to do was to use the technical jargon they had acquired previously without having formed an awareness of what justified it.

This episode illustrates the importance of exposing students to repeated opportunity to discriminate between observation and inference. One remedial encounter in one subject matter context is not nearly enough, but opportunities are available at almost every turn. Mendel's observations of nearly integral ratios of population members having different color and size characteristics must be separated from inference of the existence of discrete elements controlling inheritance. In the study of literature, analysis of the *structure* of a novel or a poem must be distinguished from an *interpretation* of the work. In the study of history, primary historical data or information cited by the historian must be separated from the historian's interpretation of the data.

A powerful exercise once employed by some of my colleagues in history was to give the students a copy of the Code of Hammurabi accompanied by the assignment: "Write a short paper addressing the following question: From this code of laws, what can you infer about how these people lived and what they held to be of value." This exercise obviously combines exposure to both processes 1 and 3.

4. *Recognizing that words are symbols for ideas and not the ideas themselves. Recognizing the necessity of using only words of prior definition, rooted in shared experience, in forming a new definition and in avoiding being misled by technical jargon.* From the didactic manner in which concepts (particularly scientific concepts) are forced on students in early schooling, it is little wonder that they acquire almost no sense of the process of operational definition and that they come to view concepts as rigid, unchanging entities with only one absolute significance that the initiated automatically "know" and that the breathless student must acquire in one intuitive gulp. It comes as a revelation and a profound relief to many students when they are allowed to see that concepts evolve; that they go through a sequence of redefinition, sharpening, and refinement; that one starts at crude, initial, intuitive levels and, profiting from insights gained in successive applications, develops the concept to final sophistication.

In my own courses, I indicate from the first day that we will operate under the precept "idea first and name afterwards" and that scientific terms acquire meaning only through the description of shared experience in words of prior definition. When students try to exhibit erudition (or take refuge from questioning) by name dropping technical terms that have not yet been defined, I and my staff go completely blank and uncomprehending. Students catch on to this game quite quickly. They cease name dropping and begin to recognize, on their own, when they do not understand the meaning of a term. Then they start drifting in to tell me of instances in which they got into trouble in a psychology, or sociology, or economics, or political science course by asking for operational meaning of technical terms. It is interesting that this is an aspect of cognitive development to which many students break through relatively quickly and easily. Unfortunately, this is not true of most other modes of abstract logical reasoning.

5. *Probing for assumptions (particularly the implicit, unarticulated assumptions) behind a line of reasoning.* In science courses, this is relatively easy to do. Idealizations, approximations, and simplifications lie close to the surface and are quite clearly articulated in most presentations. They are ignored or overlooked by the students, however, principally because explicit recognition and restatement are rarely, if ever, called for on tests or examinations. In history, humanities, and the social sciences, underlying assumptions are frequently more subtle and less clearly articulated; probing for them requires careful and self-conscious attention on the part of instructors and students.

6. *Drawing inferences from data, observations, or other evidence and recognizing when firm inferences cannot be drawn. This subsumes a number of processes such as elementary syllogistic reasoning (e.g., dealing with basic propositional, "if . . . then" statements), correlational reasoning, recognizing when relevant variables have or have not been controlled.* Separate from the analysis of another's line of reasoning is the formulation of one's own. "If . . . then" reasoning from data or information must be undertaken without prompting from an external "authority." One must be able to discern possible cause-and-effect relations in the face of statistical scatter and uncertainty. One must be aware that failure to control a significant variable vitiates the possibility of inferring a cause-and-effect relation. One must be able to discern when two alternative models, explanations, or interpretations are equally valid and cannot be discriminated on logical grounds alone.

As an illustration of the latter situation, I present a case I encounter very frequently in my own teaching. When students in a general education science course begin to respond to assignments leading them to watch events in the sky (diurnal changes in rising, setting, and elevation of the sun, waxing and waning of the moon, behavior of the stars and readily visible planets), they immediately expect these naked eye observations to allow them to "see" the "truth" they have received from authority, namely that the earth and planets revolve around the sun. When they first confront the fact that both the geo- and heliocentric models rationalize the observations equally well and that it is impossible to eliminate one in favor of the other on logical grounds at this level of observation, they are quite incredulous. They are shocked by the realization that either model might be selected provisionally on the basis of convenience, or of aesthetic or religious predilection. In their past experience, there has always been a pat answer. They have never been led to stand back and recognize that one must sometimes defer, either temporarily or permanently, to unresolvable alternatives. They have never had to wait patiently until sufficient information and evidence were accumulated to develop an answer to an important question; the answer has always been asserted (for the sake of "closure") whether the evidence was at hand or not, and the ability to discriminate decidability versus undecidability has never evolved.

An essentially parallel situation arises in the early stages of formation of the concepts of static electricity (see Sections 6.7 and 6.8). Students are very reluctant to accept the fact that, before we know anything about the microscopic constitution of matter and the role of electrical charge at that level, it is impossible to tell from observable (macroscopic) phenomena whether positive charge, negative charge, or both charges are mobile or being displaced. They wish to be told the "right answer" and fail to comprehend that any one of the three models accounts equally well for what we have observed and predicts equally well in new situations.

If attention is explicitly given, experiences such as the ones just outlined can play a powerful role in opening student minds to spontaneous assessment of what they know and what they do not know, of what can be inferred at a given juncture and what cannot.

7. *Performing hypothetico-deductive reasoning; that is, given a particular situation, applying relevant knowledge of principles and constraints and visualizing, in the abstract, the plausible outcomes that might result from various changes one can imagine to be imposed on the system.* Opportunities for such thinking abound in almost every course. Yet students are most frequently given very circumscribed questions that do not open the door to more imaginative hypothetico-deductive reasoning. The restricted situations are important and provide necessary exercises as starting points, but they should be followed by questions that impel the student to invent possible changes and pursue the plausible consequences.

8. *Discriminating between inductive and deductive reasoning; that is, being aware when an argument is being made from the particular to the general or from the general to the particular.* The concepts of "electric circuit," "electric current," and "resistance" can be induced from very simple observations made with electric batteries and arrangements of flashlight bulbs. This leads to the construction of a "model" of operation of an electric circuit. The model then forms the basis for deductions, that is, predictions of what will happen to brightness of bulbs in new configurations or when changes (such as short circuiting) are imposed on an existing configuration.

Exactly similar thinking can be developed in connection with economic models or processes. Hypothetico-deductive reasoning is intimately involved in virtually all such instances, but one should always be fully conscious of the distinction between the inductive and the deductive modes.

9. *Testing one's own line of reasoning and conclusions for internal consistency and thus developing intellectual self-reliance.* The time is long past when we could teach our students all they needed to know. The principal function of education—and higher education in particular—must be to help individuals to their own intellectual feet: to give them conceptual starting points and an awareness of what it means to learn and understand something so that they can continue to read, study, and learn as need and opportunity arise, without perpetual formal instruction.

To continue genuine learning on one's own (not just accumulating facts) requires the capacity to judge when understanding has been achieved and to draw conclusions and make inferences from acquired knowledge. Inferring, in turn, entails testing one's own thinking, and the results of such thinking, for correctness or at least for internal coherence and consistency. This is, of course, a very sophisticated level of intellectual activity, and students must first be made aware of the process and its importance. Then they need practice and help.

In science courses, they should be required to test and verify results and conclusions by checking that the results make sense in extreme or special cases that can be reasoned out simply and directly. They should be led to solve a problem in alternative ways when that is possible. Such thinking should be conducted in both quantitative and qualitative situations. In the humanities and social sciences, the checks for internal consistency are more subtle, but they are equally important and should be cultivated explicitly. Students should be helped to sense when they can be confident of the soundness, consistency, or plausibility of their own reasoning so that they can consciously dispense with the teacher and cease relying on someone else for the "right answer."

10. *Developing self-consciousness concerning one's own thinking and reasoning processes.* This is perhaps the highest and most sophisticated reasoning skill, presupposing the others that have been listed. It involves standing back and recognizing the processes one is using, deliberately invoking those most appropriate to the given circumstances, and providing the basis for conscious transfer of reasoning methods from familiar to unfamiliar contexts.

Given such awareness, one can begin to penetrate new situations by asking oneself probing questions and constructing answers. Starting with artificial, idealized, oversimplified versions of the problem, one can gradually penetrate to more realistic and complex versions. In an important sense, this is the mechanism underlying independent research and investigation.

13.3 WHY BOTHER WITH CRITICAL THINKING?

The preceding list of thinking and reasoning processes underlying the broad generic term "critical thinking" is neither complete nor exhaustive. For illustrative purposes, I have tried to isolate and describe processes and levels of awareness that appear to be bound up with clear thinking and genuine understanding in a wide variety of disciplines and to show a deep commonality in this respect among very different kinds of subject matter. I submit that these processes underlie the capacity defined by Jacques Barzun in the paragraph that heads this chapter.

Developing these intellectual skills requires extensive, sustained practice. Such practice is not possible in a space devoid of subject matter. It is only through contact with, and immersion in, rich areas of subject matter that interesting and significant experience can be generated. Although it may be possible, in principle, to generate limited aspects of such practice through artificial kinds of exercises and puzzle solving, or even through analysis of scores in sports contests, it seems a waste of time to resort to such sterile channels when all the vital disciplines of our culture lie at our disposal.

Why should we want to cultivate skills such as those I have listed? There are many obvious reasons having to do with quality of life, with professional competence, with the advance of culture and of society in general, but I particularly wish to suggest a sociopolitical reason: the education of an enlightened democratic citizenry. What capacities characterize such a citizenry?

Justice Learned Hand, the distinguished jurist of the preceding generation, argued with telling irony that we would be able to preserve civil liberties only so long as we were willing to engage in the "intolerable labor thought, that most distasteful of all our activities." John Dewey in *Democracy and Education* contends that "The opposite to thoughtful action are routine or capricious behavior. Both refuse to acknowledge responsibility for the future consequences which flow from present action."

The requirements set by Barzun, Hand, and Dewey can be broken down to more fundamental components. The sophisticated distinction between enlightened and short-range self-interest is based on hypothetico-deductive reasoning. Such reasoning is also inevitably involved in visualizing possible outcomes of decisions and policies in economic and political domains—policies on which one must exercise a vote.

There is need to discriminate between facts and inferences in the contentions with which one is surrounded. There is the necessity of making tentative judgments or decisions, and it is better that this be done in full awareness of gaps in available information than in an illusion of certainty. There is the highly desirable capacity to ask critical, probing, fruitful questions concerning situations in which one has little or no expertise. There is the need to be explicitly conscious of the boundaries of one's own knowledge and understanding of a particular problem.

Each of these capacities appears on the preceding list, and I believe that each can be cultivated and enhanced, at least to some degree, in the great majority of college students through properly designed experiences embracing a wide variety of subjects.

I hasten to emphasize that these skills *alone* are not sufficient to assure good citizenship or other desirable qualities of mind and person. Other ingredients are necessary, not the least of which are moral and ethical values, which impose their own constraints and boundary conditions on the naked processes of thinking and reasoning. Although values are not disconnected from thinking and reasoning, I believe that the educational problems they pose transcend the limits of this short essay and require discussion in their own right.

13.4 EXISTING LEVEL OF CAPACITY FOR ABSTRACT LOGICAL REASONING

In the United States some investigators have rather belatedly come to realize that much of our science curricular material, and the volume and pace with which we thrust it at our students, are badly mismatched to the existing levels of student intellectual development at virtually every age. I am convinced that the same is true in other disciplines, but the fact is less readily discerned because assignments and tests concentrate on end results and procedures rather than on reasoning and understanding.

I say that "some" have become aware of this problem because, despite the unequivocal and relentlessly accumulating statistics, many who teach in the schools, colleges, and universities remain unaware of the emerging data; others fail to see any relevance to their own teaching.

Beginning about 1971, investigators began administering elementary tasks in abstract logical reasoning (such as those pioneered by Jean Piaget [see Piaget and Inhelder (1958)] in his studies of the development of abstract reasoning capacity in children) to adolescents and adults of college age and beyond [see, for example, Chiapetta (1976); McKinnon and Renner (1971)]. The tests have centered principally on arithmetical reasoning with ratios or division and on awareness of the necessity of controlling variables in deducing cause–effect relationship.

Although the results vary significantly from one population to another (economically disadvantaged versus economically advantaged; concentrating in science and engineering versus concentrating in humanities or fine arts versus concentrating in the social sciences, etc.), the overall averages have remained essentially unchanged with increasing volume of data since the first small samples were reported in 1971, and, most suggestively, the averages do not change appreciably with increasing age beyond about 12 or 13: roughly one third of the total number of individuals tested solves the tasks correctly; roughly one third performs incorrectly but shows a partial, incipient grasp of the necessary mode of reasoning; the remaining third fails completely. In Piagetian terminology, the first group might be described as using formal patterns of reasoning, the third group as using principally concrete patterns, and the middle group as being in transition between the two modes [Arons and Karplus (1976)].

The weaknesses revealed by these two specific tasks would mean relatively little if they stood by themselves, but, in fact, these weaknesses are closely correlated with weaknesses in other modes of abstract logical reasoning such as discriminating between observation and inference; dealing with elementary syllogisms involving inclusion, exclusion, and serial ordering; recognizing gaps in available information; doing almost any kind of hypothetico-deductive reasoning.

Most of the curricular materials thrust at students in the majority of their courses at secondary and college level implicitly require well-developed reasoning capacity in the modes that have been listed in this discussion. In fact, only a small proportion of the students (less than one third) are ready for such performance. The rest, lacking the steady, supportive help and explicit exercises required, resort, in desperation, to memorization of end results and procedures. Failing to develop the processes underlying critical thinking, they fail to have experience of genuine understanding and come to believe that knowledge is inculcated by teachers and consists of recognizing juxtapositions of arcane vocabulary on multiple choice tests. (Readers familiar with the studies of William G. Perry will recognize his first category of intellectual outlook among college student. [Perry (1970)]).

13.5 CAN CAPACITY FOR ABSTRACT LOGICAL REASONING BE ENHANCED?

In our Physics Education Group at the University of Washington, we have worked intensively for some years with populations of pre- and in-service elementary school teachers and other nonscience majors ranging in age from 18 to over 30. Initially no more than about 10% were using formal patterns of reasoning. By starting with very basic, concrete observations and experiences, forming concepts out of such direct experience, going slowly, allowing students to make and rectify mistakes by confronting contradiction or inconsistency, insisting that they speak and write out their lines of reasoning and explanation, repeating the same modes of reasoning in new contexts days and weeks apart, we have been able to increase the fraction who successfully use abstract patterns of reasoning to perhaps 70 to 90%, depending on the nature of the task.

The most important practical lesson we have learned is that repetition is absolutely essential—not treading water in the same context until "mastery" is attained, but in altered and increasingly richer context, with encounters spread out over time. Quick, remedial exercises in artificial situations preceding "real" course work are virtually useless. One must patiently construct repeated encounters with the same modes of reasoning in regular course work and allow students to benefit from their mistakes. Progress becomes clearly visible in the sense that the percentage of successful students increases with each repetition.[2]

It is still a very long step from the development of specific abstract reasoning processes in one area of subject matter, such as elementary science, to more advanced levels of subject matter in the same area, not to speak of transfer to entirely different areas. What little evidence exists suggests that very little transfer occurs from experience acquired in only one discipline. I myself am strongly convinced, however (mostly by fragmentary, anecdotal evidence, and perhaps some admixture of wishful thinking), that very great progress could be effected if students were *simultaneously* exposed to such intellectual experience in entirely different disciplines. This is largely a matter of conjecture since a concerted experiment at the college level has not really been tried.

The fragmentary evidence to which I appeal comes from two disparate sources:

1. *Experience with a tightly organized core curriculum at Amherst College during the '50s and '60s.* In this curriculum, there was a very strong interaction among an English composition course, a science course, an American Studies course, and, toward the later stages, a Western Civilization course, all of which had certain attitudes, approaches and intellectual standards in common [Kennedy (1955); Arons (1978)]. Alumni of that period tend to comment very favorably, in retrospect, on the effect of that experience on their own intellectual development. (So tightly organized a curriculum was a rather special case and, as then implemented, would be possible only with a small, homogeneous student body. Judicious modifications should, however, be effective in more heterogeneous situations.)

2. *Data being reported on effects of the elementary school science curricula developed under auspices of the National Science Foundation during the '60s. The latter evidence is very indirect, but it is highly suggestive and merits a bit of discussion.* The groups that developed the new curricula worked directly with the children they sought to teach and met the latter on their own ground and at their existing verbal and conceptual starting points rather than in some never-never land of unchecked and untested hypotheses and assumptions about children and learning. Everything in these materials begins with hands-on experience and observation. Concepts are developed through induction and synthesis from this experience, with the teacher as guide and pilot rather than as verbal inculcator. Ideas are developed *first* and names are invented *afterwards*; technical terms

---

[2]On any one kind of task (e. g., arithmetical reasoning involving division, or forming a clear operational and intuitive distinction between mass and volume), the following history of progress is typical: For the first few repetitions the percentage of successful performers increases substantially, but the curve is concave downward and invariably levels off somewhere between 70 and 90% after four to six opportunities [Arons (1976); Rosenquist (1982)]. We have never been able to achieve 100% success. Approximately 15% of the students never developed, under our guidance, the capacity to perform successfully on the given task, even with further repetition and with intensive personal tutoring.

There are certain obvious questions: Were we insufficiently skillful in providing guidance and instruction to the unsuccessful 15%? Would success still be achieved over much longer periods of time? What would have happened if these individuals had received such instruction at the age of 11 or 12 instead of so much later? Are there some individuals who are intrinsically unable to develop these capacities? Are our observations, for some reason, invalid? All we can say is that we do not know the answers. We had reached a point of diminishing returns under available time and resources and were unable to press the issue further. The empirical fact is that the progress curve leveled off below 100%. We hope that answers to some of the preceding questions will begin to emerge as time goes by.

are generated operationally only *after* experience has given them sanction and meaning. [As examples, the reader might refer to programs such as *Elementary Science Study*, Webster Division McGraw-Hill Book Co., St. Louis, MO; *Science Curriculum Improvement Study* and *Science, A Process Approach*, both from Delta Education, Nashua NH.]

The essence of instruction in these programs, whether the subject matter is physical or biological, is to give the children *time*—time to explore, to test, to manipulate, to talk and argue about meaning and interpretation, to articulate hypotheses, to follow trails to dead ends and retrace steps if necessary, to make mistakes and to revise views and interpretations when guided to perceive contradictions (instead of being told, by assertion, that their idea was "right" or "wrong"), to decide when and how arithmetical calculations should be made.

Such learning is sometimes (misleadingly) called "discovery learning. " The children are, of course, not expected to be Newtons, Faradays, Agassizes, or Darwins, "discovering" the concepts and theories of science *de novo* by the age of ten. Ordinary, lively, curious children simply react positively to the opportunity to learn from perceptively *guided* experience and observation. They retain what they learn because they are synthesizing genuine experience rather than memorizing a jumble of meaningless and unfamiliar words. They know where their knowledge comes from and are able to address the "How do we know . . . ? Why do we believe . . . ?" questions.

Since the first appearance of these curricula, researchers have been comparing the achievement of children exposed to such materials with the achievement of controls. In addition to showing significantly improved command of science subject matter, children exposed to the new curricula show significantly greater progress in both verbal and numerical skills, and the effects are particularly strong among disadvantaged children [cf. Bredderman (1982); Shymansky, et al. (1983); Wellman (1978)]. In other words, this mode of instruction, when competently implemented, results in *transfer*, enhancing performance beyond the science subject matter alone.

Although there is no direct evidence of a similar kind supporting the notion that we would enhance the higher level reasoning skills of college students by undertaking the instructional effort I have been advocating, I submit that the observations of the effect of the inquiry-oriented science curricula on children are at least very encouraging. The processes involved are analogous, and the effort seems worth making.

13.6 CONSEQUENCES OF MISMATCH

We are indeed fortunate that a significant proportion of our student population, perhaps one quarter, does make the breakthrough on modes of abstract logical reasoning spontaneously. (Consider the consequences to our society if this were otherwise!) But this does not lessen the urgency of improving our performance. As pointed out earlier, there now exists a serious mismatch between curricular materials and expectations on the one hand and actual level of student intellectual development on the other. The curricular materials implicitly require abstract reasoning capacities and levels of insight and interpretation that many students have not yet attained. Neither the materials nor the most prevalent modes of instruction provide the gently paced, insistent, repetitive guidance that is necessary for helping students develop the necessary intellectual skills.

This mismatch has extensive deleterious consequences. We force a large fraction of students into blind memorization by imposing on them, particularly at high school and university levels, materials requiring abstract reasoning capacities they have not yet attained. And we proceed through these materials at a pace that precludes effective learning and understanding, even if the necessary reasoning capacities have been formed. Under such pressure, students acquire no experience of what understanding really entails. They cannot test their "knowledge" for plausible consequences or for internal consistency; they have no sense of where accepted ideas or results come from, how they are validated, or why they are to be accepted or believed. In other words, they do not have the opportunity to develop the habits of critical thinking defined earlier in this essay, and they acquire the misapprehension that knowledge resides in memorized assertions, esoteric technical terminology, and regurgitation of received "facts." Although such failure is widely prevalent in the sciences, it is by no means confined there. It pervades our entire system, including history, the humanities, and the social sciences.

One specific example of the mechanism through which an entire system becomes degraded emerges through our experience with arithmetical reasoning. When I first discovered that no more than 10% of my undergraduate nonscience majors could reason arithmetically with division, I wondered what had happened to the old word problems that were used to cultivate such reasoning from the fifth and sixth grades on. Going back to existing elementary school arithmetic texts, I found that such problems were still there, as in my own school days, and were probably significantly improved. When I questioned my university students, they began to reveal that they had never actually had to do such problems in school because the problems were "too hard." When I began working with in-service elementary school teachers and found that they themselves could not deal with such problems, the pattern was clear: an engineer would describe the system as a "degenerative feedback loop." The arithmetical reasoning disability of the future teachers had never been detected and remedied when they were at the university. They graduated, went into the schools, and passed their disability and fear to most of the children by not requiring the doing of the word problems and conveying the rationalization that they were "too hard." The children went on to the university, and so on, and so forth.

The case of arithmetical reasoning is just an especially clear and vivid illustration. The same pattern arises over and over again in other instances: in failure to master and understand the most fundamental scientific concepts (such as velocity and acceleration or the nature of floating and sinking); in poor writing and speaking of English; in incapacity to deal with historical reasoning and the concomitant blind concentration on historical "facts."

I wish to emphasize most strongly that the teachers whose incapacities I describe are *not* the ones to be blamed for this situation. The input terminals to the feedback loop of my metaphor reside in *our* hands at the colleges and universities. *We* are the ones who perpetuate the mismatch and fail to provide remediation of disabilities and enhancement of abstract reasoning capacities at the opportunities that *we* control. *We* are the ones who made the teachers as they are.

The mismatch about which I complain affects, of course, not only our future teachers but the whole of our student population outside the 25% who, in spite of the system, manage to break through spontaneously to abstract reasoning patterns. I dwell so insistently on the teachers only because of the crucial role they play in sustaining the feedback loop. Think of the prodigious impetus that might stem from altering the condition of the teachers and making the feedback *re*generative instead of *de*generative! What might we be able to achieve at university level if the mismatch between our materials and student readiness were removed?

13.7 ASCERTAINING STUDENT DIFFICULTIES

It might be helpful to point out some hard facts regarding the securing of reliable information concerning student learning difficulties and levels of abstract reasoning. What one must learn to do is ask simple, sequential questions, leading students in a deliberate Socratic fashion. After each question, one must shut up and listen carefully to the response. (It is the tendency of most inexperienced questioners to provide an answer, or to change the question, if a response is not forthcoming within one second. One must learn to wait as long as four or five seconds, and one then finds that students, having been given a chance to think, will respond in sentences and truly reveal their lines of thought.)

As the students respond to such careful questioning, one can begin to discern the errors, misconceptions, and missteps in logic that are prevalent. One learns nothing by giving students "right answers" or "lucid explanations." As a matter of fact, students do not benefit from such answers or explanations; they simply memorize them. Students are much more significantly helped when they are led to confront contradictions and inconsistencies in what they say and then spontaneously alter their statements as a result of such confrontation.

In such dialogs, two things immediately strike novice investigators. First they find that virtually all of their a priori conjectures concerning what students are and are not thinking are incorrect and that entirely unanticipated but very fundamental, plausible, and deeply rooted preconceptions, misconceptions, and misapprehensions (of which the investigator had no awareness) are revealed. Second, they discover the saving grace in all of this unanticipated complexity: the frequently voiced cliché that every individual is completely different from every other individual is patently untrue. Each kind of misconception or erroneous mode of reasoning occurs, with remarkable reproducibility, in many individuals. Some hurdles and misconceptions are very widely prevalent. When one finds an approach or insight that overcomes a particular difficulty, that approach will be helpful not to only one but to many individuals.

It must be strongly emphasized that conclusions must be based on careful and accurate listening to students. Casual extrapolation of one's own experience only leads to error. Those of us who are fortunate enough to have become competent professionals are among the 25% minority mentioned earlier. We made the breakthrough in spite of the system, not because of it. Our own learning experiences are not representative, and citing such experience rarely leads to correct insight into what transpires for the majority of learners.

13.8 TESTING

At present, deficiency in the quality of testing is one of the more serious ills of our profession. There is a large and perceptive literature on testing in virtually every discipline, but its influence has, unfortunately, not been extensive. Some cynics have even remarked on the existence of a destructive collusion between students and teachers—a collusion in which students agree to accept bad teaching provided they are given bad examinations.

It is useless to render lip service to sophisticated intellectual goals and then test only for end results, vocabulary, "facts," or "information." The real goals of a course are determined not by what we say but what we test for. Students quickly ascertain what the real requirements of a course are and orient their efforts accordingly. Their attention can be focussed on the higher intellectual processes and requirements only if these aspects are included in testing and writing and play an important role in the final grade.

It is my earnest hope that more self-conscious attention to thinking and reasoning processes on the part of faculty will lead to statistical improvement in the quality of test questions and writing assignments. Good questions are very hard to devise, and one individual runs out of inspiration. Collaborative effort could greatly increase the pool of good material and also provide the debugging that is always necessary.

13.9 SOME THOUGHTS ON FACULTY DEVELOPMENT

Given the almost universally accepted goal of enhancing the capacity for critical thinking in our students, it seems reasonable to lead faculty members to sharpen their own critical thinking about how this goal is to be attained through the use of units of subject matter in their own areas of expertise—units which they have taught and with which they feel comfortable. The problem is to get away from vague, mushy generalizations and to provide constraints that induce consideration and elaboration of very specific examples: analyzing a unit of subject matter so as to identify the thinking and reasoning processes that must be brought to bear by the student, and devising questions that lead the student to such penetration.

I suggest that useful results might stem from the organization of faculty workshops in which participants, working in pairs, come prepared with a response to something like the following assignment:

1. Select from within your area of expertise a unit of subject matter with which you are thoroughly familiar and study of which, you believe, will help a student attain a particular intellectual goal or insight or which will serve to exercise a particular reasoning process. (The unit should be as short as possible, but it should have a significant goal and not end up as a triviality.)
2. State the goal or insight involved.
3. Describe the essence of the unit of subject matter; that is, indicate how it provides a path toward the goal or insight.
4. List the various abstract reasoning capacities that the student must already possess, or must be helped to develop, in order to deal properly with the subject matter and not have to resort to memorization. (The basis for this analysis might be the list provided earlier in this paper or an appropriately modified or augmented list.)
5. Indicate the kind of help you might provide students who encounter difficulty penetrating the material (e. g., questions that help point up significant issues, or clarify concepts, or focus on assumptions that are likely to be overlooked.)
6. Indicate what writing you might ask the students to do in connection with the unit and how you would test for final mastery or understanding.

7. Indicate how you would lead the students to stand back, become conscious of the patterns of thinking and reasoning in which they had engaged, and, if possible, connect this experience with experiences they have had in other courses.

I imagine the workshop as bringing together pairs of individuals from the same and different disciplines. Each pair would present their analysis for discussion and comment by the entire group. There need be no "expert" or "authority" directing the proceedings. I would like to think that faculty members seriously interested in the intellectual development of their students would find such an exercise interesting and stimulating. They would become more conscious of commonalities across disciplinary lines while defining real differences more precisely. The necessity of presenting the essence of a specific intellectual exercise to colleagues in other disciplines would help minimize the use of jargon and would sharpen awareness of how units of subject matter can be utilized. And finally, the whole enterprise would help cultivate an instructional climate in which students clearly perceived that they were being helped to develop their own capacities for critical thinking.

Bibliography

ADAMS, H. (1918), *The Education of Henry Adams* (Houghton Mifflin, Boston (many subsequent editions)).
ADLER, C. (1958), "On the Humanization of Some Physics Problems," *Am. J. Phys.* **26**, 42.
AIKEN, C. (1953), *Collected Poems* (Oxford University Press, New York).
ANDERSON, J. R. (1980), *Cognitive Psychology and Its Implications* (Freeman, San Francisco).
ANDERSON, J. R. ed. (1981), *Cognitive Skills and Their Acquisition* (Lawrence Erlbaum Associates, Hillsdale, NJ).
ARONS, A. B. (1965), *Development of Concepts of Physics* (Addison-Wesley, Reading, MA).
ARONS, A. B. (1975), "Newton and the American Political Tradition," *Am. J. Phys.* **43**, 209.
ARONS, A. B. (1976), "Cultivating the Capacity for Formal Reasoning," *Am. J. Phys.* **44**, 834.
ARONS, A. B. (1977), *The Various Language: An Inquiry Approach to the Physical Sciences* (Oxford University Press, New York).
ARONS, A. B. (1978), "Teaching Science" in *Scholars Who Teach*, CAHN, S. M., ed. (Nelson-Hall, Chicago).
ARONS, A. B. (1979), "Basic Physics of the Semidiurnal Lunar Tide," *Am. J. Phys.* **47**, 934.
ARONS, A. B. (1981a), "Whither Do We Hurry Hence?" in *AAPT Pathways*. Proceedings of the Fiftieth Anniversary Symposium of the AAPT. (Published by the American Association of Physics Teachers.)
ARONS, A. B. (1981b), "Thinking, Reasoning and Understanding in Introductory Physics Courses," *The Physics Teacher* **19**, 166.
ARONS, A. B. (1982), "Phenomenology and Logical Reasoning in Introductory Physics Courses," *Am. J. Phys.* **50**, 13.
ARONS, A. B. (1983a), "Achieving Wider Scientific Literacy," *Daedalus* Spring 1983.
ARONS, A. B. (1983b), "Student Patterns of Thinking and Reasoning, Part One," *The Physics Teacher* **21**, 576.
ARONS, A. B. (1984a,b), "Student Patterns of Thinking and Reasoning, Parts Two and Three," *The Physics Teacher* **22**, 21, 88.
ARONS, A. B. (1984c), "Education Through Science," *J. Col. Sci. Teaching* **13**, 210.
ARONS, A. B. (1984d), "Computer-Based Instructional Dialogs in Science Courses," *Science* **224**, 1051.
ARONS, A. B. (1985), "Critical Thinking and the Baccalaureate Curriculum," *Liberal Education* **71**, 141.
ARONS, A. B. (1986), "Overcoming Conceptual Difficulties in Physical Science Through Computer-Based Socratic Dialogs," in *Designing Computer-Based Learning Materials*, WEINSTOCK, H., and BORK, A., eds. Proceedings of the NATO workshop held in San Miniato, Italy, July 1985. NATO ASI Series F: Computer and Systems Sciences, Vol. 23 (Springer-Verlag, Berlin, New York).

ARONS, A. B., and BORK, A. M. (1964), "Newton's Laws of Motion and the 17th Century Laws of Impact," *Am. J. Phys.* **32**, 313.
ARONS, A. B., and KARPLUS, R. (1976), "Implications of Accumulating Data on Levels of Intellectual Development," *Am. J. Phys.* **44**, 396.
ARONS, A. B., and PEPPARD, M. B. (1965), "Einstein's Proposal of the Photon Concept: A Translation of the Annalen der Physik Paper of 1905," *Am. J. Phys.* **33**, 367.
BARTLETT, A. A. (1976–1979), "The Exponential Function" (Parts I–IX) *The Physics Teacher* **14**, 393, 485; **15**, 37, 98, 225; **16**, 23, 92; **17**, 23.
BIRKS, J. B. (1962), *Rutherford at Manchester* (Heywood & Co. Ltd., London).
BOHR, N. (1913), "On the Constitution of Atoms and Molecules," *Phil. Mag.* **26**, 6, 1.
BORK, A. (1979), "Interactive Learning," *Am. J. Phys.* **47**, 5.
BORK, A. (1981), *Learning With Computers* (Digital Press, Bedford, MA).
BRASELL, H. (1987), "The Effect of Real-Time Laboratory Graphing on Learning Graphic Representations of Distance and Velocity," *J. Res. Sci. Teach.* **24**, 385.
BREDDERMAN, T. (1982), "Effects of Activity-Based Science in Elementary School," in *Education in the '80's: Science.* ROWE, M. B., ed. (National Education Association, Washington, DC).
BRIDGMAN, P. W. (1941), *The Nature of Thermodynamics* (Harvard University Press, Cambridge, MA).
BRIDGMAN, P. W. (1962), *A Sophisticate's Primer of Relativity* (Wesleyan University Press, Middletown, CT. Revised edition 1983).
BRUSH, S. G. (1961), "John James Waterston and the Kinetic Theory of Gases," *Am. Scientist* **49**, 202.
BRUSH, S. G. (1965), *Kinetic Theory. Vol. 1. The Nature of Gases and of Heat* (Pergamon Press, Oxford).
CASPER, B. M. and NOER, R. J. (1972), *Revolutions in Physics* (W. W. Norton & Co., New York).
CHAMPAGNE, A. B., GUNSTONE, R. F., and KLOPFER, L. E. (1985), "Instructional Consequences of Students' Knowledge about Physical Phenomena," in *Cognitive Structure and Conceptual Change*, WEST, L. H. T., and PINES, A. L., eds. (Academic Press, Orlando, FL).
CHAMPAGNE, A. B., KLOPFER, L. E., and ANDERSON, J. H. (1980), "Factors Influencing the Learning of Classical Mechanics," *Am. J. Phys.* **48**, 1074.
CHIAPPETTA, E. L. (1976), "A Review of Piagetian Studies Relevant to Science Instruction at the Secondary and College Level," *Sci. Ed.* **60**, 253.
CLEMENT, J. (1979), "Mapping a Student's Causal Conceptions from a Problem Solving Protocol," in *Cognitive Process Instruction*, LOCHHEAD, J., and CLEMENT, J., eds. (Franklin Institute Press, Philadelphia).
CLEMENT, J. (1982), "Students' Preconceptions in Introductory Mechanics," *Am. J. Phys.* **50**, 66.
CLEMENT, J. (1983), "A Conceptual Model Discussed by Galileo and Used Intuitively by Physics Students," in *Mental Models*, GENTNER, D., and STEVENS, A. L., eds. (Lawrence Erlbaum Associates, Hillsdale, NJ).
CLEMENT, J. (1986), "Overcoming Conceptual Difficulties in Mechanics." Paper presented at AAPT Summer Meeting 23–27 June, Columbus, OH.
CLEMENT, J. (1987), "Overcoming Students' Misconceptions in Physics: The Role of Anchoring Intuitions and Analogical Validity," in *Proceedings of Second International Seminar: Misconceptions and Educational Strategies in Science and Mathematics III*, NOVAK, J., ed. (Cornell University, Ithaca, NY).
CLEMENT, J., LOCHHEAD, J., and MONK, G. S. (1981), "Translation Difficulties in Learning Mathematics," *Am. Mathematical Monthly*, **88**, 286.
COHEN, I. B., ed. (1941), *B. Franklin's Experiments, a New Edition of Franklin's Experiments and Observations on Electricity*, edited, with a critical and historical introduction (Harvard University Press, Cambridge, MA). [Letter IV from this edition is reprinted in *Science* **123**, 47 (1956) in recognition of the 250th anniversary of the birth of Benjamin Franklin.]

COHEN, R., EYLON, B., and GANIEL, U. (1983), "Potential Difference and Current in Simple Electric Circuits: A Study of Students' Concepts," *Am. J. Phys.* **51**, 407.

CONANT, J. B., and NASH, L. K., eds. (1957), *Harvard Case Histories in Experimental Science* (Harvard University Press, Cambridge, MA).

CRANE, H. R. (1960), "Creative Thinking and Experimenting," *Am. J. Phys.* **28**, 437.

CRANE, H. R. (1969a), "Better Teaching with Better Problems and Exams," *Phys. Today* **22**(3).

CRANE, H. R. (1969b), "Problems for Introductory Physics. Part I," *The Physics Teacher* **7**, 371.

CRANE, H. R. (1970), "Problems for Introductory Physics. Part II," *The Physics Teacher* **8**, 182.

DI SESSA, A. (1982), "Unlearning Aristotelian Physics: A Study of Knowledge-Based Learning," *Cog. Sci.* **6**, 37.

DRIVER, R., and EASELEY, J. (1978), "Pupils and Paradigms: A Review of Literature Related to Concepts Development in Adolescent Science Students," *Studies in Science Education* **5**, 61.

EINSTEIN, A. (1905a), "Über eine die Erzeugung und Verwandlung des Lichtes betreffenden heuristischen Geischtspunkt," *Ann. d. Phys.* **4, 17**, 132. [English translation: Arons and Peppard (1965).]

EINSTEIN, A. (1905b), "Zur Elektrodynamik bewegter Körper" *Ann. d. Phys.* **4, 17**, 891. [English translation in *The Principle of Relativity: A Collection of Original Memoirs on the Special and General Theory of Relativity*. (Methuen & Co. Ltd., 1923. Reprinted by Dover Publications, Inc., New York.)]

EINSTEIN, A. (1961), *Relativity: The Special and the General Theory* (Crown Publishers, New York).

EISBERG, R. M. (1976), *Applied Mathematical Physics with Programmable Pocket Calculators* (McGraw-Hill, New York).

EISENBUD, L. (1958), "On the Classical Laws of Motion," *Am. J. Phys.* **26**, 144.

Elementary Science Study (ESS) (1968ff) Webster Division, McGraw-Hill Book Co., New York, also Delta Education, Nashua, NH. (These materials, together with other elementary science curricula, are available on a CD-ROM published under the title *Science Helper K-8 CD ROM* by PC-SIG, Sunnyvale, CA.)

ERICKSON, G., and AGUIRRE, J. (1984), "Student Conceptions About the Vector Characteristics of Three Physics Concepts," *J. Res. in Sci. Teach.* **21**(5).

ERLICHSON, H. (1977), "Work and Kinetic Energy for an Automobile Coming to a Stop," *Am. J. Phys.* **45**, 769.

EVANS, J. (1978), "Teaching Electricity with Batteries and Bulbs," *The Physics Teacher* **16**, 15.

FARADAY, M. (1839), "Identity of Electricities Derived from Different Sources," in *Experimental Researches in Electricity, Vol. I* (Taylor and Francis, London). (Reprinted by Dover Publications, New York. 1965a, p. 76.)

FARADAY, M. (1855), "On the Physical Character of the Lines of Magnetic Force," in *Experimental Researches in Electricity, Vol. III* (Taylor and Francis, London). (Reprinted by Dover Publications, New York. 1965b, p. 408.)

FEYNMAN, R. P., LEIGHTON, R. B., and SANDS, M. (1963), *The Feynman Lectures on Physics* (Addison-Wesley, Reading, MA).

FERGUSON-HESSLER, M. G. M., and DE JONG, T. (1987), "On the Quality of Knowledge in the Field of Electricity and Magnetism," *Am. J. Phys.* **55**, 492.

FREDETTE, M. H., and CLEMENT, J. J. (1981), "Student Misconceptions of an Electric Circuit: What Do They Mean?" *J. Coll. Sci. Teach.* **10**, 280.

GENTNER, D., and GENTNER, D. R. (1983), "Flowing Water or Teeming Crowds: Mental Models of Electricity," in *Mental Models*, GENTNER, D., and STEVENS. A. L., eds. (Lawrence Erlbaum Associates, Hillsdale, NJ).

GOLDBERG, F. M., and ANDERSON, J. H. (1989), "Student Difficulties with Graphical Representations of Negative Values of Velocity," *The Physics Teacher* **27**, 254.

GOLDBERG, F. M., and MCDERMOTT, L. C. (1986), "Student Difficulties in Understanding Image Formation by a Plane Mirror," *The Physics Teacher* **24**, 472.

GOLDBERG, F. M., and MCDERMOTT, L. C. (1987), "An Investigation of Student Understanding of the Real Image Formed by a Converging Lens or Concave Mirror," *Am. J. Phys.* **55**, 108.
GOLDSTEIN, M., and GOLDSTEIN, I. F. (1978), *How We Know* (Plenum Press, New York).
GUNSTONE, R. F. (1984), "Circular Motion: Some Pre-Instructional Alternative Frameworks," *Res. in Sci. Ed.* **14**, 125.
GUNSTONE, R. F. (1987), "Student Understanding in Mechanics: A Large Population Survey," *Am. J. Phys.* **55**, 691.
GUNSTONE, R. F., CHAMPAGNE, A. B., and KLOPFER, L. E. (1981), "Instruction for Understanding: A Case Study," *Australian Sci. Teachers Jour.* **27**, 27.
GUNSTONE, R. F., and WHITE, R. (1981), "Understanding Gravity," *Sci. Ed.* **65**, 291.
HABER-SCHAIM, U., et al. (1960ff), *PSSC Physics*, Several Editions 1960–date (D. C. Heath & Co., Lexington, MA).
HAKE, R. R. (1987), "Promoting Student Crossover to the Newtonian World," *Am. J. Phys.* **55**, 878.
HALLOUN, I. A., and HESTENES, D. (1985), "The Initial State of College Physics Students," *Am. J. Phys.* **53**, 1043; also "Common Sense Concepts About Motion," *Am. J. Phys.* **53**, 1056.
HALLOUN, I. A., and HESTENES, D. (1987), "Modeling Instruction in Mechanics," *Am. J. Phys.* **55**, 455.
HARNWELL, G. P., and LIVINGOOD, J. J. (1933), *Experimental Atomic Physics.* (McGraw-Hill, New York).
HEILBRON, J. L. (1979), *Electricity in the 17th and 18th Centuries. A Study of Early Modern Physics* (University of California Press, Berkeley, CA).
HESTENES, D. (1987), "Toward a Modeling Theory of Physics Instruction," *Am. J. Phys.* **55**, 440.
HOLTON, G. (1973), *Introduction to Concepts and Theories in Physical Science*, 2nd ed. revised by BRUSH, S. G. (Addison-Wesley, Reading, MA).
HOLTON, G. (1978), "Subelectrons, Presuppositions, and the Millikan-Ehrenhaft Dispute," in *The Scientific Imagination* (Cambridge University Press, Cambridge).
HOBBIE, R. K. (1973), "Teaching Exponential Growth and Decay: Examples from Medicine," *Am. J. Phys.* **41**, 389.
HOFSTADTER, D. R. (1982), "Number Numbness or Why Innumeracy May Be Just as Dangerous as Illiteracy," *Sci. Am.* **5**, 20.
HUGGINS, E. R. (1968), *Physics 1* (W. A. Benjamin, New York).
HUGHES, A. L. (1913), "On the Emission Velocities of Photo-Electrons," *Phil. Trans. Roy. Soc. London*, Series A **212**, 205.
HUMPHREY, R. D. (1951), *Georges Sorel, Prophet Without Honor* (Harvard University Press, Cambridge, MA).
IONA, M. (1979), "Teaching Electrical Resistance," *The Physics Teacher* **17**, 299.
IONA, M. (1983), "We Ought to Use the Conventional Current Direction," *The Physics Teacher* **21**, 334.
IONA, M. (1987), "Why Johnny Can't Learn Physics from Textbooks I Have Known," *Am. J. Phys.* **55**, 299.
INHELDER, B., and PIAGET, J. (1958), *The Growth of Logical Thinking from Childhood to Adolescence* (Basic Books, New York).
JOHNSTONE, A. H., and MUGHOL, A. R. (1978), "The Concept of Electrical Resistance," *Phys. Educ.* **13**, 46.
KARPLUS, R. (1977), "Science Teaching and the Development of Reasoning," *J. Res. Sci. Ed.* **14**, 169.
KARPLUS, R., KARPLUS, E., FORMISANO, M., and PAULSEN, A-C. (1979), "Proportional Reasoning and Control of Variables in Seven Countries," in *Cognitive Process Instruction*, LOCHHEAD, J., and CLEMENT, J., eds. (Franklin Institute Press, Philadelphia).
KELLER, J. B. (1987), "Newton's Second Law," *Am. J. Phys.* **55**, 1145.
KENNEDY, G. (1955), *Education at Amherst* (Harper and Brothers, New York).

KIDD, R., ARDINI, J., and ANTON, A. (1989), "Evolution of the Modern Photon," *Am. J. Phys.* **57**, 27.
KUNZ, K. S. (1971), "Visualizing Large Numbers," *Am. J. Phys.* **39**, 452.
LAMB, H. (1932), *Hydrodynamics* (Cambridge University Press, Cambridge).
LAPP, C. J. (1940), "Effectiveness of Mathematical versus Physical Solutions in Problem Solving in College Physics," *Am. J. Phys.* **8**, 241.
LARKIN, J. (1981), "Cognition of Learning Physics," *Am. J. Phys.* **49**, 534.
LARKIN, J. (1983), "The Role of Problem Representation in Physics," in *Mental Models*, GENTNER, D., and STEVENS, A. L., eds. (Lawrence Erlbaum Associates, Hillsdale, NJ).
LAWSON, A. E. (1980), "Relationships among Level of Intellectual Development, Cognitive Style, and Grades in a College Biology Course," *Sci. Ed.* **64**, 95.
LAWSON, A. E. (1982), "The Reality of General Cognitive Operations," *Sci. Ed.* **66**, 229.
LAWSON, A. E., LAWSON, D. I., and LAWSON, C. A. (1984), "Proportional Reasoning and the Linguistic Abilities Required for Hypothetico-Deductive Reasoning," *J. Res. Sci. Teach.* **21**, 119.
LAWSON, A. E., and WOLLMAN, W. T. (1975), "Physics Problems and the Process of Self-Regulation," *The Physics Teacher* **13**, 470.
LAWSON, R. A., and MCDERMOTT, L. C. (1987), "Student Understanding of the Work-Energy and Impulse-Momentum Theorems," *Am. J. Phys.* **55**, 811.
LENARD, P. (1902), "Über die Lichtelektrische Wirkung," *Ann. d. Phys.* **4, 8**, 149.
LIN, H. (1978), "Newtonian Mechanics and the Human Body: Some Estimates of Performance," *Am. J. Phys.* **46**, 15.
LIN, H. (1982), "Fundamentals of Zoological Scaling," *Am. J. Phys.* **50**, 72.
LOCHHEAD, J. (1981), "Faculty Interpretations of Simple Algebraic Statements: The Professor's Side of the Equation," *Jour. of Mathematical Behavior* Spring 1981.
MACH, E. (1893), *The Science of Mechanics* (Open Court Publishing Co., Chicago).
MAGIE, W. F. (1935), *Source Book in Physics* (McGraw-Hill, New York).
MALONEY, D. P. (1984), "Rule Governed Approaches to Physics: Newton's Third Law," *Phys. Ed.* **19**, 37.
MCCLOSKEY, M., CAMARAZZA, A., and GREEN, B. (1980), "Curvilinear Motion in the Absence of External Forces," *Science* **210**, 1139.
MCCLOSKEY, M. (1983), "Intuitive Physics," *Sci. Am.* **249**, 122.
MCDERMOTT, L. C. (1980), "Teaching Physics to Promote Cognitive Development in Academically Disadvantaged Students Aspiring to Science-Related Careers," in *Physics Teaching.* Proceedings of the GIREP conference held at the Weizmann Institute of Science, Rehovot, Israel. August 19–24, 1979. GANIEL, U., ed. (Balaban International Science Services, Jerusalem).
MCDERMOTT, L. C. (1984), "Research on Conceptual Understanding in Mechanics," *Physics Today* **37**, 24.
MCDERMOTT, L. C., PITERNICK, L. K., and ROSENQUIST, M. L. (1980), "Helping Minority Students Succeed in Science. I. Development of a Curriculum in Physics and Biology," *J. Coll. Sci. Teach.* **9**, 135.
MCDERMOTT, L. C., ROSENQUIST, M. L., and VAN ZEE, E. H. (1983), "Strategies to Improve the Performance of Minority Students in the Sciences," in *Teaching Minority Students: New Directions for Teaching and Learning*, no. 16. CONES, J. H. III, NOONAN, J. F., and JANHA, D., eds. (Jossey-Bass, San Francisco).
MCDERMOTT, L. C., ROSENQUIST, M. L., and VAN ZEE, E. H. (1987), "Student Difficulties in Connecting Graphs and Physics: Examples from Kinematics," *Am. J. Phys.* **55**, 503.
MCKINNON, J. W., and RENNER, J. W. (1971), "Are Colleges Concerned with Intellectual Development?" *Am. J. Phys.* **39**, 1047.
MEMORY, J. D. (1973), "Kinematics Problems for Joggers," *Am. J. Phys.* **41**, 1205.
MEMORY, J. D., and JENKINS, A. W., Jr. (1977), "Estimating Orders of Magnitude," *The Physics Teacher* **15**, 43.
MERRILL, J. R. (1976), *Using Computers in Physics* (Houghton Mifflin Co., Boston).
MILLIKAN, R. A. (1909), "A New Modification of the Cloud Method of Measuring the

Elementary Electrical Charge and the Most Probable Value of that Charge," *Phys. Rev.* **29**, 560.
MILLIKAN, R. A. (1911), "The Isolation of an Ion, a Precision Measurement of Its Charge, and the Correction of Stokes's Law," *Phys. Rev.* **32**, 349.
MILLIKAN, R. A. (1916), "A Direct Photoelectric Determination of Planck's Constant 'h'," *Phys. Rev.* **7**, 355.
MILLIKAN, R. A. (1917), *The Electron* (University of Chicago Press, Chicago (1st ed. 1917; 2nd ed. 1924)).
MILLIKAN, R. A. (1949), "Albert Einstein on his Seventieth Birthday," *Rev. Mod. Phys.* **21**, 343.
MINSTRELL, J. (1982), "Explaining the 'At Rest' Condition of an Object," *The Physics Teacher* **20**, 10.
MINSTRELL, J. (1984), "Teaching for the Understanding of Ideas: Forces on Moving Objects," in *1984 Yearbook of the Association for the Education of Teachers* (ERIC Clearinghouse, Ohio State University, Columbus, OH).
MORRISON, P. (1963), "Fermi Questions," *Am. J. Phys.* **31**, 626.
OKUN, L. B. (1989), "The Concept of Mass," *Phys. Today* **42:6**, 31.
OPPENHEIMER, J. R. (1963), "Communication and Comprehension of Scientific Knowledge," *Science* **142**, 1144.
OSBORNE, R. (1984), "Children's Dynamics," *The Physics Teacher* **22**, 504.
PANOFSKY, W. K. H., and PHILLIPS, M. (1962), *Classical Electricity and Magnetism*, 2nd ed. (Addison-Wesley, Reading, MA).
PENCHINA, C. M. (1978), "Pseudowork-Energy Principle," *Am J. Phys.* **46**, 295.
PERRY, W. G. (1970), *Forms of Intellectual and Ethical Development in the College Years* (Holt, Rinehart & Winston, New York).
PETERS, P. C. (1982), "Even Honors Students Have Conceptual Difficulties With Physics," *Am. J. Phys.* **50**, 501.
PETERS, P. C. (1986), "An Alternative Derivation of Relativistic Momentum," *Am. J. Phys.* **54**, 804.
PIAGET, J., and INHELDER, B. (1958), *Growth of Logical Thinking* (Basic Books, New York).
Plato Physics I (1983), (Control Data Corp., Minneapolis, MN).
REIF, F. (1981), "Teaching Problem Solving—A Scientific Approach," *The Physics Teacher* **19**, 310.
REIF, F. (1985), "Acquiring an Effective Understanding of Scientific Concepts," in *Cognitive Structure and Conceptual Change*, WEST, L. H. T., and PINES, L., eds. (Academic Press, Orlando, FL).
REIF, F., LARKIN, J. H., and BRACKETT, B. C. (1976), "Teaching General Learning and Problem Solving Skills," *Am. J. Phys.* **44**, 212.
REIF, F., and HELLER, J. I. (1982), "Knowledge Structures and Problem Solving in Physics," *Educational Psychologist* **17**, 102.
REIF, F., and HELLER, J. I. (1984), "Prescribing Effective Human Problem Solving Processes: Problem Description in Physics," *Cognition and Instruction* **1**, 177.
RESNICK, R., and HALLIDAY, D. (1977, 1985), *Physics*, 3rd and 4th eds. (John Wiley & Sons, New York).
RICHARDSON, O. W., and COMPTON, K. T. (1912), "The Photoelectric Effect," *Phil. Mag. Series 6*, **24**, 575.
ROGERS, E. M. (1960), *Physics for the Inquiring Mind* (Princeton University Press, Princeton, NJ).
ROLLER, D., and ROLLER, D. H. D. (1957), "The Development of the Concept of Electrical Charge," in *Harvard Case Histories in Experimental Science*, CONANT, J. B., and NASH, L. K., eds. (Harvard University Press, Cambridge, MA).
ROMER, R. H. (1982), "What do 'Voltmeters' Measure?: Faraday's Law in a Multiply Connected Region," *Am. J. Phys.* **50**, 1089.
ROSENQUIST, M. L. (1982), *Improving Preparation for College Physics of Minority Students Aspiring to Science-Related Careers* (Unpublished dissertation, University of Washington, Seattle, WA).

ROSENQUIST, M. L., and McDERMOTT, L. C. (1987), "A Conceptual Approach to Teaching Kinematics," *Am. J. Phys.* **55**, 407.

ROSNICK, P., and CLEMENT, J. (1980), "Learning Without Understanding: The Effect of Tutoring Strategies on Algebra Misconceptions," *Jour. of Mathematical Behavior* **3**, No. 1.

RUTHERFORD, E. (1903), "The Magnetic and Electric Deviation of the Easily Absorbed Rays from Radium," *Phil. Mag.* **5**, 6, 177.

RUTHERFORD, E. (1906), "The Mass and Velocity of the α-particles Expelled from Radium and Actinium," *Phil. Mag.* **12**, 6, 348.

RUTHERFORD, E., and ROYDS, T. (1909), "The Nature of the α-Particle from Radioactive Substances," *Phil. Mag.* **17**, 6, 281 [reprinted in Birks (1962)].

RUTHERFORD, F. J., HOLTON, G., and WATSON, F. G. (1981), *The Project Physics Course*, 3rd ed. (Holt, Rinehart and Winston, New York).

Science Curriculum Improvement Study (SCIS) (1968ff), Delta Education, Nashua, NH. (These materials, together with other elementary science curricula, are available on a CD-ROM published under the title Science Helper K-8 CDROM by PC-SIG, Sunnyvale, CA.)

SHAHN, E. (1988). "On Science Literacy," in *Educational Philosophy and Theory*. Journal of the Philosophy of Education Society of Australia. Special topic issue on Science Education. MATTHEWS, M. R., ed. (20)2, 42.

SHANKLAND, R. S. (1963), "Conversations with Albert Einstein," *Am J. Phys.* **31**, 47.

SHERWOOD, B. A. (1983), "Pseudowork and Real Work," *Am. J. Phys.* **51**, 597.

SHERWOOD, B. A., and BERNARD, W. H. (1984), "Work and Heat Transfer in the Presence of Sliding Friction," *Am. J. Phys.* **52**, 1001.

SHAMOS, M. (1959), *Great Experiments in Physics* (Henry Holt & Co., New York).

SHYMANSKY, J. A., KYLE, W. C., JR., and ALPORT, J. M. (1983), "The Effects of New Science Curricula on Student Performance," *J. Res. Sci. Teaching* **20**, 387.

ST. JOHN, M. (1980), "Thinking Like a Physicist in the Laboratory," *The Physics Teacher* **18**, 436.

STEAD, B. F., and OSBORNE, R. J. (1980), "Exploring Science Students' Concepts of Light," *Austr. Sci. Teach. Jour.* **26**, 84.

STEINBERG, M. S. (1983), "Reinventing Electricity," in *Proceedings of the International Seminar on Misconceptions in Science and Mathematics,* HELM, H., and NOVAK, J., eds. (Cornell University, Ithaca, NY).

STEINBERG, M. S. (1986), "The Origins of Force—Misconceptions and Classroom Controversy" (Paper presented at summer meeting of the American Association of Physics Teachers, Ohio State University, 26 June).

STEINBERG, M. S. (1987), "Transient Lamp Lighting with High-Tech Capacitors," *The Physics Teacher* **25**, 95.

STEINBERG, M. S., and FREDETTE, N. (1986), "Exploring and Improving Student Reasoning in Electricity with Capacitor-Controlled Bulb Lighting" (Workshop conducted at AAPT Summer Meeting, 23–27 June, Columbus, OH).

STEVENSON, H. W., SHIN-YING LEE, and STIGLER, J. W. (1986), "Mathematics Achievement of Chinese, Japanese, and American Children," *Science* **231**, 693.

STRIKE, K. A., and POSNER, G. J. (1982), "Conceptual Change and Science Teaching," *Eur. J. Sci. Ed.* **4**, 231.

THOMSON, G. P. (1956), " J. J. Thomson and The Discovery of the Electron," *Physics Today* **9:8**, 19.

THOMSON, J. J. (1897), "Cathode Rays," *Phil. Mag.* **44**, 5, 293 [Excerpts are to be found in Magie (1935) and Shamos (1959)].

THOMSON, J. J. (1899), "On the Masses of Ions in Gases at Low Pressures," *Phil. Mag.* **48**, 5, 547.

THOMSON, J. J. (1912), "Multiply Charged Atoms," *Phil. Mag.* **24**, 6, 668.

THORNTON, R. K. (1987a), "Access to College Science: Microcomputer-Based Laboratories for the Naive Science Learner," *Collegiate Microcomputer* **1**, 100.

THORNTON, R. K. (1987b), "Tools for Scientific Thinking: Microcomputer-Based Laboratories for Physics Teaching," *Phys. Ed.* **22**, 230.
TIPLER, P. A. (1976, 1982), *Physics*, 1st and 2nd eds. (Worth Publishers, New York).
TOBIAS, S. (1986), "Peer Perspectives on the Teaching of Science," *Change* **18**, 36.
TOBIAS, S. (1988), "Insiders and Outsiders," *Academic Connections.* Winter 1988. Office of Academic Affairs, The College Board, New York.
TOBIAS, S., and HAKE, R. R. (1988), "Professors as Physics Students: What Can They Teach Us?" *Am J. Phys.* **56**, 786.
TOLMAN, R. C., and STEWART, T. D. (1916), "The Electromotive Force Produced by the Acceleration of Metals," *Phys. Rev.* **8**, 97.
TOLMAN, R. C., and STEWART, T. D. (1917), "The Mass of the Electric Carrier in Copper, Silver, and Aluminum," *Phys. Rev.* **9**, 164.
TROWBRIDGE, D. E. (1988), "Applying Research Results to the Development of Computer Assisted Instruction," in *Proceedings of the Conference on Computers in Physics Instruction*, RISLEY, J. S., and REDDISH, E. F., eds. (North Carolina State University, Raleigh, NC).
TROWBRIDGE, D. E., and MCDERMOTT, L. C. (1980), "Investigation of Student Understanding of the Concept of Velocity in One Dimension," *Am. J. Phys.* **48**, 1020.
TROWBRIDGE, D. E., and MCDERMOTT, L. C. (1981), "Investigation of Student Understanding of the Concept of Acceleration in One Dimension," *Am. J. Phys.* **49**, 242.
TSANTES, E. (1974), "Note on the Tides," *Am. J. Phys.* **42**, 330.
VIENNOT, L. (1979), "Le Raisonnement Spontané en Dynamique Elementaire," *Eur. J. Sci. Ed.* **1**, 205.
WATTS, D. M. (1985), "Student Conceptions of Light: A Case Study," *Phys. Ed.* **20**, 183.
WEINSTOCK, R. (1961), "What's *F*? What's *m*? What's *a*?" *Am. J. Phys.* **29**, 698.
WEINSTOCK, H., and BORK, A., eds. (1986), *Design of Computer-Based Learning Materials*, Proceedings of the NATO Advanced Study Institute on Learning Physics and Mathematics via Computers held in San Miniato, Italy, 15–26 July 1985. NATO. ASI Series F, *Computer and Systems Sciences*, Vol. 23. (Springer Verlag, Berlin, New York).
WELLMAN, R. T. (1978), " Science: A Basic Language for Reading Development," in *What Research Says to the Science Teacher* Vol. 1., ROWE, M. B., ed. (National Science Teachers Association, Washington, DC).
WESTFALL, R. S. (1971), *Force in Newton's Physics* (American Elsevier, New York).
WHITAKER, R. J. (1983), "Aristotle Is not Dead: Student Understanding of Trajectory Motion," *Am. J. Phys.* **51**, 352.
WHITE, B. (1983), "Sources of Difficulty in Understanding Newtonian Dynamics," *Cog. Sci.* **7**, 41.
WHITE, B. (1984), "Designing Computer Games to Help Physics Students Understand Newtonian Laws of Motion," *Cog. and Instr.* **1**, 69.
WHITTAKER, E. (1951), *A History of the Theories of Aether and Electricity* (Thos. Nelson & Sons, London; also Harper Torchbook No. 531, Harper & Brothers, New York, 1960).
WILLIAMS, E. R., FALLER, J. E., and HILL, H. A. (1971), "New Experimental Test of Coulomb's Law: A Laboratory Upper Limit on the Photon Rest Mass," *Phys. Rev. Lett.* **26**, 721.
WISER, M., and CAREY, S. (1983), "When Heat and Temperature Were One," in *Mental Models*, GENTNER, D., and STEVENS, A. L., eds. (Lawrence Erlbaum Associates, Hillsdale, NJ).

Index

A

Acceleration, 28ff
  angular, 101
  centripetal, 138
  definition, 28
  due to gravity, 32
  objects not thrown backwards on, 78
  research on understanding of, 34
  tangential, 101
Action at distance, 67, 157
  Faraday's criticism of, 195
Algebraic signs, interpretation of, 27
Algebraic statements, verbal interpretation of, 14, 84, 105
Alpha particles, 241
Ampere, experiment with parallel wires, 192
Area, 1–2
  in kinematics graphs, 30
  in work and impulse graphs, 141–143
Arrows, representing different kinds of vectors, 68
Atoms, size of, 231–232
Avogadro's Number, 231, 242

B

Backwards science, 153, 170
Balmer–Rydberg formula, 241, 253
Banking a curve, 106
  unbanked curve, 103, 105, 106
Batteries and bulbs, 167ff
Because, use of word, 16
Bohr atom, 250ff

C

Cathode rays:
  in Crookes tubes, 229
  in Thomson experiment, 233ff
Cavendish, and electrical resistivity, 170
Centrifugal force, 108
Centripetal force, 103–107, 139
  supplied by colinear forces, 103ff
  supplied by noncolinear forces, 106ff
Chaos, 286
Charge, electrical, 146ff
  conservation of, 158
  corpuscle of, 239
  Franklin and, 158
  by induction, 154
  like and unlike, 146–149
  positive and negative, 149, 150, 158
  quantification of, 154
Circuit, electric, 167ff
  and current model, 177
  experiments with, 171
  operational definition, 167
  short, 171
Circular motion, 101–111. *See also* Centripetal force
  and center of mass, 108
  preconceptions regarding, 101
Clock reading, 21
Collisions:
  energy changes in, 134
  of molecules, 278
Compressibility, of liquids and solids *vs.* gasses, 277

Computers:
  and electromagnetism mnemonics, 193
  in instructional dialogues, 308
  in kinematics, 33
  and torque concept, 113
  and vector arithmetic, 92, 94
Concepts, invention of, 24, 289
Conductor, 146
  operational definition, 168
Coulomb's Law, 154
Critical thinking, 158, 230, 313ff
Current, electric, 168ff
  bulk or surface phenomenon, 176
  and circuit model, 177
  conventional *vs.* electron, 178
  induced in multiply connected region, 194
  operational definition, 168

D

Dimensions *vs.* units, 99
Division, 3–8
  arithmetical reasoning involving, 6
  verbal interpretation of, 4

E

Efficiency, of power lines, 180
Einstein:
  and the photon concept, 248ff
  and special relativity, 257ff
Electric interactions, distinguished from gravitational and magnetic, 145, 152
Electricity:
  current, 162ff
  and experiments at home, 146
  frictional, 145
  identity from different sources, 164
Electromagnetism, 188ff
  and Ampere's experiment, 192
Electron:
  as charge carrier in metals, 181
  and electric charge, 146, 154
  and electric current, 178
  and Thomson experiment, 233ff
Emf, induced in multiply connected region, 194
Empirical *vs.* theoretical, 241
Energy, 115ff
  changes in accelerating a car, 136
  changes in rolling, 132
  conservation of, 123–124, 230
  in inelastic collisions, 134
  internal, 125–126
  thermal internal, 121–122
Enthalpy, 126–128
Equalities, different meanings of, 81–83, 159
Estimating, 243–244, 283
Event, 21

F

Faraday:
  and action at a distance, 195
  and electric fluid, 166
  and energy in electric circuit, 174
  and identity of different electricities, 164
  and lines of force, 198
  and magnetic field around moving charge, 190
Faraday's Law, and induced emf in multiply connected region, 194
Field concept, 196ff
  and Maxwell's Theory, 199
Field strength:
  electric, 159
  gravitational, 159
First Law of Thermodynamics, 122, 123, 124, 126
Force, 49ff
  active *vs.* passive, 64
  "cancellation" of, 56
  centrifugal, 108
  centripetal, 101ff, 139
  diagrams (free body), 64ff, 105, 282
    and strings, 74
  distributed, 68
  fictitious, 108
  in free fall, 73
  of friction, 77, 79
  vs. impulse, 61
  linguistic aspects, 63
  normal, 76
  operational definition of, 52
  passive, 64, 79, 103
  in projectile motion, 73
  and rate of change of momentum, 118
  superposition of, 55
  and tension, 74ff
  zero-work, 12, 130
Force meter, 52–55, 58

Frames of reference:
  and fictitious forces, 78, 108
  search for absolute, 259
  and special relativity, 259
Free body (force) diagrams, 64ff
  for fluid in a container, 282
Friction, 79
  and heat, 121, 128
  and work, 128

G

Galileo:
  and "gravity," 70
  and modern science, 38–42
  and projectile motion, 94
Graphs:
  and arithmetical reasoning, 8
  interpretation in kinematics, 24ff
Gravitational interaction, distinguished from electric and magnetic, 145–146, 152
Gravitational mass, 58
Gravity:
  and air, 70–71
  contrasted with electricity and magnetism, 145–146
  and levity, 69
  linguistic aspects, 69
  terrestrial effects, 69

H

Heat, 118
  "conversion of work into," 121, 123, 129
  measuring transfer of, 120
  operational definition, 120
  in presence of sliding friction, 128
  transfer of, 120ff, 126
Horizontal, 13, 70
Hydrogen atom, 111, 253

I

Idea first, name afterwards, 23, 28, 70, 100, 111, 168, 171, 289, 315
Image formation:
  with converging lens, 223
  with plane mirror, 222
Impulse, 61, 116
Impulse–Momentum Theorem, 121
Inequalities, 106

Inertia:
  demonstrations of, 80
  law of, 49, 51, 59
Inertial mass:
  vs. gravitational mass, 58
  operational definition, 54
Interference:
  grating pattern, 219
  two-source pattern, 218
Instant, 22
Interpreting:
  algebraic expressions, 76, 84, 105
  kinematics graphs, 24ff

J

Joule heat (Joule's Law), 173–175

K

Kinematics, 20ff
  graphs, 24
  problem solving, 32
  rotational, 100
  use of calculators in, 33
Kinetic Theory, 274ff
  assumptions of, 276

L

Law of Inertia:
  hands-on observations, 60
  understanding of, 59
Laws of Motion, 50ff
  First Law, 51, 59
  Second Law, 52ff
  Third Law, 64ff
    and action at a distance, 67
    statement of, 65
Lenard, P.:
  and nature of cathode beam, 233
  and photoelectric effect, 245ff
Lenses, image formation with converging, 223
Levity, 69, 299
Linguistic aspects, 15
  and centripetal force, 104
  and concept formation, 7, 60
  and electrical charge, 147
  and force, 63
  and frictional opposition to motion, 80
  and gravitational *vs.* inertial mass, 59

Linguistic aspects *(Continued)*
  and "gravity," 69
  and kinematics, 22
  and "normal" force, 77

M

Magnetic field:
  around current carrying wire, 190
  measurement of, 200
  around moving electrostatic charge, 190
Magnetic interactions:
  with current carrying electrolyte, 191
  among current carrying wires, 191–192
  distinguished from electric and gravitational, 145, 152
Magnetic poles, 150
  like and unlike, 151
Mass, inertial, 50ff
  distinguishing between object and property, 55
  vs. gravitational mass, 58
  operational definition of, 54
  superposition of, 55
  units of, 58
  and weight, 57
Maxwell, J. C., and field theory, 199
Midnight, 13
Millikan experiment, 239
Mirrors, image formation by plane, 222
Momentum, 115ff, 139

N

Newton, and centripetal force, 138
Newton's Laws:
  First Law, 51ff
  Second Law, 52ff
  Third Law, 64ff
    and action at a distance, 67
    and electrostatic interaction, 156
    statement of, 65
    and strings, 74
Newton's rings, Young's elucidation of dark center in, 220
Noon, 13
North, 13, 151
Not the case, visualizing what is, 54, 59, 78, 79, 81, 92, 104, 112, 145, 149, 176, 177, 212, 236, 247, 278

O

Observation and inference, 42, 120, 229, 289
Oersted experiment, 188
Ohm's Law:
  and electrical resistance, 170, 175
  and freedom of carriers in metallic conduction, 176
  historical development, 176ff
Operational definition, 2, 15, 50, 52, 60, 315
  of electrical field strength, 159
  of force and mass, 52 ff
  and science literacy, 289
  of vector component, 94

P

Per, student use of, 6
Perrin, and charge in cathode beam, 234
Phenomenological reasoning, 96ff
  in Ampere's experiment, 192
  and the Bohr atom, 256
  with images, 223ff
  in thermal expansion, 282
  in Young's experiment with Newton's rings, 221
Photoelectric effect, 245ff
Photon concept, 245ff, 257
  and Bohr atom, 252
Plumb bob, 14
Polarization, electrical, 148, 152
Position, 21
  instantaneous, 22, 26
Potential difference, 172
Power, efficiency of electric power lines, 180
Preconceptions, 49
Pressure, hydrostatic, 281
Problem solving, 32, 83, 84
Projectile motion, 94ff
  superposition in, 95

R

Radian measure, 13, 99, 217
Ratio, 3–6
  reasoning, 177, 242, 277
    with lens aperture, 227
  verbal interpretation of, 4
Rays, 216

Reasoning:
  arithmetical, 6, 8, 101
  functional, 10
  phenomenological, 96
  ratio, 177, 242, 277
  reversing line of, 5, 28, 61, 207, 216, 218, 225
Redefinition, 27, 296
Reflection, specular *vs.* diffuse, 221
Relativity, special, 257ff
Resistance, electrical:
  and Ohm's Law, 175
  operational definition, 169
Reversing line of reasoning, 5, 28, 61, 207, 216, 218, 225
Rolling:
  down inclined plane, 132
  without slipping, 133
Rowland, H., and magnetic field around moving electrostatic charge, 190
Rutherford–Royds experiment, 241
Rydberg constant, 111, 253

S

Scaling, 10, 243, 277
Science literacy, 231, 238, 240, 241, 248, 274, 288ff
  definition, 289
Sense perception, effects eluding direct, 65, 67, 81, 215, 282
Shock wave, 212
Simultaneity:
  distant, 258
  local, 258
  in single frame of reference, 257
Slowness, 23
South, 13, 151
Special relativity, 257ff
Spiraling back, 8, 57, 60, 68, 98, 100, 101, 103, 104, 106, 110, 113, 115, 118, 138, 156, 158, 181, 183, 193, 208, 217, 229, 236, 238, 244, 247, 251, 257, 282
Strings, 74ff
  massless, 75
Superposition:
  of electrical fields, 160
  of forces, 55
  of inertial masses, 55
  of magnetic effects, 189
  in projectile motion, 95
  of waves, 205

System, 115
  deformable, 130

T

Temperature, 118
Tension, 74ff
Thermal:
  expansion, 282
  interaction, 119
Thomson, J. J., and nature of cathode rays, 233ff
Tides, 109ff
Time, 21
Tolman–Stewart experiment, 181
Top of flight, 31
Torque, 111ff
Trigonometry, 13, 99
Two-body problem, 108

U

Units, *vs.* dimensions, 99

V

Vacuum:
  free fall in, 71
  meaning of, 71
Vectors, 91ff
  addition of, 91
  arithmetic of, 91
  arrows representing different, 68
  components of, 93
  definition, 92
  subtraction of, 91
Velocity, 20, 52
  angular, 100
  average, 23
  graphs, 30
  instantaneous, 23, 26
  research on understanding of, 34
  tangential, 101
Vertical, 13, 70, 106
Volume, 3

W

Wave:
  fronts:
    in one dimension, 208–215
    in two dimensions, 216
  phenomena:

Wave (Continued)
  phenomena (Continued)
    distinguishing particle and propagation velocities, 201
    graphical representation of pulse shapes, 202–203
    propagation velocities, 208ff
    transient effects, 215
  pulses:
    generation of, 203–204
    reflection of, 204ff
  trains, periodic and sinusoidal, 217
  velocity:
    in fluid, 210
    in shallow water, 213
    on string, 208
Weight:
  and force on supporting object, 71
  and mass, 57
  units of, 58
Weightlessness, 72
Work, 115, 122
  operational definition, 124
  in presence of sliding friction, 128
  pseudo-, 123
  real, 123
Work-Kinetic Energy Theorem, 115, 121